Sweet Potato Pest Management

About the Book and Editors

Sweet potato, *Ipomoea batatas* (L.) Lam., is the seventh most important food crop world-wide. Of all root and tuber crops, it is second in importance to the white potato, *Solanum tuberosum* L. Despite its importance globally, sweet potato pest management has not received the research attention it deserves. The present book draws on the work of researchers from around the world to summarize the state-of-the-art research on pest management in this important crop. The contributors address issues of crop protection and biology of pests, such as sweet potato weevils, wireworms, cucumber beetles, flea beetles, vine borers, plant-parasitic nematodes, and aphid and whitefly vectors and their associated disease disorders. They also present new data on the biological control of pests and on biotechnological and conventional methods for developing cultivars with resistance to pests. Additionally, the researchers look beyond the field to address regulatory issues and the socioeconomic problems of pest management.

Richard K. Jansson is an associate professor of entomology at the University of Florida, Institute of Food and Agricultural Sciences, Tropical Research and Education Center in Homestead. Kandukuri V. Raman is a senior scientist in entomology, project leader in pest management, and acting chairman of the Entomology and Nematology Department at the International Potato Center, Lima, Peru.

Studies in Insect Biology
Michael D. Breed, Series Editor

Sweet Potato Pest Management: A Global Perspective, edited by Richard K. Jansson and Kandukuri V. Raman

Diversity in the Genus Apis, edited by Deborah Roan Smith

Sweet Potato Pest Management

A Global Perspective

EDITED BY

Richard K. Jansson
and Kandukuri V. Raman

Routledge
Taylor & Francis Group

NEW YORK AND LONDON

First published in paperback 2024

First published 1991 by Westview Press, Inc.

Published 2019 by Routledge
605 Third Avenue, New York, NY 10158

and by Routledge
4 Park Square, Milton Park, Abingdon, Oxon OX14 4RN

Routledge is an imprint of the Taylor & Francis Group, an informa business

Copyright © 1991, 2019, 2024 by Taylor & Francis

Library of Congress Cataloging-in-Publication Data
Sweet potato pest management : a global perspective / edited by R.K.
   Jansson and K.V. Raman.
      p. cm. — (Westview studies in insect biology)
   Includes index.
   ISBN 0-8133-7825-7
   1. Sweet potatoes—Diseases and pests—Control. 2. Insect pests—
Control. I. Jansson, R.K. (Richard K.). II. Raman, K.V.
III. Series.
SB608.S98S94 1991
635'.2297—dc20   90-12373
                                                                          CIP

Publisher's Note
The publisher has gone to great lengths to ensure the quality of this reprint but points out that some imperfections in the original copies may be apparent.

ISBN: 978-0-367-30481-2 (pbk)
ISBN: 978-0-367-28935-5 (hbk)
ISBN: 978-0-429-30810-9 (ebk)

DOI: 10.1201/9780429308109

CONTENTS

FOREWORD

Sweet potato is one of the world's major crops, especially in developing countries, where it ranks third in value of production and fifth in calorie contribution to human diets. In the light of such importance, one might think that the knowledge base for sweet potato would be highly developed or that, at least, active research programs would be under way to underpin sweet potato production. Alas, neither is true. In fact, the knowledge base for sweet potato is quite poor, and sweet potato research is woefully inadequate to match the importance of the crop. Indeed, just a few years ago the Technical Advisory Committee of the Consultative Group on International Agricultural Research (CGIAR) concluded that sweet potato research was underfunded by a ratio of eight times in relation to its global importance. Thus, the sweet potato has long been neglected as far as research is concerned.

Despite the general lack of research on sweet potato, breeding of new varieties has been carried out successfully at several places including North Carolina State University, Louisiana State University, the United States Department of Agriculture laboratory in Charleston, South Carolina, the Asian Vegetable Research and Development Center (AVRDC) in Taiwan, the International Institute of Tropical Agriculture (IITA) in Nigeria, and more recently by the International Potato Center (CIP) in Peru. Some national programs in developing countries have begun breeding sweet potato. The People's Republic of China, the world's largest producer of sweet potato, has several active breeding programs.

In a time when sweet potato research is generally lacking and the knowledge base for the crop is mostly weak, it was a great pleasure to learn of this new book, *Sweet Potato Pest Management: A Global Perspective*, by Richard K. Jansson and Kandukuri V. Raman. Sweet potato is well known as a hardy crop that performs better under marginal or low-input conditions than most food crops. In particular, sweet potato is more drought-tolerant and less demanding of soil nutrients than most other crops. However, sweet potato does have severe insect and disease problems that limit the crop's performance in many places.

This book deals with insects and other pests of sweet potato, in the context of a pest management perspective. The major pests of sweet potato are surveyed and the available knowledge about the pests is analyzed. The book contains an impressive global perspective. The authors point to the lack of knowledge about sweet potato pests in developing countries and make a strong case for increased research on the crop.

As a former root crop scientist, I was pleased to see a book of such quality on sweet potato, which has been neglected for so long. I commend the book to scientists, extensionists, and decision makers in agricultural research who carry out agricultural programs in both developing and industrialized countries. The book provides a basis for planning and carrying out research and supporting activities on integrated pest management strategies for sweet potato in many locations.

Donald L. Plucknett
Consultative Group on International Agricultural Research
The World Bank
Washington, D.C.

ACKNOWLEDGMENTS

This book, *Sweet Potato Pest Management: A Global Perspective*, evolved from an international conference, *International Conference on Sweet Potato Pest Management*, that was held in Miami, Florida on 18-22 June 1989. This conference was the first meeting to bring together researchers and extension personnel from various land-grant universities, international and national research centers, government agricultural agencies, and industry to exchange information and develop collaborative projects on sweet potato pest management world-wide. In excess of 80 scientists from over 25 countries participated in this conference.

Although the sweet potato, *Ipomoea batatas* (L.) Lam., is the seventh most important crop world-wide, sweet potato pest management has been a neglected research topic and has been seriously underfunded. This lack of attention, especially in developing countries, can be attributed to several factors including a lack of information on the major constraints of sweet potato pest management, a lack of trained research and technology transfer specialists, inadequate facilities and transportation to conduct the needed research, and in some countries, the low status of the crop.

This book compiles the most current information on the major insect, nematode, and virus pests of sweet potato world-wide. Most of the book reviews the biology and management approaches for sweet potato weevils, *Cylas* spp. and *Euscepes postfasciatus* (Fairmaire), which are the most important biotic constraints to sweet potato production in the world. In addition, certain selected pests on which there is considerable information, such as wireworms, cucumber beetles, flea beetles, vine borers, plant-parasitic nematodes, and virus vectors, are discussed. Other major insect pests, such as the sweet potato moth, *Herse convolvuli* L., sweet potato butterfly, *Acraea acerata* Hew., striped sweet potato weevil, *Alcidodes dentipes* (Oliver), various tortoise beetles, *Aspidomorpha* sp., sweet potato clearwing moth, *Synathedon dasysceles* Bradley, and several minor insect pests, can be important regionally; however, because of a lack of research, little information is available on most of these pests. For this reason, these pests were not included in the present book. In addition to examining field problems, this book also addresses regulatory and socioeconomic aspects of pest management programs on sweet potato.

We thank the many sponsors of the conference for their support: University of Florida, I.F.A.S., Office of the Dean for Research, Gainesville, Florida; International Programs, Gainesville, Florida; International Potato Center, Lima, Peru; AgriSense, Fresno, California; Monsanto Co., St. Louis, Missouri; Helena Chemical Co., Memphis, Tennessee; Rhone-Poulenc Chemical Co., Monmouth Junction, New Jersey; Dow Chemical U.S.A., Midland, Michigan; Biosis, Inc., Palo Alto, California; Alger Farms, Homestead, Florida; IT Supply, Homestead, Florida; S&M Farm Supply, Inc., Homestead, Florida; V&S Growers, Miami, Florida; Cuban Agricola, Miami, Florida; Guilarte Brothers Farms Corp., Miami, Florida; and South Florida Potato Growers Exchange, Homestead, Florida.

We thank B. Towry, C.M. Mannion, L.J. Mason, S.H. Lecrone, D.R. Seal, A.G.B. Hunsberger, R. Lance, T. Bailey, A. Albinana, and H. Valenzuela for helping to make the conference a success. This conference would not have been possible without the support of O.S. Malamud, to whom we are most grateful. We also thank D. Davis, Program Manager, U.S. Department of Agriculture, Tropical/Subtropical Agriculture Program, Caribbean Basin Administrative Group, for support of research on sweet potato pest management that led to the development of this conference. Lastly, we thank T. Bailey for help in preparing the final draft of the book and H. Huseman for drafting the cover of the book.

Each chapter was peer reviewed by at least two authorities that work on the subject. We thank the following reviewers for their time and effort: M. Altieri, R. Anderson, T.C. Baker, C.S. Barfield, K.R. Barker, W.J. Bell, F.D. Bennett, K.J. Brown, H.T. Burke, J.L. Capinera, T. Casey, R.B. Chalfant, F. Cisneros, W.E. Clark, J.A. Coffelt, W.W. Collins, S. Finch, R.E. Foster, J.E. Funderburk, J.R. Fuxa, R.R. Gaugler, P. Gregory, D.G. Hall, A.M. Hammond, K.F. Haynes, R.R. Heath, M.G. Hutton, A. Jones, C.I. Kado, S.J. Kays, A.J. Keaster, G.G. Kennedy, J.B. Kring (deceased), P.J. Landolt, B. Leonhardt, R.E. Litz, T.P. Mack, F.W. Martin, L.J. Mason, L. McDonough, R. McSorley, D. Midmore, E.R. Mitchell, C. O'Brien, T.M. Perring, C.E. Pray, F.I. Proshold, R. Rodriguez-Kabana, W.E. Roelofs, D. Roff, J.M. Schalk, C.M. Smith, N.S. Talekar, W.M. Tingey,

I.K. Vasil, J.K. Waage, S.E. Webb, L.T. Wilson, G.W. Wolfe, and G. Zehnder. Their independent reviews helped to improve the quality of this publication.

This book is dedicated to all international research center scientists, such as those supported by the Consultative Group for International Agricultural Research (CGIAR), who despite the hardships associated with being located in violent and politically unstable environments continue to pursue and transfer knowledge to the developing world so that the less fortunate may become subsistent.

Richard K. Jansson
Kandukuri V. Raman

1. Sweet Potato Pest Management: A Global Overview

Richard K. Jansson
Tropical Research and Education Center
University of Florida, I.F.A.S.
Homestead, Florida 33031 U.S.A.

Kandukuri V. Raman
International Potato Center
P.O. Box 5969
Lima, Peru

Sweet potato, *Ipomoea batatas* (L.) Lam., originated in or near northwestern South America around 8,000-6,000 B.C. (Austin 1988). The countries of Guatemala, Colombia, Ecuador, and Peru have the greatest diversity in sweet potato germplasm (Austin 1983). Secondary centers of genetic variability are Papua New Guinea, the Philippines, and parts of Africa (Yen 1982). The plant was first discovered and cultivated by Proto-Chibhuan, Chibhuan, or Chibhuan-influenced people around 3,000 B.C. (Austin 1988, O'Brien 1972). By approximately 2,500 B.C., sweet potato had spread throughout most of the Neotropics (Austin 1988, O'Brien 1972), with the exception of the temperate zones of the New World (Brand 1971). Fossil records indicate that the plant was present in New Zealand between 500 B.C. and 1419 A.D. (Yen 1974). It was present on Easter Island between 1426 and 1626 and on Hawaii between 1545 and 1785 (Yen 1974, 1982). European explorers introduced the crop into Africa and India by the early 1500's, China by 1594, Taiwan by 1597, and Japan by 1698 (Yen 1974, 1982). More complete descriptions of the history and dispersal of sweet potato world-wide are available (Austin 1988, Brand 1971, Yen 1974, 1982).

Sweet potato ranks seventh among all food crops world-wide, with an annual production of 115 million metric tons (FAO 1984). Of the root and tuber crops, sweet potato ranks third in acreage (7.9 million ha) behind white potato, *Solanum tuberosum* L., (20.1 million ha) and cassava, *Manihot esculentum* Crantz, (14.8 million ha) (FAO 1984). It is grown in more than 100 countries (Figure 1.1), and among the world's root and tuber crops, it is second in importance to the white potato (Horton 1987). Horton (1988a) published production, consumption, and trade statistics for sweet potato world-wide. Developing countries account for 98% of the world production (Gregory *et al.* 1990, Horton 1988a). Approximately 92% of the world's sweet potato is produced in Asia and the Pacific Islands; 89% of this is grown in China (Horton 1988b). Since 1961, sweet potato production has increased by 20% in developing countries (Gregory *et al.* 1990). In China, sweet potato production has increased by 41% since 1961. In all other Asian countries, excluding China, production has only increased by 2.5%.

Sweet potato is an important staple food in areas of subsistence farming and is a drought-tolerant crop (Bouwkamp 1985b). Sweet potato production adapts well to both low and high technology input agricultural systems. Adequate to high yields are produced in areas with high levels of inputs and technology, such as the United States and Japan, as well as in many lesser developed tropical countries that use labor-intensive, low technology systems primarily for subsistence farming.

Sweet potato is used as a staple food, vegetable (both fleshy roots, tender leaves, and petioles), snack food, animal feed, for industrial starch extraction and fermentation, and for various processed products (Bouwkamp 1985a, Kays 1985, Lin *et al.* 1985, Sakamoto & Bouwkamp 1985). Use of sweet potato for human consumption has declined, whereas its use as an animal feed and a raw material for the manufacture of industrial products has increased (Gregory *et al.* 1990). Two of the principal factors inhibiting greater use of sweet potato in food systems are the limited knowledge of its nutritional value and its low cultural status in some regions of the world (Anonymous 1989). Among the ten leading food crops, sweet potato ranks third in terms of calories produced per square meter (Bouwkamp 1985a). Sweet potato is high in nutritional value; with the exception of protein and niacin, sweet potato provides over 90% of the nutrients per calorie required for most people (Food and Nutrition Board 1980, Watt & Merrill 1975). Roots are a valuable source of carbohydrates, vitamins (provides 100% of the Recommended Daily Allowance [RDA] for

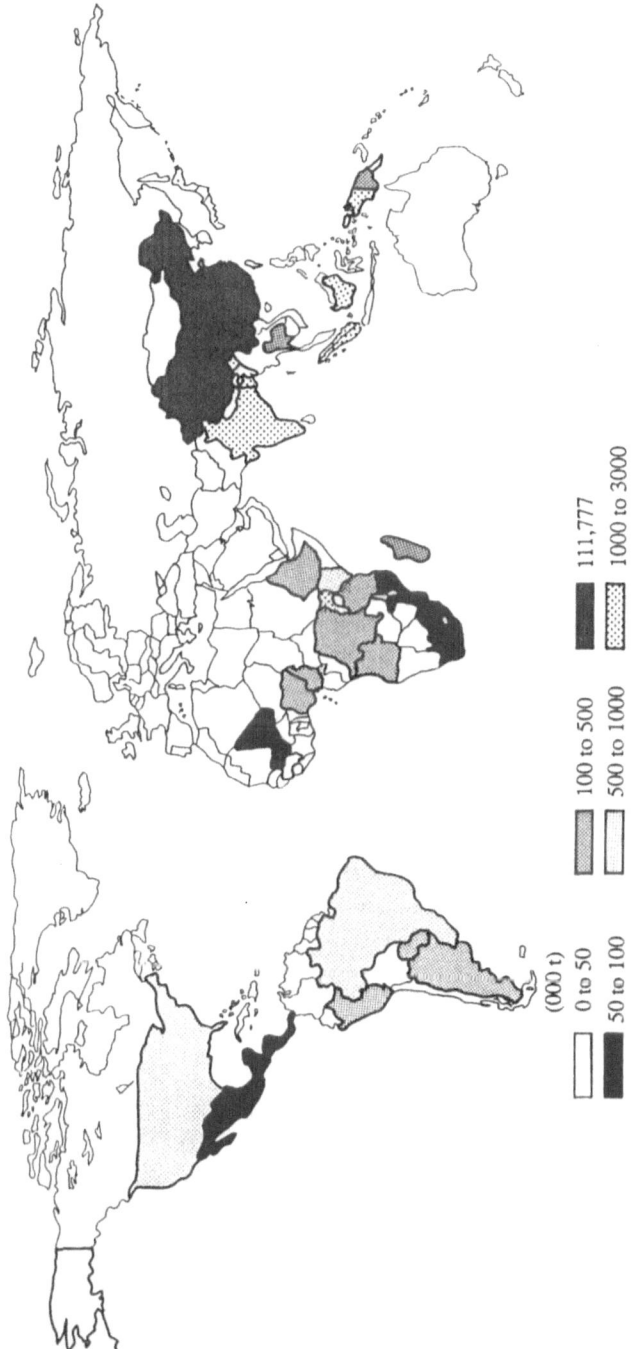

Fig. 1.1. Global distribution of sweet potato production (1987-1989; x 1,000 metric tons).

Vitamin A and 49% of the RDA for Vitamin C), and minerals (provides 10% of the RDA for iron and 15% of the RDA for potassium) (Anonymous 1980, Food and Nutrition Board 1980).

SWEET POTATO PRODUCTION CONSTRAINTS

There are several constraints that limit sweet potato production world-wide. These include pre- and post-harvest losses from insects (especially sweet potato weevils, *Cylas* spp.), nematodes, diseases, and rodents; storage methods, costs, and losses; lack of processed products; transportation problems; high marketing costs; unstable supplies and prices; environmental factors, such as drought, soil fertility, flooding, and soil structure; and a lack of improved cultivars and planting material (Horton 1989, Lin *et al.* 1985, also see Chapter 22). Of these, post-harvest factors have been found to be most limiting. For example, in storage, insect, disease, and rodent pest problems are prominent; in marketing, problems of supplying the market at a reasonable cost are most important; and in consumer demand, the lack of varieties with desirable eating and/or processing quality, and the lack of a diversity of processed products are most limiting (Anonymous 1989). A recent poll of sweet potato researchers in Asia found that the most limiting field production problem was crop loss due to insects followed in decreasing order by plant viruses, environmental factors, nematodes, available germplasm, and fungi and bacteria (Figure 1.2) (Horton 1989). This concurs with a recent, more extensive survey presented in this book (see Chapter 22).

ARTHROPOD PESTS OF SWEET POTATO

Losses due to insect feeding, especially sweet potato weevils, may often reach 60 to 100% because most sweet potatoes are produced in low input agricultural systems (Chalfant *et al.* 1990). The major insect pests of sweet potato in Java (Franssen 1934), Africa (Terry 1976), China (Hua & Zhou 1984), and the Caribbean (Suah 1981) have been described. West (1977) listed 100 insect and three mite species attacking sweet potato world-wide. The majority of these were leaf feeders (58), followed by stem and vine feeders (32), root feeders (9), and flower feeders (4). The biology and management of the major insect pests world-wide have been reported (Chalfant *et al.* 1990, Hill 1975, Schalk & Jones 1985, Ward 1978). Talekar (unpublished) listed

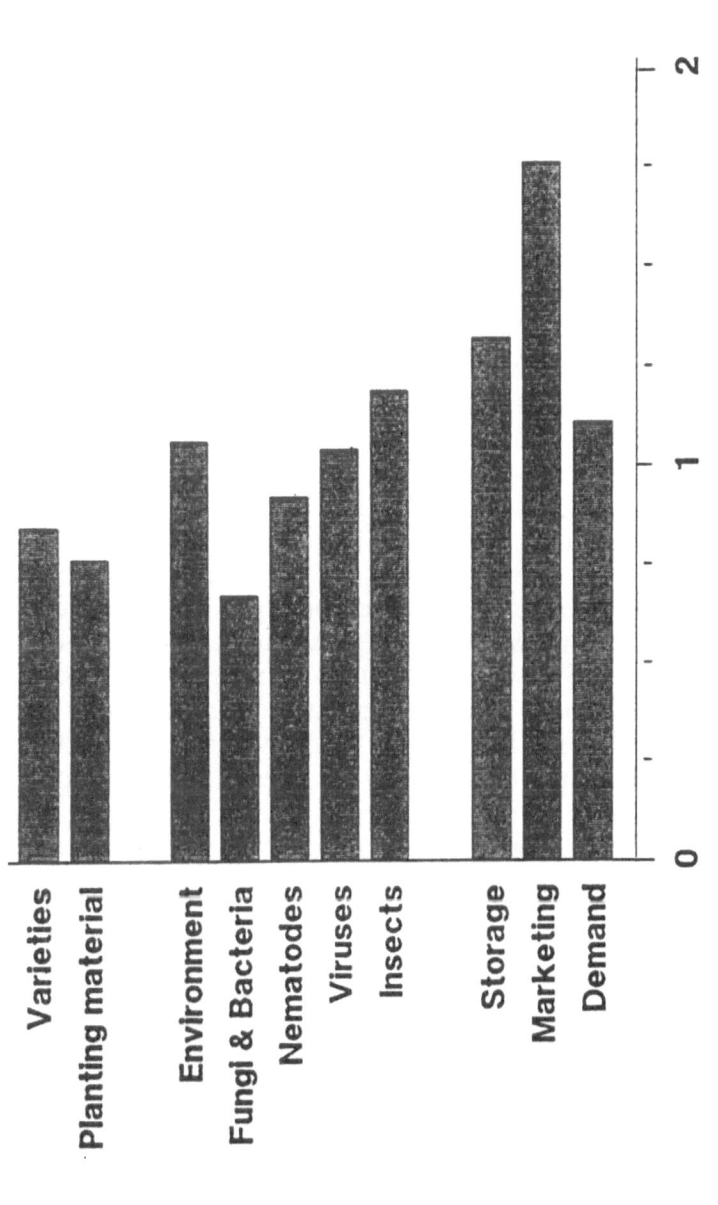

Fig. 1.2. The major constraints to sweet potato production in Asia (severity scores: 0, not present; 1, of little practical importance; 2, somewhat important; 3, very important; from Horton 1989; reprinted with permission from the International Potato Center, Lima, Peru).

280 insect species and 18 mite species attacking sweet potato in the field and in storage around the world, of which the weevils, *C. formicarius* (Fabricius) and *C. puncticollis* (Boheman), are most damaging. All plant parts are susceptible to damage by insects and mites. Although the plant is attacked by a large number of pest species, few cause significant crop loss. For example, many foliar feeders do not cause yield reductions because of a compensatory ability of the plant to tolerate high levels of defoliation (Chalfant *et al.* 1990).

NEMATODE AND DISEASE PESTS OF SWEET POTATO

Losses due to nematodes, fungi, bacteria, viruses, and other disease organisms can also be devastating. The major bacterial and fungal diseases of sweet potato and their control were recently reviewed (Clark 1988, Clark & Moyer 1988, Moyer 1982, 1985). Virus and nematode problems have also been reviewed previously (Chiu *et al.* 1982, Hua & Zhou 1984, Jatala 1989, Jatala & Raman 1988, Moyer 1988, Moyer & Salazar 1989, Rossel & Thottapilly 1988, Terry 1976, 1982). In general, disease problems vary in different parts of the world, and may also vary within regions of a given country (e.g., China [Lu *et al.* 1989]). In some countries, total yield loss may occur from these pests (Palomar *et al.* 1989), whereas in other countries, disease pests are not considered limiting to sweet potato production (Nayar & Rajendran 1989, Siddique & Rashid 1989). However, in general, little is known about the status and severity of disease and nematode problems in developing countries. As mentioned previously, losses from fungi and bacteria were considered less important than those from insects and viruses (Horton 1989).

SWEET POTATO PEST MANAGEMENT CONSTRAINTS

Historically, sweet potato pest management has been a neglected research topic world-wide, especially in the developing world. Its lack of attention is due to several factors. The most concentrated research on this topic has been in the United States, where the crop is of minor importance (0.6 million MT; $160 million; USDA 1989). In regions of the world where the crop is a major staple and losses due to insects are high, funding for integrated pest management (IPM) research is minimal. World-wide, the crop has been seriously underfunded (Gregory *et al.* 1990, TAC 1985). In addition, certain IPM methods, such as improved germplasm, biological control agents, etc., are not available in developing countries. For this reason, approaches for managing pests of sweet potato have relied on the use of remedial measures on a pest-by-pest basis and, in most cases, management has relied exclusively on cultural practices. This is most apparent with the management of sweet potato weevils, *Cylas* spp. and *Euscepes postfasciatus* (Fairmaire).

Also, few specialists are present in the National Agricultural Research Systems (NARS) of developing countries, and these specialists often lack an understanding of IPM. Many of these countries lack trained entomologists, nematologists, virologists, and plant pathologists. Also, mechanisms for technology transfer, such as training, on-farm research, and human resource development, are limited.

Concepts and practices of sweet potato IPM have begun to mature, especially during the last five years. Rapid progress has been made in developing management strategies for use in IPM. In order to use these approaches in developing countries, the role of sweet potato in farming and cultural systems must be better understood. More recently, it has become clear that in addition to farmers, the perspectives and needs of consumers, processors, and market agents must also be considered in developing research priorities in sweet potato IPM. We must recognize that research on sweet potato IPM focuses on a commodity in a food system, and not only on a crop in a production system. In many regions of the world, sweet potato is considered a low-status crop. The way in which the crop will be used is clearly linked to its image, and much remains to be done with regard to public awareness of its potential.

Sweet potato pest management, however, is still in its infancy. If we are to improve pest management programs on sweet potato, we must better understand the biology of insect, mite, nematode, and disease pests that attack sweet potato, and select management approaches that are compatible with the limitations of the agricultural systems in which the crop is grown. This book reviews the status of current knowledge on insect, nematode, and virus pests of sweet potato world-wide, including their biologies, distributions, damage, sampling approaches, management programs, and the future of IPM in sweet potato in developed and developing countries.

We review the status of research on the biology and management of only the most important insect and nematode pests that attack sweet potato world-wide. These include sweet potato weevils, *C. formicarius*, *C. puncticollis*, *C. brunneus* (Fabricius), and *E.*

postfasciatus; sweet potato vine borers, *Omphisa anastomasalis* Guenee and *Megastes grandalis* Guenee; wireworms, *Conoderus* spp.; cucumber beetles, *Diabrotica* spp.; flea beetles, *Systena* spp.; virus vectors; and plant-parasitic nematodes. In Chapters 2 through 9 and Chapters 11 through 13, the current status of the biology, taxonomy, and control of the most important sweet potato weevils, *Cylas* spp., are presented. In Chapter 10, the potential of *Agrobacterium*-mediated gene transfer to confer resistance of sweet potato to pests is described. In Chapters 14 through 19, the biologies and management approaches for other important insect, virus, and nematode pests are reviewed. In Chapters 20 and 21, the history and future outlook of pest resistance breeding programs are described. In Chapter 22, pest management programs for sweet potato are discussed from a social science perspective. Lastly, in the closing chapter, the current perspectives and future outlook of sweet potato pest management are presented.

We recognize that not all pest problems were included in this treatise, and that other insect, mite, and disease pests may be important on a regional level; however, because these pests are only of regional importance and/or little published information is available, they were not included in this book. Others have reviewed some of the insect pests of lesser importance (Chalfant *et al.* 1990, Schalk & Jones 1985). Information on disease pests is also available elsewhere (Clark 1988, Clark & Moyer 1988, Moyer 1985, 1988, Rossel & Thottapilly 1988), and was not included in the present treatise.

ACKNOWLEDGMENTS

We thank J.L. Capinera, C.S. Barfield, D.L. Plucknett, P. Gregory, K. Brown, and L.J. Mason for critically reviewing this manuscript. This is Florida Agricultural Experiment Station Journal Series No. R-01028.

REFERENCES

Anonymous. 1980. Sweet potato quality. U.S.D.A. Southern Cooperative Series Bulletin No. 249. Russell Research Center, Athens, Georgia.

Anonymous. 1989. An assessment of CIP's programs: achievements, impact, and constraints. International Potato Center, Lima, Peru.

Austin, D.F. 1983. Variability in sweet potatoes in America. Proc. Amer. Soc. Hort. Sci. Trop. Reg. 27(B):15-26.

Austin, D.F. 1988. The taxonomy, evolution, and genetic diversity of sweet potatoes and related wild species, pp. 27-60. *In* Exploration, maintenance, and utilization of sweet potato genetic resources. International Potato Center, Lima, Peru.

Bouwkamp, J.C. 1985a. Introduction - part 1, pp. 3-7. *In* J.C. Bouwkamp (ed.). Sweet potato products: a natural resource for the tropics. CRC Press, Boca Raton, Florida.

Bouwkamp, J.C. 1985b. Production requirements, pp. 9-33. *In* J.C. Bouwkamp (ed.). Sweet potato products: a natural resource for the tropics. CRC Press, Boca Raton, Florida.

Brand, D.D. 1971. The sweet potato: an exercise in methodology, pp. 343-365. *In* C.L. Riley, J.C. Kelley, C.W. Pennington & R.L. Rands (eds.). Man across the sea. Univ. Texas Press, Austin.

Chalfant, R.B., R.K. Jansson, D.R. Seal & J.M Schalk. 1990. Ecology and management of sweet potato insects. Annu. Rev. Entomol. 35:157-180.

Chiu, R.J., C.H. Liao & M.L. Chung. 1982. Sweet potato virus diseases and meristem culture as a means of disease control in Taiwan, pp. 169-176. *In* R.L. Villlareal & T.D. Griggs (eds.). Sweet potato: Proceedings of the 1st international symposium. Asian Vegetable Research and Development Center, Shanhua, Tainan, Taiwan, Republic of China.

Clark, C.A. 1988. Principal bacterial and fungal diseases of sweet potato and their control, pp. 275-289. *In* Exploration, maintenance, and utilization of sweet potato genetic resources. International Potato Center, Lima, Peru.

Clark, C.A. & J.W. Moyer. 1988. Compendium of sweet potato diseases. Amer. Phytopathol. Soc. Press, St. Paul, Minnesota.

FAO. 1984. FAO production yearbook, 1983. Food and Agriculture Organization, Rome.

Food and Nutrition Board. 1980. Recommended daily allowances. National Academy of Sciences, National Research Council, Washington, D.C.

Franssen, C.J.H. 1934. Insect pests of the sweet potato crop in Java. Korte Meded. Inst. Peziekt., Buitenzorg 10:205-225 (Translation published, 1986. Asian Vegetable Research and Development Center, Shanhua, Taiwan).

Gregory, P., M. Iwanaga & D.E. Horton. 1990. Sweet potato: global issues, pp. 462-468. *In* R.H. Howeler (ed.). Proceedings of the 8th symposium of the International Society of Tropical Root Crops. International Center of Tropical Agriculture (CIAT), Bangkok, Thailand.

Hill, D.S. 1975. Agricultural insect pests of the tropics and their control. Cambridge Univ. Press, London.

Horton, D.E. 1987. Potatoes: production, marketing, and programs for developing countries. Westview Press, Boulder, Colorado.

Horton, D.E. 1988a. Underground crops. Long-term trends in production of roots and tubers. Winrock International, Morrilton, Arkansas.

Horton, D.E. 1988b. World patterns and trends in sweet potato production. Trop. Agric. 65:268-270.

Horton, D.E. 1989. Constraints to sweet potato production and use, pp. 219-223. *In* Improvement of sweet potato (*Ipomoea batatas*) in Asia. International Potato Center, Lima, Peru.

Hua, C. & X.T. Zhou (eds.). 1984. Sweet potato cultivation in China. Shanghai Science and Technology Press, Shanghai, People's Republic of China (In Chinese).

Jatala, P. 1989. Important nematode parasites of sweet potatoes and their control, pp. 213-218. *In* Improvement of sweet potato (*Ipomoea batatas*) in Asia. International Potato Center, Lima, Peru.

Jatala, P. & K.V. Raman. 1988. Major insect and nematode pests of sweet potatoes and recommendations for transfer of pest free germplasm, pp. 319-321. *In* Exploration, maintenance, and utilization of sweet potato genetic resources. International Potato Center, Lima, Peru.

Kays, S.J. 1985. Formulated sweet potato products, pp. 205-218. *In* J.C. Bouwkamp (ed.). Sweet potato products: a natural resouce for the tropics. CRC Press, Boca Raton, Florida.

Lin, S.S.M., C.C. Peet, D.-M. Chen & H.-F. Lo. 1985. Sweet potato production and utilization in Asia and the Pacific, pp. 139-148. *In* J.C. Bouwkamp (ed.). Sweet potato products: a natural resource for the tropics. CRC Press, Boca Raton, Florida.

Lu, S.Y., Q.H. Xue, D.P. Zhang & B.F. Song. 1989. Sweet potato production and research in China, pp. 21-30. *In* Improvement of sweet potato (*Ipomoea batatas*) in Asia. International Potato Center, Lima, Peru.

Moyer, J.W. 1982. Postharvest disease management for sweet potatoes, pp. 177-184. *In* R.L. Villlareal & T.D. Griggs (eds.). Sweet potato: Proceedings of the 1st international symposium. Asian Vegetable Research and Development Center, Shanhua, Tainan, Taiwan, Republic of China.

Moyer, J.W. 1985. Major disease pests, pp. 35-57. *In* J.C. Bouwkamp (ed.). Sweet potato products: a natural resource for the tropics. CRC Press, Boca Raton, Florida.

Moyer, J.W. 1988. Principal virus diseases of sweet potatoes, their control and eradication. *In* Exploration, maintenance, and utilization of sweet potato genetic resources. International Potato Center, Lima, Peru.

Moyer, J.W. & L.F. Salazar. 1989. Virus and viruslike diseases of sweet potato. Plant Dis. 73:451-455.

Nayar, G.G. & P.G. Rajendran. 1989. Sweet potato production, utilization and constraints in India, pp. 31-42. *In* Improvement of sweet potato (*Ipomoea batatas*) in Asia. International Potato Center, Lima, Peru.

O'Brien, P.J. 1972. The sweet potato; its origin and dispersal. Amer. Anthropol. 74:342-365.

Palomar, M.K., E.F. Bulayog & T. Van Den. 1989. Sweet potato research and development in the Philippines, pp. 79-85. *In* Improvement of sweet potato (*Ipomoea batatas*) in Asia. International Potato Center, Lima, Peru.

Rossel, H.W. & G. Thottapilly. 1988. Complex virus diseases of sweet potato. *In* Exploration, maintenance, and utilization of sweet potato genetic resources. International Potato Center, Lima, Peru.

Sakamoto, S. & J.C. Bouwkamp. 1985. Industrial products from sweet potatoes, pp. 219-234. *In* J.C. Bouwkamp (ed.). Sweet potato products: a natural resource for the tropics. CRC Press, Boca Raton, Florida.

Schalk, J.M. & A. Jones. 1985. Major insect pests, pp. 59-78. *In* J.C. Bouwkamp (ed.). Sweet potato products: a natural resource for the tropics. CRC Press, Boca Raton, Florida.

Siddique, A. & M. Rashid. 1989. Present status and future prospects of sweet potatoes in Bangladesh, pp. 7-20. *In* Improvement of sweet potato (*Ipomoea batatas*) in Asia. International Potato Center, Lima, Peru.

Suah, J.R.R. 1981. Major pests of root crops in the Caribbean, *In* Short course on integrated pest management of tropical crops. Faculty of Agriculture, University of the West Indies, St. Augustine, Trinidad.

TAC (Technical Advisory Committee, Consultative Group on International Agricultural Research). 1985. TAC review of CGIAR priorities and future strategies. FAO, Rome.

Terry, E.R. 1976. A review of the major pests and diseases of sweet potato with special reference to plant exchange and introduction. Afr. Plant Prot. 1:117-133.

Terry, E.R. 1982. Sweet potato (*Ipomoea batatas*) virus diseases and their control, pp. 161-168. *In* R.L. Villlareal & T.D. Griggs (eds.). Sweet potato: Proceedings of the 1st international symposium. Asian Vegetable Research and Development Center, Shanhua, Tainan, Taiwan, Republic of China.

USDA. 1989. Crop production 1988 summary. U.S. Department of Agriculture, National Agricultural Statistics Service and Agricultural Statistics Board, Washington, D.C.

Ward, A. 1978. Sweet potato insect pests, pp. 80-96. *In* Pest control in tropical root crops. PANS Manual No. 4. Centre Overseas Pest Research, London.

Watt, B.K. & A.L. Merrill. 1975. Composition of foods. U.S. Department of Agriculture Handbook No. 8, Washington, D.C.

West, S.A. 1977. Studies on the biology and ecology of the sweet potato stem borer *Megastes grandalis* Guen. in Trinidad. M.S. Thesis. University of the West Indies, Trinidad.

Yen, D.E. 1974. The sweet potato and Oceana. Bishop Mus. Bull., Honolulu 236:1-389.

Yen, D.E. 1982. Sweet potato in historical perspective, pp. 17-30. *In* R.L. Villareal & T.D. Griggs (eds.). Sweet potato. Proceedings of the 1st international symposium. Asian Vegetable Research and Development Center, Shanhua, Taiwan, Republic of China.

2. The Origin and Dispersal of the Pest Species of *Cylas* with a Key to the Pest Species Groups of the World

G. William Wolfe [1]
Department of Entomology
Rutgers, The State University of New Jersey
New Brunswick, New Jersey 08903 U.S.A.

Sweet potato, *Ipomoea batatas* (L.) Lam., is grown circumglobally in tropical and sub-tropical latitudes and constitutes one of the seven most important crops on a worldwide basis (Chalfant *et al.* 1990, FAO 1984, Jones 1970). In terms of production and nutrition, sweet potato is particularly important to subsistence farmers in developing tropical countries (DeVries *et al.* 1967, Hahn *et al.* 1984).

Numerous insects have been recorded as pests of sweet potato, including various members of Lepidoptera, Thysanoptera, Orthoptera, and Hemiptera (Chalfant *et al.* 1990, Kay 1973, Talekar 1988, Wyniger 1962). Among Coleoptera, members of Elateridae, Scarabaeidae, Chrysomelidae, and Curculionidae attack sweet potato (Pollard 1984, Wyniger 1962). Although importance of pest species varies from region to region, weevils *sensu lato* constitute the most important insect threat on a worldwide basis and certain species of *Cylas* particularly are problematic. *Cylas* is most diverse in Africa and all known or suspected pest species of *Cylas* occur in Africa and/or Madagascar. Only one pest species occurs extensively outside of Africa/Madagascar, the circumglobally distributed *C. formicarius* (Fabricius).

The overall objectives of this chapter are to provide an overview of the systematics of the genus, to highlight the most important literature pertaining to systematics of *Cylas*, and to point out systematic data that have a bearing on control of these important pest species. Specifically,

[1] Current address: Division of Mathematics and Science, Dobbs Hall, Reinhardt College, Waleska, Georgia 30183 U.S.A.

this chapter will: 1) discuss the total number of species names that have been applied to various pest populations and assign each species name to a species group; 2) sort out nomenclatural confusion, especially regarding *C. formicarius sensu lato*; 3) provide a key (with illustrations) for identification of each pest species group; 4) indicate the distribution of the pest species group members, their possible centers of origin, and/or aspects of dispersal; and 5) discuss a phylogeny of the pest species in the context of the genus.

At first glance, the relationships between insect taxonomy (the process of naming and classifying insects), insect systematics (the study of different kinds of relationships among insects), and historical zoogeography (temporally based study of geographic distribution of insects) with the development of management programs for insect pests might not be obvious. However, all scientists need to be able to consistently identify, name, and communicate information for each species (Rosen 1978a).

These kinds of systematic data allow scientists to more optimally target and evaluate a variety of types of ongoing sweet potato research programs, such as the interpretation of pheromone studies, the evaluation of resistant sweet potato plants, determination of insect-host plant relationships, and surveys for natural enemies associated with pest species.

LEVEL OF ANALYSIS

Specimens of *Cylas* were borrowed for study from major museums in North America, Europe, Africa, India, and Southeast Asia. Although many species now are defined adequately, several species groups remain enigmatic due to high degrees of intraspecific variation, subtle differences between species, a paucity of specimens for some species, and/or nomenclatural problems. Because it is not possible to resolve each of these issues completely and uniformly for every pest species, the discussion below emphasizes the species group level of analysis.

Table 2.1 lists all species names placed in *Cylas* and indicates centers of distribution for each species. Undescribed species are not included. Most undescribed species are related to *C. cyanescens* Boheman; however, some populations currently assigned to the pest species *C. puncticollis* (Boheman) may be undescribed, cryptic species.

Table 2.1. List of all names placed in *Cylas* with center of distribution for each.

Species	Distribution
Pest species	
C. formicarius (Fabricius)	Circumglobal
C. turcipennis (Boheman)[a]	S.E. Asia\Indonesia
C. elegantulus (Summers)	U.S.A.\Caribbean
C. brunneus (Fabricius)	Africa
C. angustatus (Labram and Imhoff)	Africa
C. femoralis (Faust)	Africa
C. puncticollis puncticollis (Boheman)	Africa
C. puncticollis opacus Voss	Africa
C. nigrocoerulans Fairmaire	Madagascar
C. compressus Hartmann	Africa
C. hovanus Hustache	Madagascar
Species inconclusively designated as pests	
C. vanderplasi Voss	Africa
C. cyanescens Boheman	Africa
Non-pest species	
C. laevigatus Fahraeus	Africa
C. pumilus Marshall	Africa
C. rufescens Fairmaire	Madagascar
C. aeneus Hustache	Africa
C. longicollis Chevrolat	Africa
C. glabripennis Hartmann	Africa
C. nitens Hustache	Africa
C. semipunctatus Fahraeus	Africa
C. freyi Voss	Africa
C. curtipennis Fairmaire	Africa
C. robustus Fairmaire	Africa
C. impunctatus Faust	India
C. submetallicus Desbrochers	India
C. coimbitores Subramanian	India
C. rufipes Faust	Southeast Asia
C. laevicollis Chevrolat	Southeast Asia

[a] Currently accepted junior synonyms are indented.

Species Criteria

Species are defined as lineages consistent with the evolutionary species concept as outlined in Wiley (1981). Although this approach circumvents inconsistencies associated with reproductive isolation and the biological species concept *sensu* Mayr (1970), it is still necessary to discern "character gaps" sufficient to justify the designation of a population(s) as a species level lineage (Rosen 1978b). Erwin's (1970) approach for species recognition is used herein in this regard: an array of specimens displaying a multidimensional continuum of characters but still separable from other specimens by a distinct gap. To distinguish pest species, specimens were first separated by differences in aedeagal and head structure. A search was then made for correlation between these and other characters, such as antennal shape, femoral shape, punctation, etc.. Allopatric populations were validated as species level lineages if the gap between them was similar to the difference between presumed closely related sympatric species. A more complete discussion of this overall approach is in Wolfe and Roughley (1990).

Phylogeny

The Hennigian approach to phylogenetics was used (see Wiley 1981) to determine relationships. For this analysis, *Aporhina* Lacordaire, (an apionid *sensu* Wagner (1910)), was used as the outgroup to polarize characters.

A single fully dichotomous phylogeny for all species placed in *Cylas* cannot be produced. In fact, computer-assisted analysis using PAUP, (version 2.4, Swofford (1985)), revealed between 15 and 100 equally parsimonius trees depending on subtle changes in assumptions concerning polarity, addition or deletion of characters, and whether the characters were ordered or unordered. Even though the relationships among many species of *Cylas* remain unknown, analysis of all equally parsimonius trees revealed several extremely interesting communalities regarding the pest species groups (see below).

Distribution

Analysis of distributions always reveals region(s) where certain species are not known to occur; however, it is very difficult to prove that a species definitely is absent from a region.

Obviously, the more thoroughly an area is collected, the more reasonable it is to accept absence of a species as valid. Beyond the obvious, a hypothesis of absence is enhanced if information on microhabitat "preference" is available and it is known that a particular species in question co-occurs in the same microhabitat with other related species, in at least some regions. Collectively, these kinds of data are important because then even if the species in question is not captured, the discovery of at least some related species suggests that the proper microhabitat was examined.

For example, representatives of the *C. formicarius* and *C. puncticollis* group co-occur in sweet potato fields (and even individual tubers) in parts of South Africa. However, *C. formicarius* currently is known from only one locality in Central Africa, despite the fact that *C. brunneus* (Fabricius) and *C. puncticollis* are reported from numerous localities (including sweet potato fields) throughout Central Africa. Therefore, it appears that *C. formicarius* is rare to absent in many parts of Africa, and where it does occur, it was probably secondarily introduced. This is discussed further below.

TAXONOMIC OVERVIEW

Family Level Classification

Supra-specific classification of *Cylas* has been confused since Fabricius (1792, 1798, 1801) assigned males of *C. formicarius* and *C. brunneus* to *Brentus* and two females of *C. formicarius* to *Attelabus*. The name *Cylas* was created by Latrielle (1802) for *Brentus brunneus*; he placed *Cylas* between *Brentus* and *Attelabus* in his discussions. More recently, most weevil taxonomists have assigned *Cylas* to either Brentidae (Blatchley & Leng 1916, LeConte & Horn 1876, Morimoto 1976), Apionidae (Crowson 1955, Kissinger 1968, O'Brien & Wibmer 1982, Pierce 1918, 1940, Wagner 1910), or even in its own family, Cyladidae (Brues *et al.* 1954). Although this issue is not resolved conclusively, it is obvious that *Cylas* along with plesiomorphic ("primitive") brentids and apionids form a complex of somewhat similar and presumably closely related forms. I currently accept the conclusions of Crowson (1955) and Kissinger (1968) and assign *Cylas* to Apionidae.

Genus-Subgenus Classification

Rafinesque (1815) proposed the name *Cylanus* as a new name for *Cylas*. *Cylanus* certainly never was used widely and Pierce (1940) placed it as a junior synonym of *Cylas*.

Pierce (1940) described a new genus, *Protocylas*, for members of *Cylas* with globose body shapes and short, blunt snouts. Subramanian (1957) briefly discussed the subgenus and described a new species. However, body shape of members of *Cylas* grades from elongate and slender to moderately globose to distinctly globose. Kissinger (1968) recognized this and placed *Protocylas* as a junior synonym of *Cylas*.

Species Level Classification

Nineteen names were placed in *Cylas* by Wagner (1910). Since then, ten other species and/or subspecies have been described (see Table 2.1) and there currently are at least four undescribed species. However, a number of the available names are junior synonyms. All in all, it is reasonable to estimate that at the conclusion of the systematic revision of *Cylas* there will be approximately 25 valid species of *Cylas*.

It specifically has been speculated that eleven species of *Cylas* attack attack sweet potato: *C. brunneus*, *C. compressus* Hartmann, *C. cyanescens*, *C. elegantulus* (Summers), *C. femoralis* (Faust), *C. formicarius*, *C. nigrocoerulans* Fairmaire, *C. puncticollis*, *C. puncticollis opacus* Voss, *C. turcipennis* (Boheman), and *C. vanderplasi* Voss (Burgeon 1936, Pierce 1918, 1940, Risbec 1947, G.W. Wolfe, unpubl. data). However, there currently is adequate evidence to apply only nine of the above names to pest populations and those nine names are divided into three pest species groups.

Cylas formicarius **Group**. *Cylas formicarius*, *C. elegantulus*, and *C. turcipennis* are assigned to this group. These three species names have been variously split (e.g., Pierce 1918, 1940) or synonymized (Kemner 1924, Kissinger 1968, LeConte & Horn 1876). The larva and pupa of *C. formicarius* from the United States were described by Pierce (1918) and compared to specimens from India, Indonesia, and the South Pacific by Pierce (1940).

Pierce (1940) particularly advocated using the name *C. formicarius elegantulus* (Summers) for specimens from the New World, Madagascar, and the South Pacific. He used the name *C. formicarius formicarius* (Fabricius) for specimens from India, but was not even sure if the nominate subspecies attacked sweet potato. He applied the name *C. turcipennis* primarily to specimens from Indonesia and the Philippines.

Pierce (1940) separated specimens of *C. turcipennis* from *C. formicarius elegantulus* mainly by the more greenish elytra, longer antennal club, and minor differences in male genitalic structure of the former. He admitted, however, that his sample size was too small for making worldwide comparisons (one specimen of *C. formicarius*, seven specimens of *C. turcipennis*, and 70 specimens of *C. formicarius elegantulus*), that large amounts of intraspecific variation existed, and that appropriate type material had not been examined.

Pierce (1940) was apparently unaware that Kemner (1924) compared the types of *C. formicarius* and *C. turcipennis* and placed *C. turcipennis* as a junior synonym of *C. formicarius*. However, Kemner continued to recognize *C. elegantulus* (Summers). Champion (1914 (in Hustache [1924])) independently reached the same conclusion.

I also examined the types of *C. formicarius* and *C. turcipennis* and compared both of the former types to representative New World specimens (because the type series of *C. elegantulus* is lost). Considering the amount of variation in the several thousand specimens that have been examined, it is appropriate to follow LeConte & Horn (1876) and recognize the name *C. elegantulus* as a junior synonym of *C. formicarius*. Likewise, I agree with Kemner and Champion that the name *C. turcipennis* is a junior synonym of *C. formicarius*.

The only other study related to this subject is the interesting karyological study by Hung (1985). Hung found differences among New World and Old World specimens of *C. formicarius* based on male sex chromosomes. Unfortunately, his sample sizes were too small (18 specimens from Louisiana, 97 specimens from Hawaii, and 25 specimens from Taiwan) for definitive conclusions, especially considering the amount of chromosomal variation that he observed.

Cylas brunneus **Group.** *Cylas brunneus*, *C. femoralis*, and *C. angustatus* (Labram and Imhoff) belong in this group. Although the type of *C. brunneus* cannot be located, *C. brunneus* was described from Senegal as a brownish, cylindrical species. The type was probably a female because the antennae were described as moniliform. Generally, the name *C. brunneus* is applied to specimens with widely separated eyes and only moderately long antennal clubs (club equal in length to segments 1-9). *Cylas femoralis* is extremely close to my concept of *C. brunneus*, and these names may be synonymous.

Cylas angustatus was placed as a junior synonym of *C. brunneus* by Pierce (1940) and Marshall (1953). Based on the brief original description and illustration of an apparent female and the fact that *C.*

angustatus was described from Senegal, the conclusions of Pierce (1940) and Marshall (1953) are appropriate.

Cylas puncticollis **Group.** *C. puncticollis, C. puncticollis opacus, C. nigrocoerulans, C. compressus,* and *C. hovanus* Hustache are placed in the *C. puncticollis* group. Voss (1966) described *C. puncticollis opacus* from Kano, Nigeria. I have not been able to locate and examine the type of *C. puncticollis opacus*; however, I have examined numerous specimens of *C. puncticollis* from West Africa (including Nigeria) and thus far I have not found a consistent set of characters to separate populations at the subspecies level. All members of this species group are uniformly black with the eyes dorsally narrowly separated in males. Conclusive identification requires genitalic examination.

Only *C. puncticollis* and *C. compressus* are definitely associated with sweet potato (Risbec 1947, G.W. Wolfe, unpubl. data). Risbec (1947) suspected that *Cylas nigrocoerulans* was a pest of sweet potato in Madagascar, probably because of its extreme similarity to *C. puncticollis*. I agree with Risbec on that point; furthermore, I include *C. hovanus* herein as a probable pest of sweet potato because of its extreme similarity to *C. compressus*. Based on this pattern of overall similarity, two distinct species pairs can be recognized in the *C. puncticollis* groups: 1) *C. puncticollis* and *C. nigrocoerulans*, and 2) *C. compressus* and *C. hovanus*.

Interestingly, within the first species pair, *C. puncticollis* is recorded only from continental Africa while *C. nigrocoerulans* is considered endemic to Madagascar. In the second pair, the distribution of *C. compressus* is centered in East Africa while *C. hovanus* is recorded only from Madagascar. For now, distribution is the best way for non-taxonomists to separate the extremely similar species within each species pair. From a taxonomic standpoint, *C. puncticollis* is the most problematic species within this species group and some populations currently assigned to *C. puncticollis* may represent undescribed cryptic species.

Other Species of *Cylas*. Risbec (1947) indicated that at Bambey, Senegal, specimens of *C. cyanescens* were observed feeding on leaves of sweet potato and that damage was so severe that it was difficult to "conserve" the storage roots. However, he did not state that specimens of *C. cyanescens* were from storage roots. Furthermore, he noted that some fields had been abandoned for several years and that there were numerous wild species of *Ipomoea* in the region. For these reasons, the presumed pest status *C. cyanescens* is not conclusive.

Risbec (1947) indicated that *C. vanderplasi* (another globose shaped species) was "without doubt" another "enemy" of sweet potato because it was very similar to *C. cyanescens*. In fact, these species are not "similar" (at least not compared to the level of similarity that exists between members of the *C. puncticollis* species pair and the *C. compressus* species pair) and the pest status of *C. vanderplasi* is also far from conclusive.

Risbec (1947) included *C. longicollis* Chevrolat and *C. aeneus* Hustache in his article on pest species. However, he did not discuss *C. aeneus* and only indicated that the biology of *C. longicollis* was unknown.

Lastly, specimens of the *C. laevigatus* Fahraeus group, which have not been implicated as pests of sweet potato, could possibly complicate identification of pest species because they occassionally occur on sweet potato. This is taken into account in Table 2.2 which separates all pest species groups from other members of *Cylas*.

IDENTIFICATION OF PEST SPECIES GROUPS

Pierce (1918), Hustache (1924), Burgeon (1936), Risbec (1947), and Voss (1966) provided keys to separate at least some of the pest species discussed above. Some of those keys were regional in scope while others need to be updated because they include a number of species names considered herein as synonyms. Furthermore, the authors were not aware of the amount of intraspecific variation in some of the species.

Although a conclusive key for identification of pest species cannot be constructed, the key provided in Table 2.2 is useful for separating species groups. A conservative approach was used by emphasizing a species group level of analysis; thus, the key accounts for nomenclatural problems and intra-specific variation that complicates identification. Also, the key separates the pest species groups from all non-pest species. Additionally, the key is useful worldwide and it is thorougly illustrated. Lastly, it relies mostly on non-genitalic characters for separating groups. This should prove useful to non-taxonomists.

It should be noted that a number of the species groups designated herein might prove to be monotypic (e.g., the *C. formicarius* and *C. brunneus* groups). Even the *C. puncticollis* group, which definitely is not monotypic, may be simplified by synonymizing the names within each of the proposed species pairs.

Also, separation of specimens of *C. cyanescens* and *C. vanderplasi* from the *C. puncticollis*, *C. brunneus*, and *C. formicarius* pest species groups is relatively easy based on femoral length and body

Table 2.2. Key for separating pest species from non-pest species of *Cylas*.

---

1. Hind femora usually projecting beyond apex of elytra (Figures 2.1A-B), or, snout very short and blunt (sometimes with eyes vertically elongate), or both (similar to Figure 2.1A); abdomen usually distinctly globose or oval shaped (specimens keying here are not currently verified pest species; adults feed on foliage; immature stages probably occur in seed pods). **Approximately 20 species key to this couplet, including *C. cyanescens* and *C. glabripennis*.**
1'. Hind femora not projecting or only slightly projecting beyond elytral apex; abdomen always elongate and cylindrical (Figures 2.2 and 2.3); snout never extremely short and blunt. 2

2. Size from tip of snout to apex of elytra 3.7 mm or less; antennae not distinctly sexually dimorphic; length of male antennal club at most equal to only five of preceding segments (Figure 2.4C); not pest species. *C. laevigatus* group
2'. Size from tip of snout to elytral apex greater than 3.8 mm; antennae distinctly sexually dimorphic (Figures 2.3B,C); length of male antennal club equal to or greater than combined length of all preceding segments (Figure 2.3C). 3

3. Eyes close together in dorsal view; distance between eyes about 1/6 of minimum width of snout (see Figure 2.1C). 4
3'. Eyes widely separated in dorsal view; distance between eyes subequal to minimum width of snout (see Figure 2.1D). 6

4. Pronotum in lateral view more distinctly arched, posterior constriction evident (Figure 2.4B); color uniformly blackish, sometimes with a bluish sheen but never with a distinctly shiny, copperish sheen; members of the *C. puncticollis* group key to this couplet. 5
4'. Pronotum in lateral view less noticeably arched, posterior constriction scarcely if at all evident (Figure 2.4A); color blackish, often with an evident copperish or brassy sheen; not verified pest species. *C. longicollis* group

5. Sclerite of set II of aedeagus small, barely larger than one sclerite of sclerite set III (Figure 2.5B); sclerite of set II positioned dorsal to sclerite set III.*C. compressus* or
C. hovanus
5'. Sclerite of set II of aedeagus distinctly larger than one sclerite of sclerite set III (Figure 2.5A); sclerite of set II positioned posteriorly to sclerite set III. *C. puncticollis*
or *C. nigrocoerulans*

6. Antennal club approximately equal in length to the scape (Figure 2.2C); internal sac of the aedeagus with two sets of sclerites (Figure 2.5C). *C. brunneus* group
6'. Antennal club longer than scape (Figure 2.3C); internal sac of aedeagus with one set of sclerites (Figure 2.5D). *C. formicarius* group

---

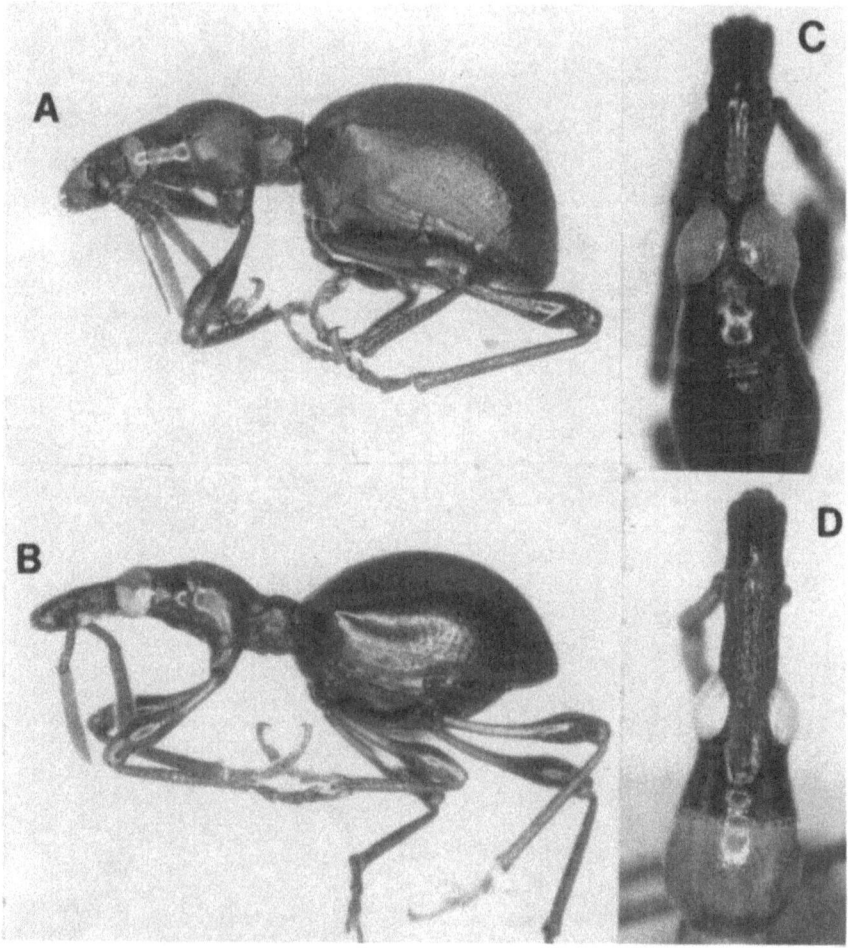

Fig. 2.1. Basic body structure/shape and coloration of non-pest species (A and B), and dorsal view of head (C and D) of *C. formicarius* and *C. puncticollis* groups: (A) *C. cyanescens* from Mali, 12x; (B) *C. glabripennis* (similar to *C. vanderplasi*) from Uganda, 18x; (C) *C. puncticollis* from Nigeria, 25x; and (D) *C. formicarius* from India, 25x.

shape. However, it should be noted that some species keying to this couplet have relatively short femora; but individuals of those species always have extremely short blunt snouts and sometimes have vertically elongate eyes (as in Figure 2.1A).

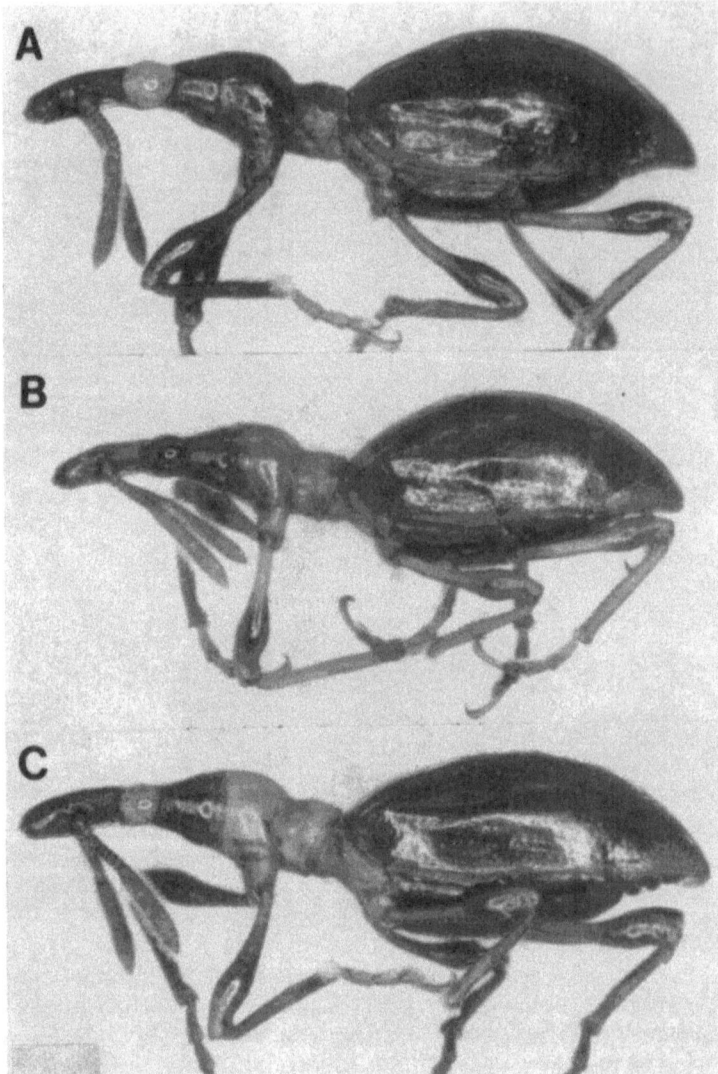

Fig. 2.2. Basic body structure/shape and coloration of representatives of the *C. brunneus* group (all males): (A) dark specimen from Nigeria, 18x (occasionally confused with members of the *C. puncticollis* group, compare with Figure 2.3A); (B) intermediate, moderately bicolored specimen from Nigeria, 18x; and (C) bicolored specimen from Burundi, 18x (in Africa), often confused with members of the *C. formicarius* group (compare with Figures 2.3B and 2.3C).

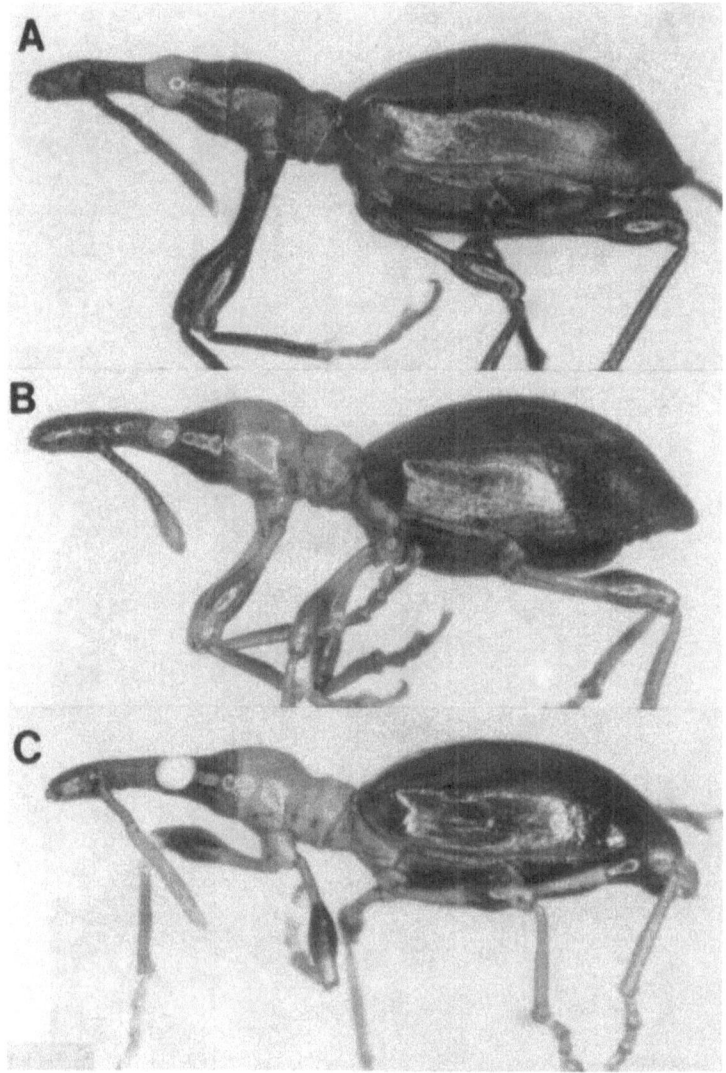

Fig. 2.3. Basic body structure/shape and coloration of representatives of the
C. formicarius and *C. puncticollis* groups: (A) *C. puncticollis* from
Nigeria, 13x; (B) female *C. formicarius* from Hawaii, 18x; and (C) male
C. formicarius from India, 15x.

26

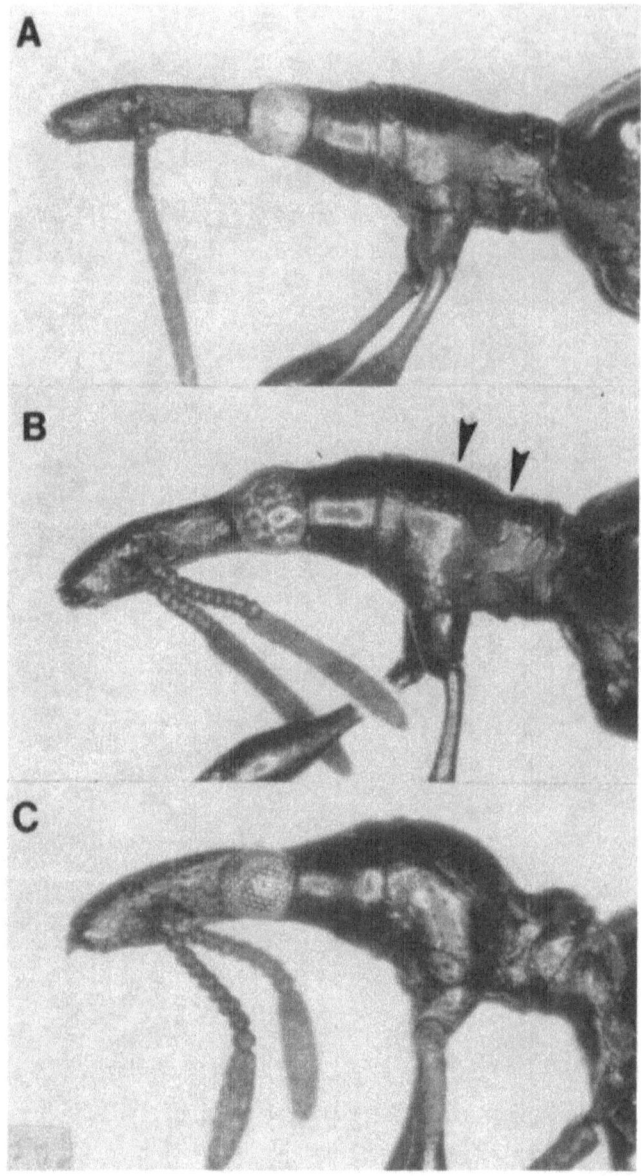

Fig. 2.4. Head and pronotal structure (all males): (A) *C. longicollis* group from Chad, 20x; (B) *C. puncticollis* group from Burundi, 22x, left arrow indicates area of pronotum that is slightly more arched than similar area in Figure 2.4A; right arrow indicates posterior constriction that is more pronounced than in Figure 2.4A; (C) *C. laevigatus* group from Zaire, 40x, notice the small antennal club (compare with Figures 2.2 and 2.3).

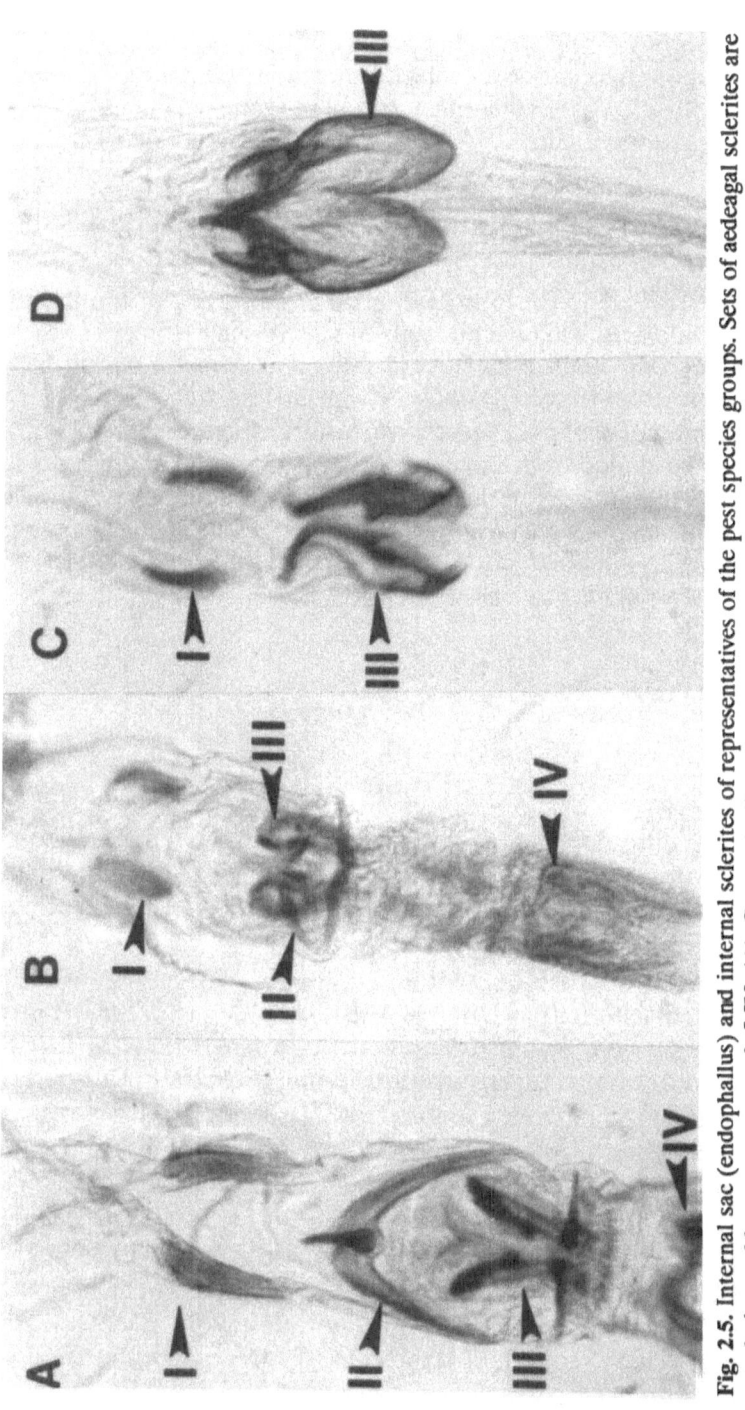

Fig. 2.5. Internal sac (endophallus) and internal sclerites of representatives of the pest species groups. Sets of aedeagal sclerites are designated by roman numerals I-IV: (A) *C. puncticollis* group, *C. puncticollis* group, *C. puncticollis* and *C. nigrocoerulans*; (B) *C. puncticollis* group, *C. compressus* and *C. hovanus*; note reduced size of sclerite II and its position relative to sclerite III; (C) *C. brunneus* group; note absence of sclerites II and IV; and (D) *C. formicarius* group; note absence of sclerites I, II, and IV.

Third, although uniformly black specimens of the *C. brunneus* group (Figure 2.2A) are sometimes confused with specimens of the *C. puncticollis* group (Figure 2.3A), the widely separated eyes (in dorsal view) of the former (as in Figure 2.1D) compared to the narrowly separated eyes in the latter (Figure 2.1C) are always diagnostic. Completely confident separation of bicolored specimens of the *C. brunneus* group (Figure 2.2C) from those of the *C. formicarius* group (Figures 2.3B,C) requires examination of the sclerites of the internal sac of the aedeagus. There is one pair of sclerites in members of the *C. formicarius* group (Figure 2.5D) and two pairs in members of the *C. brunneus* group (Figure 2.5C). Four pairs of aedeagal sclerites are present in members of the *C. puncticollis* group (Figures 2.5A,B).

Finally, specimens of *C. puncticollis* and *C. nigrocoerulans* can be separated confidently from those of *C. compressus* and *C. hovanus* by the distinctly larger size of sclerite II in the former (e.g., compare Figures 2.5A and 2.5B). As stated above, species within each species pair are best separated by distribution.

DISTRIBUTION

Cylas sensu lato

Most species of *Cylas* are confined to Africa-Madagascar (at least 20 species), with a secondary center of diversity in India-Southeast Asia (approximately six species) (see Table 2.1). All undescribed species that I am aware of occur only in Africa (G.W. Wolfe, unpubl. data).

Cylas puncticollis **Group.** Within continental Africa, members of this group collectively have the widest distribution. Specimens were examined from 22°N, south to South Africa and Madagascar (Figure 2.6B). There are notable gaps in distribution in central-southwestern Africa, especially in Angola, Zambia, and Zimbabwe, and in northern Africa, especially in Chad, Niger, and Egypt. It is likely that these gaps are the result of insufficient collecting.

Cylas brunneus **Group.** Members of the *C. brunneus* group are known from West and Central Africa (Figure 2.6A). Although the exact southern limit is not known, they appear to be absent in South Africa and Madagascar. Similarly, the northern limit is not clearly established. I have collected specimens of this species group in association with members of the *C. puncticollis* group in Nigeria, Rwanda, Burundi, and Kenya.

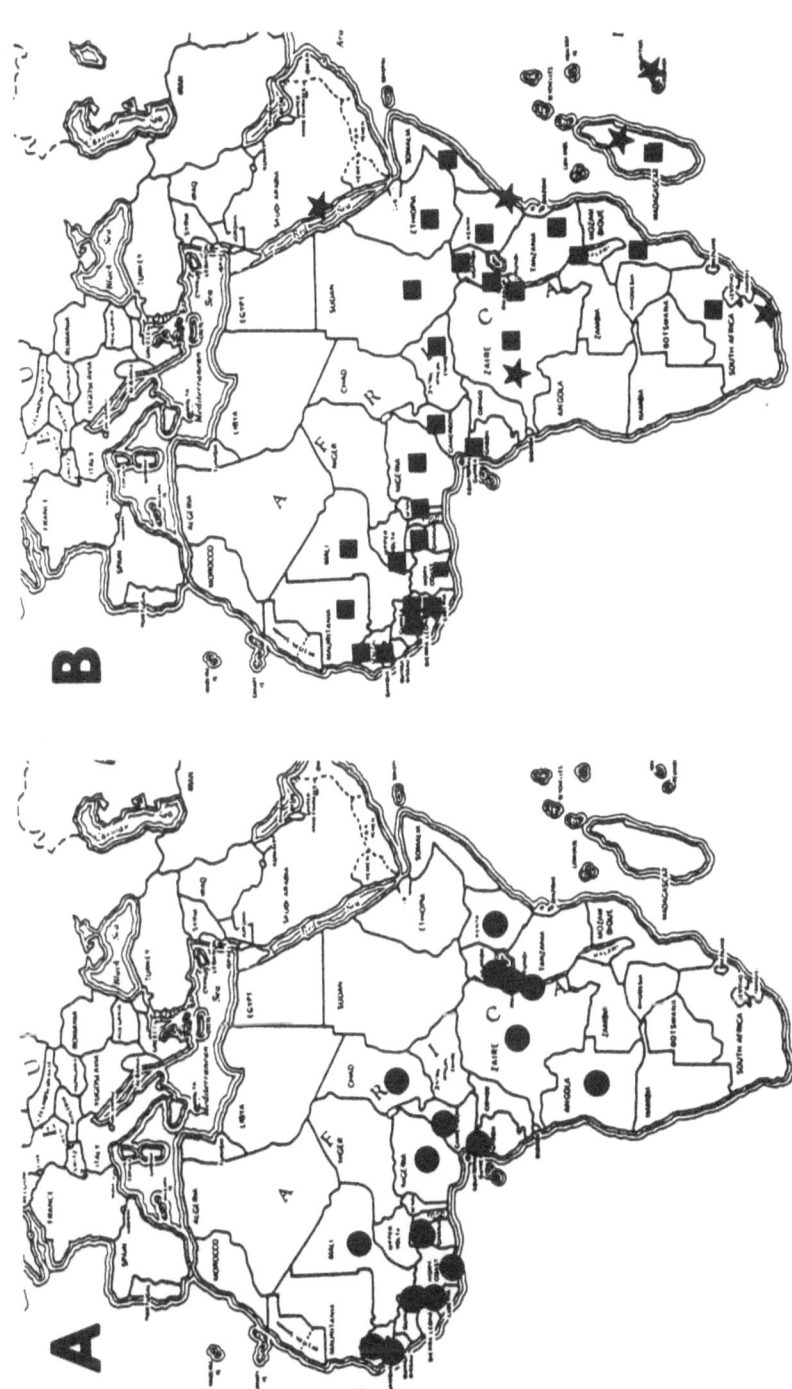

Fig. 2.6. Distribution of pest species groups in and proximate to Africa. Symbols indicate countries in which occurrence is verified by taxonomic analysis: (A) *C. brunneus* group (circles); and (B) *C. puncticollis* group (squares) and *C. formicarius* group (stars).

Cylas formicarius **Group.** As stated above, this is the only species of *Cylas* that occurs circumglobally. However, in continental Africa, this species appears to be rare because substantial numbers of specimens of *C. formicarius* have only been found in eastern Kenya and South Africa (Figure 2.6B). One specimen is known from "Congo da Lemba"; a location that is unclear, but may be located in Central Africa, possibly Zaire or Angola (B. Parker & G.W. Wolfe, unpubl. data).

The apparent rarity of members of the *C. formicarius* group in Africa is probably a real phenomenon and is based on analysis of thousands of museum specimens and extensive field collecting. Efforts to collect specimens of *C. formicarius* even included the use of synthetic sex pheromone (see Chapters 5 and 6) for trapping adult males in southern Ethiopia and western Kenya (B. Parker, pers. comm.). For reasons outlined below, the known occurrences of this species in Africa and the New World are regarded as secondary, agriculturally-related introductions.

PHYLOGENY

Phylogenetic systematics demonstrates genealogical (ancestor-descendent) relationships among organisms (or groups of organisms) through analysis of character transformation series where "advanced" (=apomorphic) characters are distinguished from "primitive" (=plesiomorphic) characters. Through this kind of analysis, monophyletic groups are postulated when members share certain apomorphic characters. A group is considered monophyletic if it includes the common ancestor and all its descendants (see Wiley 1981).

Cylas as a Monophyletic Unit

There are two apomorphic characters shared by all *Cylas* that clearly indicate the genus is monophyletic (i.e., all species are derived from the same common ancestor). Those apomorphies are a large, dorsoventrally elongate, mesothoracic pit (Figure 2.7A) (also noted by Kissinger [1968]), and specialized pits located in the antennal scrobe or groove (Figures 2.7B,C). Other characteristics which occur in all members of *Cylas* but are either plesiomorphic or are difficult to evaluate phylogenetically are a constriction at the posterior end of the prothorax (Figures 2.1-2.4), the 10-segmented antennae, abdomen

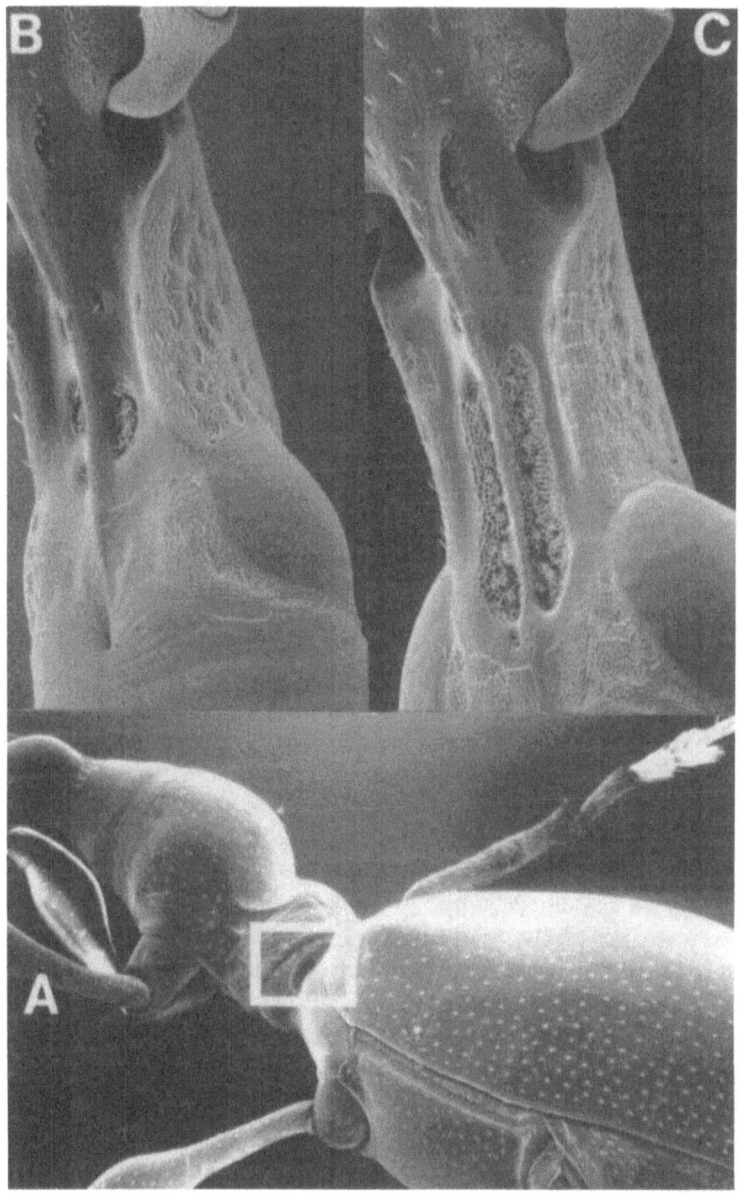

Fig. 2.7. Scanning electron micrography of apomorphies shared by all members of *Cylas*: (A) mesothoracic pit (highlighted in white box) in male of *C. brunneus* from Nigeria, 30x; (B) and (C) specialized pits in antennal scrobe (or groove); (B) female of *C. formicarius* from Texas, 100x; and (C) male of *C. formicarius* from Texas, 100x.

with suture between sternum-1 and sternum-2 reduced, and sternum-2 equal in length to sterna-3 and 4 combined.

Characters Used for Species Group Phylogeny

Three primary character systems were used to generate the species group phylogeny: head structure, body shape and hind femoral length, and aedeagal characters.

Head Structure. The most important characteristic involving head structure is the distance between the eyes in dorsal view. In outgroup representatives, members of the *C. formicarius* and *C. brunneus* groups, and a number of non-pest species, the distance between the eyes is sub-equal to the minimum width of the snout (Figure 2.1D); that condition is considered plesiomorphic. The apomorphic state occurs when the eyes are extremely closely placed (Figure 2.1C), as in members of the *C. puncticollis* group and several non-pest species.

Body Shape/Hind Femora Length. Body shape varies from elongate and slender (Figures 2.2 and 2.3) to oval and globose (Figures 2.1A,B). The more globose body shape is primarily a function of change in elytral-abdominal structure which together become shorter and more inflated. In more globose specimens, the hind femora usually extend to varying degrees past the abdominal-elytral apex.

In members of the *C. formicarius* and *C. brunneus* groups and outgroup specimens, the body shape is slender and the posterior end of the hind femora ends immediately anterior to the apex of the abdomen; therefore, that character state is considered plesiomorphic. A slightly more apomorphic condition is evident in members of the *C. puncticollis* group where, although the overall body shape is slender, the posterior end of the hind femora extends just past the elytral apex. The most apomorphic condition occurs in a number of non-pest species with a distinctly oval and inflated elytral-abdominal shape and hind femora that project well beyond the elytral apex (Figures 2.1A,B).

Genitalic Characters. The best reviews of male weevil genitalic structures and associated terminology are Clark (1977, 1982) and Hamilton (1979). Analysis of genitalic parts for identification and phylogeny centers on characteristics of small sclerites found in the internal sac (endophallus) of the median lobe. There are up to four sets of sclerites (numbered I-IV, Figure 2.5). Sclerite sets I, III, and IV consist of a pair of sclerites each. Sclerite set II is a single curved sclerite that is curved so that it is posteriorly convex. All four sets of sclerites are evident in members of the *C. puncticollis* group (Figures

2.5A,B) and most non-pest species. The fewest internal sclerites are found in members of the *C. formicarius* (one set of sclerites, Figure 2.5D) and *C. brunneus* groups (two sets of sclerites) (Figure 2.5C).

Since no internal sclerites were evident in the outgroup, the occurrence of them in members of *Cylas* is considered apomorphic. Furthermore, I recognize a morphocline with the fewest sclerites (in specimens of the *C. formicarius* group) being least apomorphic and the highest number of sclerites (in members of the *C. puncticollis* group and most globose-shaped species) being the most apomorphic. In this sense, specimens of the *C. brunneus* group are somewhat intermediate.

Species/Species Group Relationships

Although a species level phylogeny cannot be dichotomously resolved fully in *Cylas*, certain patterns regarding the pest species groups are evident. Of primary importance is the observation that the *C. formicarius* and *C. brunneus* groups are very plesiomorphic. Most evidence suggests that the *C. formicarius* group is the sister group to all other members of the genus. However, some evidence suggests that the *C. formicarius* and *C. brunneus* groups are sister groups which together form the sister lineage to all other species.

Members of the *C. puncticollis* group are more closely related to the more globose, short-snouted, non-pest species (Figure 2.8) of *Cylas* than to either the *C. brunneus* or *C. formicarius* groups.

HISTORICAL ZOOGEOGRAPHY

Probably the two most frequently asked taxonomic questions concerning pest species of *Cylas* pertain to *C. formicarius* and address the same issue. The questions are: 1) Where did *C. formicarius* originally evolve? and 2) Is *C. formicarius* native to the New World?

These questions can only be answered if species distributions and pertinent historical geological facts are analyzed in relation to phylogenetic evidence. Crowson (1981) stated that curculionoid fossils are known from the late Jurassic period (approximately 150 million years before present [mybp]). At that time, India, Madagascar, Africa, South America, Antarctica, and Australia together formed Gondwanaland (Cracraft 1974). Subsequently, during the Cretaceous period, South America, Antartica, and Australia had separated from Africa, India, and Madagascar. As recently as the Upper Cretaceous

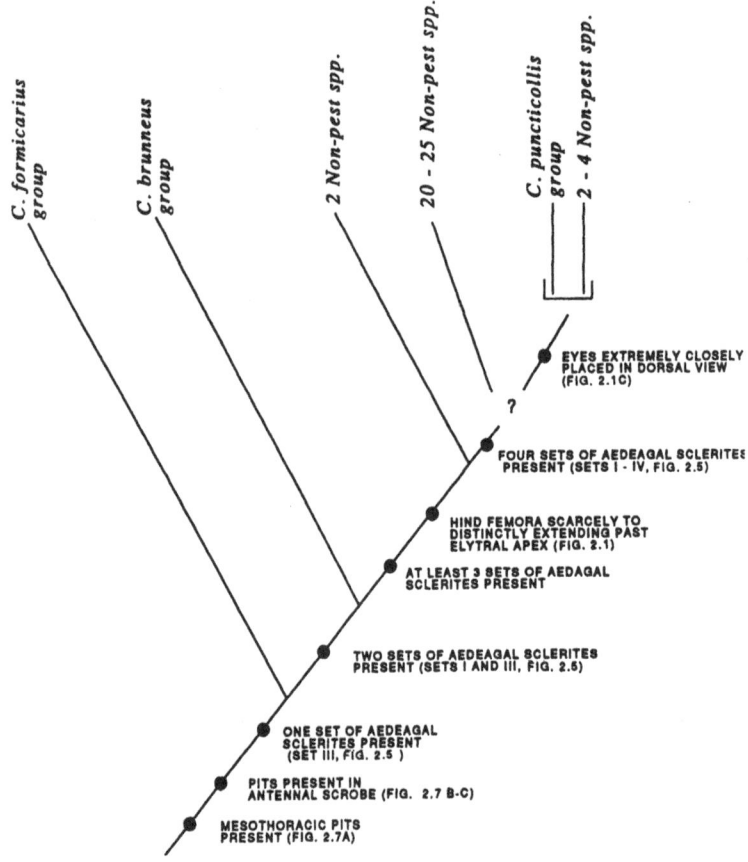

Fig. 2.8. Phylogenetic relationships of the pest species groups of *Cylas*.

period (80-90 million years before present), these latter three land masses were still almost contiguous. I hypothesize that the common ancestor to *Cylas* evolved in Africa/India by about the mid-Jurassic and had a distribution spanning Africa/India by the Mid- to Upper Cretaceous period (Figure 2.9A).

If the *C. formicarius* group is the most plesiomorphic lineage of *Cylas*, then one must assume that when India and Africa were separated during the late Cretaceous period, two populations of the *Cylas* ancestral stock were isolated: one in Africa and the other in India (Figure 2.9B). The Indian population represented the ancestor to *C. formicarius* and the African population represented the ancestor to all other species of *Cylas*, including members of the *C. brunneus* and *C. puncticollis* groups. Since Madagascar and India may have remained

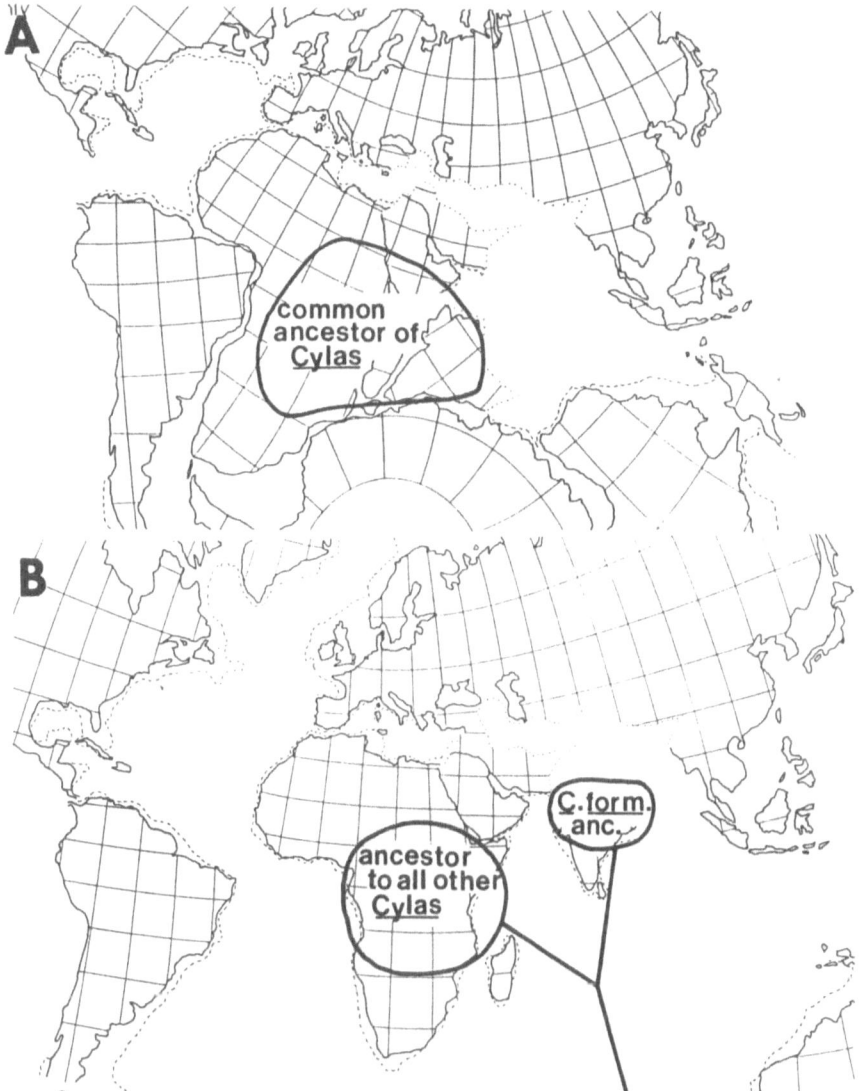

Fig. 2.9. Illustration of the hypothesized effect of continental drift on distribution and evolution of the common ancestor of *Cylas* assuming *C. formicarius* group is the sister group to all other members of *Cylas* (see text for alternative explanation): (A) hypothesized distribution of common ancestor, mid to late Mesozoic (approximately 150 million years before present); and (B) divergence of *C. formicarius* group ancestor from main stock as a result of separation of India from Africa in late Mesozoic to early Cenozoic (approximately 100 to 80 million years before present).

proximate after they separated from Africa, it is possible that *C. formicarius* originated and then dispersed throughout both regions before they separated entirely in the early Cenozoic period (Cracraft 1974).

If the *C. formicarius* and *C. brunneus* groups are sister lineages, a vicariant hypothesis is still reasonable if it is assumed that some diversification within *Cylas* occurred before India/Madagascar separated from Africa. At the very least, the common ancestors to the *C. formicarius* and *C. brunneus* lineage, the species of the *C. puncticollis* group, and most other more globose shaped, non-pest species already existed. Under this set of assumptions, the separation of India from Africa simultaneously isolated populations of several lineages. While this latter hypothesis is more complicated, it does provide a non-dispersal mechanism for the occurrence of the non-pest species that are indigenous to India/Southeast Asia.

Under both phylogenetic hypotheses, *C. formicarius* may have dispersed to southeast Asia or even portions of Indonesia and the South Pacific, after India collided with Asia in the early to mid-Cenozoic period. However, neither phylogenetic/zoogeographic hypothesis suggests that *C. formicarius* is native to Africa, portions of southeastern North America, or the Neotropics. Further evidence for this conclusion is discussed below.

In summary, there is evidence that there were two centers of origin for pest species of *Cylas*: one in Africa (*C. puncticollis* group, *C. brunneus* group and numerous non-pest species) and a second in India (*C. formicarius* and a few non-pest species).

DISTRIBUTIONS OF *Cylas formicarius* AND *Ipomoea batatas*

The evidence that *I. batatas* originated in northwestern South America is substantial if not overwhelming (Austin 1978, 1988, Austin *et al.* 1990). The crop was dispersed to New Zealand and portions of Polynesia before the arrival of Europeans (500 BC - 1616) and was established on all continents by 1674. Austin *et al.* (1990) speculated that since *C. formicarius* was an Old World species and *I. batatas* originated in the New World, that the weevil-plant association was secondarily established as the crop was dispersed. Evidence provided herein (see above) indicates that the first contact between *I. batatas* and *C. formicarius* must have occurred in India (possibly, but less likely in Southeast Asia). According to Austin (1988), sweet potato was present in India around 1509.

If the above insect phylogenetic-zoogeographic hypotheses are true, then the occurrence of *C. formicarius* in Africa and the New World must be the result of accidental, rather recent introductions (probably via infested storage roots).

Information on patterns of Indian immigration and known occurrence of *C. formicarius* are consistent with this phylogenetically- and zoogeographically-based hypothesis of secondary introduction into Africa and the New World.

Africa-India

Contact between Indian and North African civilizations extends as far back as 3,000 BC. However, since the sweet potato was not introduced into India until the early 1500's (Austin 1988), introduction of *C. formicarius* into Africa from India prior to 1500 was not likely. Furthermore, as recently as 1873 there were less than 7,000 Indians in East Africa-Zanzibar (Gregory 1971).

However, between 1896 and 1922, the British began major railroad construction projects in Kenya and Uganda. For this reason, approximately 39,000 indentured Indian servants were brought to Africa as laborers between 1896 and 1922. In South Africa, the first Indians arrived about 1860; however, by 1912, the population had grown to 150,000; 120,000 of them were concentrated in the Natal Province.

The fact that most verified records of *C. formicarius* are from eastern Kenya and the Natal region of South Africa (despite significant attempts to find this species elsewhere [B. Parker & G.W. Wolfe, unpubl. data]) is consistent with the above information on immigration patterns. Also, although *C. formicarius* was described from India in 1798, the oldest specimen record from Africa that I examined was collected in 1914 from Natal Province, South Africa.

In summary, the distribution of *C. formicarius* in Africa is sporadic and correlated with regions significantly influenced by late 19th Century immigration from India. Thus far, I have examined and identified specimens of the *C. puncticollis* and the *C. brunneus* group from more than 60 localities in Central Africa and found *C. formicarius* only once (Congo da Lemba). This is true despite the fact that where members of *C. formicarius* group do occur, they usually occur in association with those of *C. puncticollis* or *C. brunneus* groups. This information, in light of phylogenetic/zoogeographic data, suggests that members of the *C. formicarius* group are not native to Africa.

38

It is likely that members of the *C. formicarius* group already are established all along the east coast of Africa by dispersal north along the coast of South Africa into Mozambique and south into Tanzania from Kenya. Based on analysis of available material, *C. formicarius* has not spread widely into the interior of Central Africa, perhaps because it is not able to adapt to regional conditions, cannot compete with native pest species, or simply because it has not had sufficient time.

New World-India

According to Austin (1988), the sweet potato was introduced as a crop into North America at least by 1648. However, a vexing dilemma has been that the weevil was not recorded in North America until 1875.

The answer to this dilemma appears to be correlated with immigration patterns in South America and the Caribbean. In the early 1600's, Guyana was occupied by the Dutch who soon established an extensive system of sugar plantations that were supported primarily by African slaves. The British gained control of Guyana in 1781 and abolished slavery in 1838. To compensate, plantation owners switched to an indentured labor system and brought 31,628 Portuguese, 14,000 Chinese, and 238,960 Indians to Guyana between 1838 and 1917. Trinidad also received thousands of immigrants from India (Parry *et al.* 1987, Singh 1988).

It is likely that *C. formicarius* was introduced into the New World by infested storage roots brought by indentured Indians, probably in the mid-nineteenth century. The first occurrence of *C. formicarius* in the United States in New Orleans in 1875 was probably an introduction from the West Indies in the Caribbean.

CONCLUSIONS

Individuals of nine species of *Cylas* are considered pests of sweet potato. These nine species are classified into one of three monophyletic species groups: the *C. formicarius* group, the *C. brunneus* group, and the *C. puncticollis* group. These three species groups can be separated by the key in Table 2.2. Completely confident identification requires male genitalic comparisons.

The most intractable systematic problems in *Cylas* involve members of the *C. formicarius* and *C. puncticollis* groups. It is unlikely that conventional taxonomic procedures will sort out all of the problems. Therefore, future systematic work on pest species will have to include comprehensive morphometric and multivariate approaches, biochemical systematic studies, karyological investigations, and perhaps DNA analysis for more conclusive answers.

The *C. formicarius* group or the *C. formicarius* and the *C. brunneus* group(s) represent the sister lineage to all other species of *Cylas*. The *C. puncticollis* group is more closely related to non-pest, globose-shaped species than to either the *C. formicarius* or *C. brunneus* groups. *Cylas formicarius* is not considered native to the New World or Africa. This conclusion is based on phylogenetic, zoogeographic, and extant distributional evidence, as well as on immigration patterns of indentured Indians from the mid 1800's to about 1920.

In the future, the above information will become increasingly important to pest management specialists because accurate identification is the first step in agricultural research on pest species of *Cylas*. Furthermore, distributional and phylogenetic evidence may help plant breeders target development and subsequent distribution of resistant plant varieties. Finally, based on predicted centers of origin/evolution, one can predict that parasitoids, predators, and other natural enemies of *C. formicarius* will be most abundant/diverse in India (possibly southeast Asia) and that the greatest diversity of natural enemies for members of the *C. brunneus* and *C. puncticollis* groups probably occurs in Central Africa.

ACKNOWLEDGMENTS

This article is New Jersey Experiment Station Publication No. F-08410-01-90 supported by state funds. This research was supported by an US-AID special constraints grant (86-CRSC-2-2798). I thank R.K. Jansson for his comments and suggested improvements on early drafts of this chapter, and D. Whitehead for initially pointing out to me the taxonomic problems in *Cylas* and their relevance to control procedures. B. Parker, S.K. Hahn, M. Alvarez, and R. Oberprieler provided valuable assistance and/or advice on collecting specimens of *Cylas*. Many hours of curation of *Cylas* specimens by S. Oygur and S. Kung are gratefully acknowledged. I also thank C. Bevere for preparing the manuscript.

40

REFERENCES

Austin, D.F. 1978. The *Ipomoea batatas* complex - I. Taxonomy. Bull. Torrey Bot. Club. 105:114-129.

Austin, D.F. 1988. The taxonomy, evolution, and genetic diversity of sweet potatoes and related wild species, pp. 27-60. *In* Exploration, maintenance, and utilization of sweet potato genetic resources. International Potato Center, Lima, Peru.

Austin, D.F., R.K. Jansson & G.W. Wolfe. 1990. Convolvulaceae and *Cylas*: a proposed hypothesis on the origins of this plant:insect relationship. Trop. Agric. (In press).

Blatchley, W.S. & C.W. Leng. 1916. Rhynchophora or weevils of Northeastern America. The Nature Publishing Co., Indianapolis.

Brues, C.T., A.L. Melander & F.M. Carpenter. 1954. Classification of insects, keys to the living and extinct families of insects, and to living families of other terrestrial arthropods. Cambridge, London.

Burgeon, L. 1936. Les *Cylas* du Congo Belge (Col.: Curculionidae). Rev. Zool. Bot. Africaines 23:421-518.

Chalfant, R.B., R.K. Jansson, D.R. Seal & J.M. Shalk. 1990. Ecology and management of sweet potato insects. Annu. Rev. Entomol. 35:157-180.

Champion, G.C. 1914. Dans la faune des Iles Seychelles. Trans. Linn. Soc. London. 16:450. (*In* Hustache [1924]).

Clark, W.E. 1977. Male genitalia of some Curculionoidea (Coleoptera): musculature and discussion of function. Coleop. Bull. 31:101-115.

Clark, W.E. 1982. Classification of the weevil tribe Lignyodini (Coleoptera, Curculionoidea, Tychiinae), with revision of the genus *Plocetes*. Trans. Am. Entomol. Soc. 108:11-151.

Cracraft, J. 1974. Continental drift and vertebrate distribution. Annu. Rev. Ecol. Syst. 5:215-261.

Crowson, R.A. 1955. The natural classification of the families of Coleoptera. Nathaniel Lloyd, London. (Reprinted 1967, E. W. Classey Ltd., Middlesex, England).

Crowson, R.A. 1981. The biology of the Coleoptera. Academic Press, London.

DeVries, J.D., J.D. Ferwerda & M. Flach. 1967. Choice of food crops in relation to actual and potential production in the tropics. Neth. J. Agric. Sci. 15:241-248.

Erwin, T.L. 1970. A reclassification of bombadier beetles and a taxonomic revision of the North and Middle American species (Carabidae: Brachinida). Quaest. Entomol. 6:4-215.

Fabricius, J.C. 1792. Entomologia systematica emendata et aucta. Secundum classes, ordines, genera, species, adjectis synonimis, locis, observationibus, descritionibus. Vol. 1. Hafninae.

Fabricius, J.C. 1798. Supplementum entomoligiae systematicae. Hafninae.

Fabricius, J.C. 1801. Systema eleutheratorum secundum ordines, genera, species: adjectis synonymis, locis, observationibus. Vol. 2. Kiliae.

FAO. 1984. FAO production Yearbook 1983. Food and Agricultural Organization, Rome.

Gregory, R.G. 1971. India and East Africa, a history of race relations within the British empire. Clarendon Press, Oxford.

Hamilton, R.W. 1979. Taxonomic use of endophallic structures in some Attelabidae and Rhynchitidae of America, North of Mexico (Coleoptera: Curculionoidea), with notes on nomenclature. Ann. Entomol. Soc. Am. 72:29-34.

Hung, A. 1985. Chromosomal polymorphism in the sweet potato weevil, *Cylas formicarius*. Cytologia 50:769-772.

Hustache, A. 1924. Synopsis des Curculionides de Madagascar. Bull. l'Acad. Malagache. Nouvelle serie 7:275-277.

Jones, A. 1970. The sweet potato - today and tomorrow, pp. 3-6. *In* D. L. Plucknett (ed.). Proceedings of the 2nd symposium for the International Society of Tropical Root Crops. Vol. 1., University of Hawaii, Honolulu.

Kay, D.E. 1973. Root crops. Crop and production digest. No. 2. Tropical Products Institute, Foreign Commonwealth Office, United Kingdom.

Kemner, N.A. 1924. Der batatenkafer (*Cylas formicarius* [F.]) auf Java und den benachbarten Inseln Ostindiens. Z. Angew. Entomol. 10:398-435.

Kissinger, D.G. 1968. Curculionidae subfamily Apioninae of North and Central America with reviews of the world genera of Apioninae and world genera of *Apion* Herbst (Coleoptera). S. Lancaster, Massachusetts (Privately published).

Latrielle, P.A. 1802. Histoire naturelle, generale et particuliere, des crustaces et des insectes. Ouvrage faisant suite a l'Histoire naturelle generale et particuliere, composee par Leclerc de Buffon, et redigee par C. S. Sonnini, Member de Plusieurs Societes Savantes, Vol. 3. Paris.

LeConte, J.L. & G.H. Horn. 1876. The Rhynchophora of America, north of Mexico. Proc. Am. Phil. Soc. 15:1-455.

Marshall, G.A.K. 1953. On a collection of Curculionidae (Coleoptera) from Angola. Companhia de Diamantes de Angola Servicos Culturais Lisbon 16:99-119.

Mayr, E. 1970. Populations, species and evolution. Harvard University Press, Cambridge, Massachusetts.

Morimoto, K. 1976. Notes on the family characters of Apionidae and Brentidae (Coleoptera), with a key to two related families. Kontyu, Tokyo 44:469-476.

O'Brien, C.W. & G.J. Wibmer. 1982. Annotated checklist of the weevils. *In* Curculionidae *sensu lato* of North America, Central America, and the West Indies (Coleoptera: Curculionidae). American Entomology Institute, Ann Arbor, Michigan.

Parry, J.H., P.M. Sherlock & A.P. Maingol. 1987. A short history of the West Indies. St. Martin's Press, New York.

Pierce, W.D. 1918. Weevils which affect Irish potato, sweet potato, and yam. J. Agric. Res. 12:601-612.

Pierce, W.D. 1940. Studies of the sweet potato weevils of the subfamily Cyladinae. Bull. S. Calif. Acad. Sci. 39:205-228.

Pollard, G. 1984. Insect pests of sweet potato in the Caribbean with particular reference to the borer (*Megastes grandalis* Guen.), pp. 147-152. *In* D. Dolly (ed.). Root crops in the Caribbean. Proceedings Caribbean regional workshop on tropical root crops. Faculty of Agriculture, University of the West Indies, St. Augustine, Trinidad.

Rafinesque, C.S. 1815. Analyse de la natura ou tableau de l'univers et des corpo organises. Palermo.

Risbec, J. 1947. Les charancons (*Cylas*) nuisbles aux potatoes douces. Agron. Trop. 2:375-398.

Rosen, D.E. 1978a. The importance of cryptic species and specific identifications as related to biological control, pp. 23-35. *In* J.A. Romberger, R.H. Foole, L. Knutson & P. Lentz (eds.). Biosystematics in agriculture. J. Wiley, New York.

Rosen, D.E. 1978b. Vicariant patterns and historical explanation in biogeography. System. Zool. 27:159-188.

Singh, C. 1988. Guyana, politics in a plantation society. Praeger, New York.

Subramanian, T.R. 1957. Description and life history of a new weevil of the genus *Protocylas* from Coimbatore. Indian J. Entomol. 19:204-213.

Swofford, D.L. 1985. PAUP, phylogenetic analysis using parsimony, version 2.2. Privately Published Manual. Urbana, Ilinois

Talekar, N.S. 1988. Insect pests of sweet potato in the tropics. *In* Proceedings of the 11th International Congress of Plant Protection. Manila, Philippines (In press).

Voss, E. 1966. Über athiopische und madagassische Apioninen (Col.), vorweigend aus den Sammlimgen des Museums G. Frey. Entomol. Arb. Aus dem Mus. G. Frey. Bond 17:13-325.

Wagner, H. 1910. Curculionidae: Apioninae. Pars 6. Coleopterorum Catalogues. S. Schenking (ed.). W. Junk, Berlin.

Wiley, E.O. 1981. Phylogenetics: the theory and practice of phylogenetic systematics. J. Wiley, New York.

Wolfe, G.W. & R.E. Roughley. 1990. A taxonomic, phylogenetic, and zoogeographic analysis of *Laccomis* DesGozis (Coleoptera: Dytiscidae) with the description of a new tribe of Hydroporinae. Quaest. Entomol. (In press).

Wyninger, R. 1962. Pests of crops in warm climates and their control. Verlag Fur Recht Und Gesellschaft Ag., Basel, Switzerland.

3. Associations Between the Plant Family Convolvulaceae and *Cylas* Weevils

Daniel F. Austin
Department of Biological Sciences
Florida Atlantic University
Boca Raton, Florida 33431 U.S.A.

The sweetpotato weevil, *Cylas formicarius* (Fabricius), has been associated for certain with the sweet potato, *Ipomoea batatas* (L.) Lam., since the nineteenth century. The early history of both organisms is obscure, but they probably have been together since the 1500's. To date, the weevil has been reported on 35 species of Convolvulaceae distributed among seven genera in six tribes. *Cylas formicarius* is a recent arrival in the New World, having originated in the Old World. By contrast, the sweet potato originated in the New World, and arrived recently in the Old World. This short period of association between sweet potato and *C. formicarius* may reflect a recent adaptation by the weevil. This chapter presents probable origins of the association between *C. formicarius* and Convolvulaceae and phylogenies for the *Ipomoea* sections and the tribes within the Convolvulaceae.

SWEET POTATO AND WEEVILS -- NATIVITY

The sweet potato is the species most often associated with the weevils of the genus *Cylas*. Damage caused on this food crop is so extensive that these weevils have become known as the "sweet potato weevils." Before dealing with the specific relationships between the weevils and other members of the Convolvulaceae, it is important to establish the geographic regions of origins of the sweet potato and *Cylas*.

All current data indicate that the sweet potato originated in or near northwestern South America (Austin 1988). The species was first cultivated in Peru as early as 8,000 B.C (Engel 1970, Ugent & Peterson

1988, Ugent *et al.* 1981, 1983, Yen 1974), even though one author thought it was not cultivated until 3,000 B.C. (O'Brien 1972). Prior to the arrival of Columbus in the New World, the sweet potato had been dispersed with people throughout the tropics, but not into the temperate zones of either North or South America (Brand 1971).

We have a fairly complete record of the disperal of the crop into various parts of the world from the Americas (Austin 1988). We know, for example, that the sweet potato reached Europe with the Spanish by 1493; that the Portuguese took it to Africa and India by the first decade of the 1500's; and that it was in China by 1594. Perhaps because the crop was not as important to the northern Europeans, it was not recorded in North America until 1648. Yen (1974, 1982) showed that the cultigen was definitely present in New Zealand between 500 B.C. and 1419 A.D., and that the crop had probably also been taken to the other two corners of Polynesia before Columbus.

Thus, in pre-Columbian times, the crop species was found in the American tropics and the Polynesian triangle. It was not until well within the historic period of the 1490's and later that the species was introduced from the Americas into Europe, Africa, and continental Asia.

Several species of *Cylas* weevils are considered pests of sweet potato in various parts of the world. The most damaging species is *C. formicarius*. Distribution maps (Anonymous 1970, Sorensen 1984) indicate that the weevils are present throughout the Old World tropics. The maps also show that weevils are absent from large parts of the New World, particularly most of South America.

Cladistic analysis of all the *Cylas* species showed their probable phylogenetic relationships (Austin *et al.* 1990). This analysis along with other data led Wolfe (see Chapter 2) to conclude that *C. formicarius* is a primitive species. Wolfe (Chapter 2) further believes that the weevil originated on the Indian subcontinent. He hypothesized that the genus was dispersed from the Indian region into other parts of the Old World, particularly Africa. The origin of the genus was long enough ago that numerous species evolved in the Old World. No species, however, is native to the New World. Much more recently, the weevil was carried by man into the New World.

According to Newell (1917) and Cockerham *et al.* (1954), *C. formicarius* was first recorded in the United States in 1875 (Louisiana) and 1878 (Florida). Weevils are thought to have been brought to the United States from the West Indies, probably Cuba. The arrival of the weevils in the West Indies has apparently not been documented.

Austin *et al.* (1990) concluded that the sweet potato was not associated with the weevil until after Europeans began to spread both around the world. Sweet potatoes were taken from the New World to the Old World where they became associated with the weevils. Both were spread from the point of contact as they were taken from port to port around the world. As the crop was established, so too was the weevil. Native and introduced alternate host plants probably aided in the establishment of the insect.

Cylas formicarius could not have been involved with the sweet potato until after the Europeans began to spread both around the world; therefore, the weevil must have originally been associated with some other host plant(s). A survey of the Convolvulaceae to identify the species fed on by weevils might provide information on the plant(s) that have a long historical association with the weevil. To accomplish this survey, it is necessary to introduce the classification of the family.

CLASSIFICATION OF THE CONVOLVULACEAE

Species of *Ipomoea* are known to be important host plants of *C. formicarius* (Austin *et al.* 1990). This plant genus is also the largest alliance of species within the family Convolvulaceae, and contains several hundred species. The largest numbers of species are concentrated in the New World tropics and tropical Africa, with smaller centers of diversity in Asia and Australia.

Following classifications proposed for Melasia (the area between Malaysia and New Guinea including the Philippines) (Ooststroom 1953) and Africa (Verdcourt 1957), Austin (1975, 1979, 1980) treated the American taxa within *Ipomoea*. Preliminary estimates of numbers of species are given below. Three subgenera, 13 sections, and 17 series are recognized (Austin 1979, 1980).

The thirteen sections represent several different pollination syndromes. Past classification has been confounded, in part, by misinterpretation of frequent convergence due to co-evolution with pollinators. For example, bird pollination has arisen independently in at least three lineages and moth pollination in at least two (Austin *et al.* 1990). Cladistic analysis led to a probable phylogeny (Figure 3.1).

There are three groups of tribes within the Convolvulaceae (Figure 3.2). Weevils reportedly use members of each of these three groups; six tribes are used by weevils (Table 3.1), but the groups of most importance include the tribes Argyreieae, Ipomoeeae, and Merremieae. Of these, Ipomoeeae is most important to weevils.

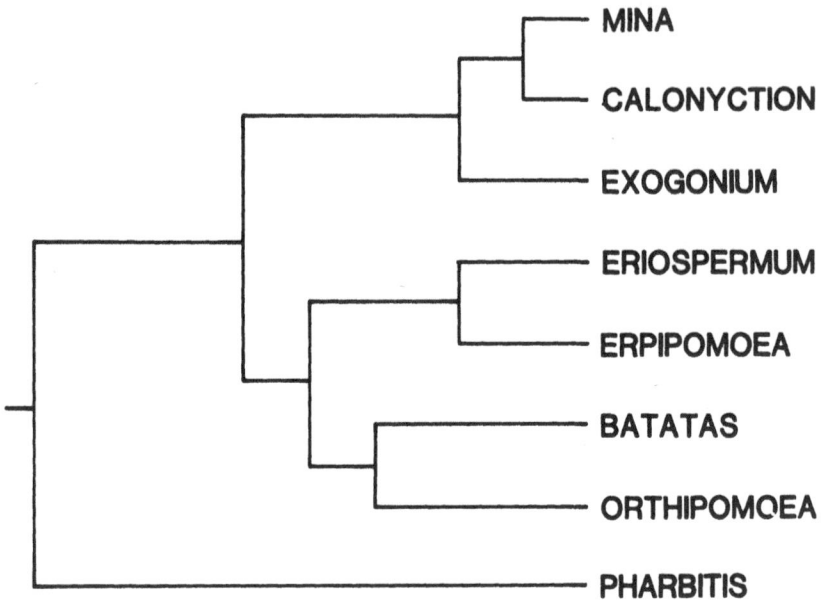

Fig. 3.1. Probable phylogeny of the sections within the genus *Ipomoea*.

Cylas formicarius: CONVOLVULACEAE ASSOCIATION

There are 24 published reports on the occurrence of *C. formicarius* on members of the Convolvulaceae (Austin *et al.* 1990). In total, the weevil has been reported on 35 plant species (Table 3.2); a revised and corrected host plant list was recently published (Austin *et al.* 1990).

Of the thirteen recognized sections of *Ipomoea*, five are confined to Africa or Asia. No further comment will be made on these sections because no weevils have been found on them. This probably does not mean that weevils are not using them; it probably reflects a lack of data. Eight sections are found in the Americas and weevils have been reported on six of these. Weevils have also been recorded on an Old World species in a seventh section (section *Eriospermum*).

Not only is the weevil found on seven sections within all three subgenera of *Ipomoea*, but it also occurs on six other genera (*Calystegia*, *Cuscuta*, *Dichondra*, *Jacquemontia*, *Merremia*, and *Stictocardia*) in five other tribes of Convolvulaceae. This wide usage of plant genera shows that there is little specificity between *Cylas* and its hosts within the Convolvulaceae.

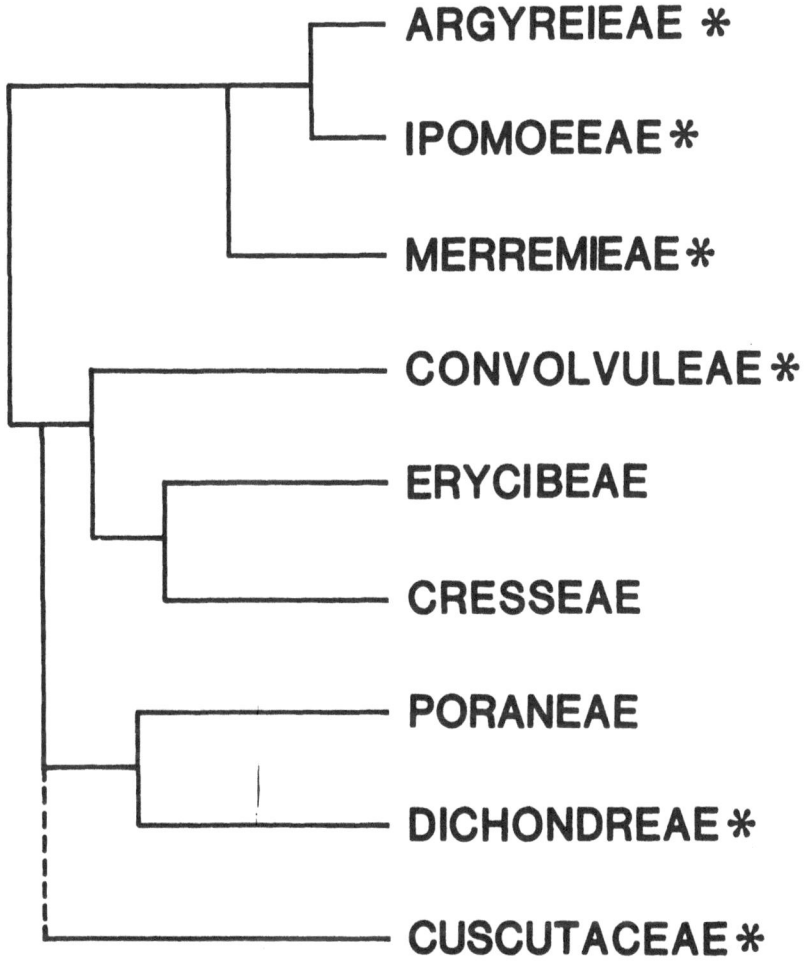

Fig. 3.2. Probable phylogeny of the tribes within the plant family Convolvulaceae. Asterisk (*) denotes that the tribe is used by *C. formicarius*.

Such limited data, involving 35 out of well over 1,000 species, preclude many generalizations. There appears to be a wide acceptance of Convolvulaceae as host-plants, and the data suggest a recent shift by the weevil to cultivated sweet potato. If there was an original close relationship between *Cylas* and any particular member(s) of the genus *Ipomoea* or other genera, it seems to have been complicated by recent historical events. More surveying for potential host plants of *Cylas* spp. is needed to draw firm conclusions.

Table 3.1. A listing of the tribes within Convolvulaceae used by C. *formicarius.*

Tribe	Description, subgenus, and section
Ipomoeae	Contains three or four genera and over 500 species.
	I. *Ipomoea* Linnaeus (1753) subgenus *Ipomoea*; the subgenus has approximately 12 species in the Americas and about 15 in East Africa.
	I.1. Section *Pharbitis* (Choisy) Grisebach (1864); at least a dozen species in the New World, some of which have become pantropical through introduction.
	II. *Ipomoea* subgenus *Quamoclit* (Moench) Clarke (1883)
	II.1. Section *Batatas* (Choisy) Grisebach (1864); twelve species and some apparent hybrids (Austin 1977a, 1978, 1983, 1988), originally American. Most are now introduced into the tropics around the world. One species is Asian (Indian and Pacific Oceans). Three Australia taxa were tentatively placed in this section; however, these probably belong to another section (D. Austin & R. Johnson, unpubl. data).
	II.2. Section *Mina* (Cervantes) Grisebach (1864); this section contains 13 American species (O'Donell 1959); some of these are now pantropical through introduction.
	II.3. Section *Calonyction* (Choisy) Grisebach (1864); this American group contains four species (Gunn 1972).
	II.4. Section *Orthipomoea* Choisy (1845); this section contains series in the Old World and in the Americas; there are approximately 200 species in the section.
	III. *Ipomoea* subgenus *Eriospermum* (H. Hallier) Verdcourt (1963)
	III.1. Section *Eriospermum*; this section has approximately 200 species in the Americas and 50 or more in East Africa (Verdcourt 1963, Austin 1977b).
	III.2. Section *Erpipomoea* Choisy (1838); pantropical, comprising about 20 species.
Argyreieae	Contains about four genera.
Merremieae	Contains about six genera.
Convolvuleae	Contains about six genera.
Dichondreae	Contains four genera.
Cuscuteae	Contains over 100 species.

Table 3.2. Species of Convolvulaceae reported as host plants of *C. formicarius*.

Genus/subgenus	Section and description	Host plant species
Ipomoea subgenus *Ipomoea*	section *Pharbitis*: American species; spread into Old World tropics.	*I. hederacea, I. indica, I. nil, I. purpurea*
Ipomoea subgenus *Quamoclit*	section *Batatas*: American species; spread into Old World tropics.	*I. batatas, I. cordato-triloba, I. lacunosa, I. triloba*
	section *Mina*: American species; spread into Old World tropics.	*I. quamoclit, I. hederifolia*
	section *Calonyction*: American species; spread into Old World tropics.	*I. alba, I. turbinata*
	section *Orthipomoea*: An Asian/African species is the one species known to be used.	*I. sinensis*
Ipomoea subgenus *Eriospermum*	section *Eriospermum*: American species; one species spread into Old World tropics. Species native to Old World tropics have not been reported to be hosts of *C. formicarius*.	*I. horsfalliae, I. macrorhiza, I. pandurata, I. sagittata, I. setosa, I. tuboides*
	section *Erpipomoea*: Old World and pantropical species; some spread into New World. Old World species; some spread into New World.	*I. aquatica, I. barlerioides, I. cairica, I. imperati, I. obscura, I. pes-caprae, I. sepiaria, I. wrightii*
Stictocardia	American species; spread into Old World.	*S. tiliifolia*
Merremia	American species; one spread into Old World.	*M. dissecta*
Jacquemontia	Probably American species, but amphi-Pacific.	*J. curtissii, J. tamnifolia*
Calystegia	*C. formicarius* found on only American species.	*C. sepium, C. soldanella*
Dichondra	Cosmopolitan species; *C. formicarius* found on American species only.	*D. carolinensis*
Cuscuta		*Cuscuta* sp.

Of the thirteen taxonomic groups on which weevils have been recorded, all but three are of American origin (Table 3.2). These American taxa will not be considered further since the weevils are known to have originated in the Old World. The other three predominantly Old World groups are *Ipomoea* section *Erpipomoea*, *Ipomoea* section *Orthipomoea* and the genus *Stictocardia*. *Cylas formicarius* fed on *Stictocardia* under laboratory conditions (Sutherland 1986), probably indicating a forced relationship. One species of *Ipomoea* section *Orthipomoea* in Asia is used by *C. formicarius*, seven (or eight) species in *Ipomoea* section *Erpipomoea* are also used by *C. formicarius*.

Regarding section *Erpipomoea*, it is pertinent to note that Boyden (1922) recorded that "....*I. pes-caprae*....[is] preferred even to the sweet potato...." Boyden (1927) later speculated that in the southeastern United States, the same species was the major reservoir of *C. formicarius* and enhanced its spread along the coasts. This species is still a suitable food plant in southern Florida (R.K. Jansson, unpubl. data) and in coastal North Carolina (K.A. Sorenson, unpubl. data). Not all reports, however, indicate such a strong preference for *I. pes-caprae* (Kemner 1924).

The goat's-foot or railroad vine, *I. pes-caprae*, is a pantropical member of section *Erpipomoea*. Its subspecies *pes-caprae* is native to the Indian ocean region, where the weevils are thought to have originated (see Chapter 2). Moreover, this is the only host plant of *C. formicarius* that was pantropical in pre-Columbian times. Accordingly, Austin *et al.* (1990) hypothesized that the most likely plant taxon to have a long-term association with *C. formicarius* is *Ipomoea* section *Erpipomoea*.

Because all of the species of *Cylas* originated in the Old World, *Ipomoea* section *Erpipomoea* should be searched for possible relationships between insects and plants. Current data suggest that other taxonomic groups are less likely to have been primary participants in the original association. No traits have been found, morphological or otherwise, that attract the weevil to some of the species and not to others (Figures 3.3 and 3.4). Phytochemical cues are probably involved and this is a promising research topic (see Chapter 12).

Austin *et al.* (1990) hypothesized that the insects evolved with some of the Old World members of the family and later shifted to the introduced sweet potato. Why the weevil has been found mostly on plant species of American origin is still unknown.

Fig. 3.3. Examples of the *Ipomoea* sections on which *C. formicarius* has been found feeding or reproducing: (A) *I. indica* (section *Pharbitis*); (B) *I. batatas* (section *Batatas*); (C) *I. quamoclit* (section *Mina*); (D) *I. alba* and *I. turbinata* (section *Calonyction*); (E) *I. pandurata* (section *Eriospermum*); and (F) *I. pes-caprae* (section *Erpipomoea*).

ACKNOWLEDGMENTS

Studies were supported by grants from the Florida Atlantic University Division of Sponsored Research, the Bache Fund, the Society of Sigma Xi, P.L. 480 funds through the Smithsonian Institution (F.R. Fosberg, principal investigator), the National Science

54

Fig. 3.4. Examples of the other genera on which *C. formicarius* has been found feeding or reproducing: (A) *Stictocardia tiliifolia*; (B) *Merremia dissecta*; (C) *Jacquemontia tamnifolia*; (D) *Calystegia soldanella*; (E) *Dichondra carolinensis*; and (F) *Cuscuta sandwichensis*.

Foundation (INT 78-23341, G.T. Prance, principal investigator), sabbaticals awarded by Florida Atlantic University during the academic years 1979-1980, and 1989-1990, the Food and Agricultural Organization (FAO Project No. 82/75 and No. 84/92), the International Potato Center, Lima, Peru, and the U. S. Department of Agriculture, Experiment, Georgia.

S. Austin, F. Posin and P.N. Honychurch gave technical assistance. R.K. Jansson, A. Jones, L. Rolston, J. Schalk, and G.W. Wolfe helped in a variety of ways. M. Stewart read this paper for the author at the International Conference on Sweet Potato Pest Management.

REFERENCES

Anonymous. 1970. Distribution maps of pests, *Cylas formicarius* (F.). Series A., Map No. 278. Commonwealth Institute of Entomology, London.

Austin, D.F. 1975. Typification of the New World subdivisions of *Ipomoea*. Taxon 24:107-110.

Austin, D.F. 1977a. Hybrid polyploids in *Ipomoea* section *Batatas*. J. Heredity 68:259-260.

Austin, D.F. 1977b. Realignment of the species placed in *Exogonium* (Convolvulaceae). Ann. Mo. Bot. Garden 64:330-339.

Austin, D.F. 1978. The *Ipomoea batatas* complex - I. Taxonomy. Bull. Torrey Bot. Club 105:114-129.

Austin, D.F. 1979. An infrageneric classification for *Ipomoea* (Convolvulaceae). Taxon 28:359-361.

Austin, D.F. 1980. Additional comments on infrageneric taxa in *Ipomoea* (Convolvulaceae). Taxon 29:501-502.

Austin, D.F. 1983. Variability in sweet potato in America. Proc. Am. Soc. Hort. Sci., Trop. Region 27(B):15-26.

Austin, D.F. 1988. The taxonomy, evolution and genetic diversity of sweet potatoes and related wild species, pp. 27-60. *In* Exploration, maintenance, and utilization of sweet potato genetic resources, International Potato Center, Lima, Peru.

Austin, D.F., R.K. Jansson & G.W. Wolfe. 1990. Convolvulaceae and *Cylas*: a proposed hypothesis on the origins of this plant: insect relationship. Trop. Agric. (In press).

Boyden, B.L. 1922. The sweet potato weevil in Florida. Fla. State Plant Bd. Quart. Bull. 6:76-87.

Boyden, B.L. 1927. Sweet potato weevil eradication in Florida and Georgia. The Monthly Bull. Fla. State Plant Bd. 12:17-53.

Brand, D.D. 1971. The sweet potato, an exercise in methodology, pp. 343-365. *In* C.L. Riley (ed.). Man across the sea. University of Texas Press, Austin.

Choisy, J.D. 1838. De Convolvulaceis dissertatio secunda. Mem. Société Physique de Genéve 8:121-164.

56

Clarke, C.B. 1883. Convolvulaceae, *In* J.D. Hooker (ed.). Flora of British India, Vol. 4. London.

Cockerham K.L., O.T. Deen, M.B. Christian & L.D. Newsom. 1954. The biology of the sweet potato weevil. La. Agric. Exp. Stn. Bull. No. 483.

Engel, F.A. 1970. Explorations of the Chilca Canyon, Peru. Current Anthropol. 11:55-58.

Grisebach, A.H.R. 1864. Convolvulaceae, pp. 446-473. *In* Flora of the British West Indian islands. Lovell, Reeve, and Co., London.

Gunn, C.R. 1972. Moonflowers, *Ipomoea* section *Calonyction*, in temperate North America. Brittonia 24:150-168.

Jansson, R.K., A.G.B. Hunsberger, S.H. Lecrone, D.F. Austin & G.W. Wolfe. 1989. *Ipomoea hederifolia* L., a new host record for the sweetpotato weevil, *Cylas formicarius elegantulus* (Coleoptera: Curculionidae). Fla. Entomol. 72:551-553.

Kemner, N.A. 1924. Der batatenkäfer (*Cylas formicarius* [F.]) auf Java und den benachbarten Inseln Ostindiens. (The sweet potato weevil, *Cylas formicarius* [F.], on Java and the neighboring islands of the East Indies). Z. Angew. Entomol. 10:398-435.

Linnaeus, C. 1753. Species plantarum. Laurentii Salvii, Vienna. Facsimile edition. 1957. 2 Vols. The Ray Society, London.

Newell, W. 1917. Sweet potato root weevil. Fla. State Plant Bd. Quart. Bull. 2:81-100.

O'Brien, P.J. 1972. The sweet potato; its origin and dispersal. Amer. Anthropol. 74:342-365.

O'Donell, C.A. 1959. Las especies americanas de *Ipomoea* L. sect. *Quamoclit* (Moench) Griseb. Lilloa 29:19-86.

Ooststroom, S.J. van. 1953. Convolvulaceae, pp. 388-512. *In* C.G.G.J. van Steenis (ed.). Flora Malesiana, Ser. 1., Vol. 4, No. 4. Utrecht.

Sorensen, K.A. 1984. Impact of the sweetpotato weevil in the southeast, pp. 1-16. *In* M.A. Mullen & K.A. Sorensen (eds.). Proceedings of a sweetpotato weevil workshop. Department of Entomology, North Carolina State University, Raleigh.

Sutherland, J.A. 1986. A review of the biology and control of the sweetpotato weevil *Cylas formicarius* (Fab.). Trop. Pest Manage. 32:304-315.

Ugent, D.S. & L.W. Peterson. 1988. Archeological remains of potato and sweet potato in Peru. CIP Circular 16(3): 1-10.

Ugent, D.S., S. Porozski & T. Porozski. 1981. Prehistoric remains of the sweet potato from the Casma Valley of Peru. Phytologia 49:401-405.

Ugent, D.S., S. Porozski & T. Porozski. 1983. Restos arqueológicos de tubérculos de papas y camotes del Valle de Casma en el Perú. Boletín de Lima 25:1-17.

Verdcourt, B. 1957. Typification of the subdivisions of *Ipomoea* L. (Convolvulaceae) with particular regard to the East African species. Taxon 6:150-152.

Verdcourt, B. 1963. Convolvulaceae, pp. 162. *In* C.E. Hubbard & E. Milne-Redhead (eds.). Flora of tropical East Africa. Crown Agents for Overseas Governments and Administrations, London.

Yen, D.E. 1974. The sweet potato and Oceania. Bishop Mus. Bull., Honolulu 236:1-389.

Yen, D.E. 1982. Sweet potato in historical perspective, pp. 17-30. *In* R.L. Villareal & T.D. Griggs (eds.). Sweet potato, Proceedings of the 1st international symposium. Asian Vegetable Research and Development Center, Tainan, Taiwan, Republic of China.

4. Growth of Sweet Potato in Relation to Attack by Sweet Potato Weevils

Stephen K. O'Hair
Tropical Research and Education Center
University of Florida, I.F.A.S.
Homestead, Florida 33031 U.S.A.

In the humid tropics, sweet potato, *Ipomoea batatas* (L.) Lam., is a perennial species grown as a crop. The photosynthetic potential of sweet potato in terms of root calorie production ha^{-1} day^{-1} as outlined by De Vries *et al.* (1967) is easily attainable under experimental conditions (Wilson 1982). In Japan, a maximum crop growth rate of 207 to 271 kg ha^{-1} day^{-1} has been reported (Agata 1982, Agata & Takeda 1982), which is greater than the expected maximum potential reported by De Vries *et al.* (1967). Thus, it is likely that under optimal conditions, sweet potato is among the most productive of root and tuber crops.

Sweet potato weevils, *Cylas* spp. and *Euscepes postfasciatus* (Fairmaire), attack nearly all plant parts and immatures can develop successfully on mature stems as well as on storage roots (Vasquez & Gapasin 1980). Thus, few plant parts appear to be immune from weevil attack.

The relationship between sweet potato plant growth and weevil attack is complex. There are critical periods during the growth of the plant when it is more vulnerable to attack by these weevils (e.g., late in the season when storage roots are more available), and this relationship is affected by the population dynamics of these weevils. In the field, populations of *C. formicarius* increase exponentially at a rate of about one weevil per plant per day, with most of the increase in population density occurring late in the growing season when storage roots are more available (Jansson *et al.* 1990). At this time, a greater percentage of the weevil population is found in the storage roots than in the vines of most cultivars (Jansson *et al.* 1990). Also, several overlapping generations of these weevils occur during the growth of the sweet potato crop. *Cylas formicarius* requires between 22 to 78 days to develop from the egg to adult emergence depending on temperature (Chalfant *et al.* 1990 and references therein). Between 27 and 30°C, only about 33 days are needed to complete development (Chalfant *et al.* 1990 and references therein). Female weevils begin laying eggs about four days after adult emergence and continue to lay eggs for over 75 days thereafter (Jansson & Hunsberger 1991). Lifetime fecundities per female have been estimated at between 56 and 256 eggs (Chalfant *et al.* 1990 and references therein, Jansson & Hunsberger 1991). It is apparent, that with such great potential for rapid population growth, sweet potato plants are extremely vulnerable to this insect.

Several characteristics of the sweet potato plant affect its growth and concomitant attack by these weevils. These include anatomical characteristics of the plant, growth phase, cultural factors, cultivar or genotypic variations, and environmental factors. Most of these factors are interrelated, but they can be discussed independently. Yet, one may not be considered to be more important than the other(s) when considering relationships with these weevils until weak links in the chain of events are clearly identified. The relationship between each of these characteristics and weevil attack is discussed below.

ANATOMY

In general, information on the growth of specific anatomical parts of the plant and weevil attack is minimal. Weevils feed on storage roots, stems, and leaves (Hahn & Leuschner 1982). Little is known about the relationship between weevil attack and variation in leaf or stem anatomy. It was suggested that the age of plants influences the preference of weevils for certain plant parts (Hahn & Leuschner 1982).

Examination of feeding habits of larvae indicated that older stem tissue was preferred by larvae, and vascular bundles were bypassed; larvae remained primarily in pithy tissues (Leuschner 1979). Similarly, Pole (1988) observed that thick vines were preferred over thin vines. On plants with enlarged storage roots, the storage root, especially the periderm of the root, was the preferred plant part for oviposition (Cockerham *et al.* 1954, Nottingham *et al.* 1987). However, because leaves appear to be an essential part of the diet of *Cylas* weevils (Hahn & Leuschner 1982, Nottingham *et al.* 1988), growth patterns of all plant parts may affect the population dynamics of these weevils.

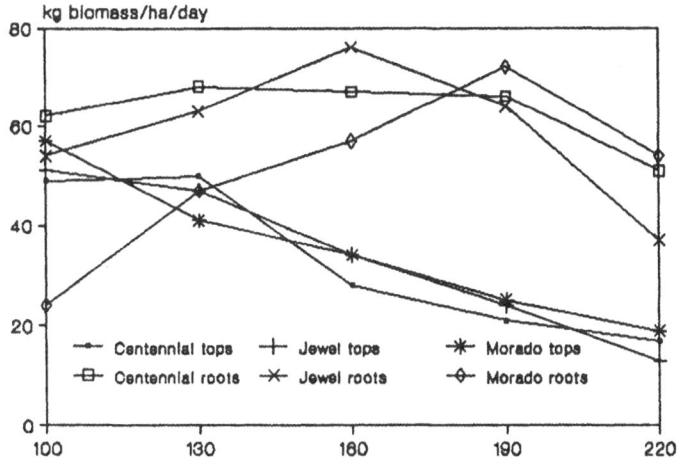

Fig. 4.1. Growth of sweet potato in southern Florida.

PLANT GROWTH CYCLE

Growth of sweet potato follows a typical sigmoidal curve, with leaf and stem production occurring faster than storage root production during the first 100 days of growth (Agata 1982). By approximately 100 days after planting (DAP), the maximum leaf area is reached, with further increases in plant biomass being attributed to storage root enlargement (Figure 4.1) (Agata 1982, O'Hair *et al.* 1986). Significant reductions in photosynthetic activity occur as leaves age; thus, when leaf production peaks, a decline in whole plant photosynthetic assimilation occurs (Agata & Takeda 1982, Hozyo 1982). At this time, leaf and stem damage from weevil attack could result in greater decreases in plant productivity than during earlier stages of growth, because stem and leaf production tends to peak and then decline around 100 DAP (O'Hair *et al.* 1986) and leaves produce less photosynthate as they age (Agata & Takeda 1982, Hozyo 1982). There may also be a loss of leaf area from direct feeding on the leaves and damage to the lower parts of the stems due to adult feeding and larval tunneling, both of which result in reduced plant growth.

Storage Root Initiation

Storage root initiation and storage root growth are considered to be two separate developmental phases. Starch deposition is first visible in root cortical tissue as early as 8 DAP (Indira & Kurian 1977) or it may be delayed until 56 DAP (Wilson 1982). Storage root formation is visible as early as 28 DAP and is associated with accumulation of pigments in the roots (Wilson & Lowe 1973). The first 20 days of plant growth are especially critical in determining total storage root number (Wilson 1977b, Wilson 1982). This phase of storage root growth appears to have little effect on leaf and stem growth under normal conditions. However, if weevil population densities are high, there could be serious consequences because weevils oviposit on young stems in the absence of fleshy roots (Hahn & Leuschner 1982).

As in most of the growth phases, there is genetically- and environmentally-controlled variability in timing of storage root initiation (Ramanujam & Indira 1979). In addition, early storage root initiation is not necessarily synonymous with early harvestable yield. Thus, selection for early yield potential at the storage root initiation stage may be premature and may not help to escape weevil attack and subsequent population buildup because the peak in storage root enlargement (i.e., maturity) is not correlated to the time when storage roots are identifiable.

Storage Root Enlargement

Severe damage from weevils has been observed as early as 90 DAP (Ramanujam & Indira 1979). Consequently, the period between storage root initiation and harvest is critical in relation to weevil population dynamics. The rate of storage root bulking may reach a maximum as early as 90 DAP (Ramanujam & Indira 1979). Weevil infestation can become serious if harvest is delayed (Nawale 1981, Sutherland 1986). Sutherland (1986) observed a linear increase in storage root damage by weevils starting between 80 and 100 DAP. Thus, selecting for early and rapid storage root growth may be invaluable in areas with the potential for rapid weevil population buildup.

Most commonly, the number of storage roots formed reaches its maximum between 56 and 84 DAP; by 49 DAP, at least 80% of the storage roots can be identified (Wilson 1982, Wilson & Lowe 1973). However, for some cultivars, up to 112 days may be needed for all storage roots to form. Storage root enlargement was linear for the period between 56 and 112 DAP, with some cultivar variation as to timing. During this linear phase, 76 to 96% of storage root enlargement was completed. Nevertheless, the environment can significantly alter growth patterns. Chen and Yang (1980) observed a linear pattern of storage root enlargement up to approximately 200 DAP in Mississippi. O'Hair et al. (1986) noted that the optimum efficiency in terms of dry matter production of storage roots over space and time occurred between 100 and 190 DAP, depending upon the location where the plants were grown. During this period, storage root dry matter production peaked at several locations in Florida on or before 160 DAP with a maximum of 119 kg ha^{-1} day^{-1}. Thus, once the efficiency of storage root production peaks, roots should be harvested to reduce the potential for severe weevil damage.

Lowe and Wilson (1974) observed that storage root stalk length was at or near its maximum by 56 DAP. This characteristic determines how deeply storage roots develop below the soil surface. Selection for longer stalk lengths could reduce accessibility of storage roots to weevils (Burdeos & Gapasin 1980).

Longitudinal and lateral root enlargement are different events, with longitudinal growth being completed by 112 DAP (Lowe & Wilson 1974). The rate of lateral enlargement was greatest between 112 and 168 DAP. Thus, the highest yielding cultivars will have a longer linear phase of storage root enlargement and greater growth laterally. Cultivars that produce globular-shaped roots differ in root enlargement patterns from those that produce narrow, fusiform roots (Wilson 1977a). Cultivars that produced globular-shaped roots had the greatest increase in storage root diameter between 112 to 168 DAP. This coincided with a corresponding increase in total weight. Storage roots of cultivars that produced narrow, fusiform roots continued to lengthen to a small extent during the same period; and their total weight did not increase as rapidly as those of cultivars with globular-shaped roots. Therefore, selection for cultivars with long, thin roots, which are less likely to develop close to the soil surface where weevils could oviposit, will be less productive, but also less prone to weevil damage than cultivars with globular-shaped roots.

Storage Root Maturity

There is no set method of establishing maturity, since sweet potato roots continue to enlarge throughout the life of the plant. By 120 DAP, rates of cell enlargement, proliferation and starch deposition have peaked (Hahn & Hozyo 1984). Thus, storage roots may be considered to be mature when these events happen. This coincides with the end of the linear phase of storage root enlargement mentioned before.

CULTURAL EFFECTS

Sweet potato is at a disadvantage in relation to weevil attack in the tropics and sub-tropics because of crop continuity. In many tropical and sub-tropical regions, the crop may be present at various growth stages in various fields most, if not all, of the time. Cultural practices can be manipulated to change the growth of the plant and concomitantly reduce the potential for weevil attack. However, the planting options may be restricted because of environmental constraints. In regions with a long dry season, such as in parts of Africa, the availability of not only adequate soil moisture for planting, but also adequate planting material determine when the grower will plant the field (Hahn 1984).

Cropping Systems

One of the main factors which affects sweet potato growth is the crop production system chosen by the grower. Mixed cropping, which is common in the tropics, can have a substantial affect on sweet potato growth. Yields of plants grown in mixed cropping systems are usually below those of plants grown in monoculture (Singh *et al.* 1984). This is due, in part, to the shading effect of upper story crops on the lower story sweet potato crop. Shading of greater than 25% full sun results in longer vine length and slower growth during the first few weeks in the field (Roberts-Nkrumah *et al.* 1986a). Thus, in weevil-infested areas, damage could be more devastating to shaded plants due to the reduced ability of the plant to compensate for weevil feeding. Shading may also delay storage root formation such that little root enlargement occurs (Roberts-Nkrumah *et al.* 1986b). As with other traits, it appears that there is genetic variability for shade tolerance and root enlargement in sweet potato. Thus, the relationship between sweet potato plant growth and weevil attack probably differs between plants grown in monoculture and those grown in mixed plantings. Studies are needed to determine the effects of mixed plantings on weevil attack to shade-tolerant and shade-intolerant sweet potato plants.

Propagules and Planting

Propagation Systems. In the tropics, sweet potatoes are normally grown from vine-tip cuttings harvested from fields that are ready for harvest. Ray *et al.* (1983) suggested that this practice enhances the severity of weevil infestation, and he concluded that growers should obtain planting material as slips from selected storage roots. Although the reduction in weevil population density due to the use of slips was minor in terms of the overall density, yields were increased with the use of slips rather than vine-tip cuttings. Thus, the selection of pest-free cuttings alone may not have a major effect on reducing weevil damage, but in combination with other control measures, it may be a significant component, especially because plant productivity might increase. Further investigation is needed before absolute recommendations could be made.

Propagule Types. There are two choices for propagules in sweet potato production other than vine-tip cuttings: rooted slips produced in special growing beds and cut portions of the storage roots (Kim *et al.* 1983). The use of slips is most common in temperate climates where storage roots are stored in protective structures during the cooler months. The cut portions of storage roots have been utilized experimentally, with hopes of reducing labor costs incurred in the production of the slips and vine-tip cuttings. Results from trials utilizing portions of the storage roots as propagules have been quite variable (Hsi 1980, Kim *et al.* 1983).

Advantages and disadvantages are implicit for each propagule type for crop growth and for the spread of weevils within a region. For example, use of portions of the storage root as propagules, as an alternative to vine cuttings, should only involve selection and use of weevil-free roots. Kim and Woo (1972) suggested that a crop produced from portions of the root could be harvested 20 to 30 days earlier than a crop produced from vine cuttings. This might be attributed to the more vigorous vine growth observed in the former and might lead to reduced weevil problems. At harvest, vines from the former weighed more than four times than those from vine cuttings.

Propagule Quality. As with most vegetatively-propagated crops, the quality of the planting material affects growth qualities of the resultant plant (Bryan 1983, Martin 1984, Nzima & Del Rosario 1982). The use of healthy planting material maximizes storage root formation and early leaf area development (Huett 1982, Wilson 1982). Typically, growers in the tropics are not concerned with the health of the plant when selecting planting material. For this reason, systemic diseases are perpetuated from one planting to the next, resulting in a decline in the growth and yield potential of the crop. In areas with a distinct dry season, growers have access to propagules consisting mostly of older vine parts (Hahn 1984). The use of such propagules results in slower plant growth, lower yields, a lengthened growing season, and a prolonged time period for weevil attack to occur.

Propagule Orientation. The orientation, length, and planting depth of vine cuttings and slips can vary, and consequently affect plant growth, weevil feeding, and the distribution of weevils within the plant. In addition, the presence or absence of a shoot apex on vine cuttings may affect subsequent growth (Hall 1987). Removal of the shoot apex resulted in bushier plants. Studies on the effects of orientation and planting depth of vine cuttings and/or slips on productivity have produced variable results. Increasing the length of the cutting, minimizing the number of nodes planted below the ground to less than 5, and horizontal orientation of the planted cuttings resulted in higher yields in some clones (Chen *et al.* 1982, Hall 1986, Hall 1987). These higher yields might be due to faster or earlier growth of storage roots. Vertical planting versus horizontal planting tended to result in fewer jumbo roots (Chen *et al.* 1982) which may result in less exposure of roots to the soil surface. Lowe and Wilson (1974) observed that 80 to 100% of the storage root production was on the first four nodes below the soil surface. However, they indicated that there may be some genetic variation in this trait, since one cultivar produced 70% of its storage roots on the first four nodes. Thus, genotypic variation may be an important factor influencing the response of the plants to stem orientation and planting depths (Chen *et al.* 1982). Additional studies are needed to determine the genotypic and propagule orientation interactions in relation to the depth of storage root formation, because this would have a major effect on weevil oviposition.

Planting Density. Planting densities can vary considerably. Planting at higher densities resulted in delayed storage root enlargement (Bouwkamp & Scott 1980). Thus, in areas with high weevil populations, wider plant spacing may help reduce the length of time that the crop is in the field and concomitantly help to reduce weevil population densities. Alternatively, high planting densities also result in smaller roots and reduced root exposure at the soil surface (Hahn 1984).

Fertilization

Sweet potato usually responds to nitrogen (N) fertilization (Badillo & Lugo 1976, Talleyrand & Lugo 1976). Under continuous cropping conditions in Papua New Guinea, Bourke (1985) found that N fertilization had a greater influence on growth and productivity than did potassium (K). It is generally accepted that approximately 100 kg N ha^{-1} is optimal for high storage root yields (Anonymous 1970). Storage root initiation was influenced by fertility level with differences being detected as early as 28 DAP (Acock & Garner 1987). High amounts of slow-release fertilizer (19N-3P-10K) delayed storage root formation by 9 to 18 days. Wilson (1973) found that increasing N above optimal levels delayed storage root development by three weeks or more. As expected, vine biomass and growth increased with an increase in the level of N applied (Badillo & Lugo 1976, Bourke 1977). Similarly, delayed applications of N, beyond 60 DAP, also resulted in increased vine growth at the expense of storage root production (Bartolini 1982). As with the other growth characteristics, delayed storage root development due to improper fertilization might increase the likelihood of weevil attack. Cultivars vary somewhat in their response to N (Bourke 1977), but under conditions of high N fertilization, the trend towards enhanced vine growth is common to all (Badillo & Lugo 1976, Bourke 1977). However, Bourke (1985) concluded that the beneficial effects of N fertilization were realized in greater leaf area duration, which in turn improved other growth and yield parameters. In contrast, Pushkaran et al. (1976) found a negative correlation between leaf area at harvest and yield. However, the differences were related to genetic variation, which was not considered in their analysis. Other genetic factors, such as delayed maturity, could have caused the negative correlation between yield and leaf area at harvest.

During an evaluation of K on growth and yield of sweet potato, Bautista and Santiago (1981) found that soil type was more important than the level of fertilization on sweet potato growth and yield, with silt and sandy loams being preferable to clay loams. Since similar studies with N fertilization have not been reported, the relative importance of soil type in relation to soil fertility level can not be discussed.

Irrigation and Water Table

Although the sweet potato is native to the American tropics, it is grown over a wide range of environments (Hahn & Hozyo 1984). These areas include some with high amounts of rainfall. Nevertheless, sweet potato can not tolerate waterlogged soil conditions. The effects of wet conditions are recognized at a relatively early growth stage, whereby such conditions retard storage root formation (Wilson 1982). Drought also retards storage root development and encourages the formation of soil cracks, thereby making the storage roots more accessible to weevils (Hahn & Leuschner 1982). During extended dry periods, supplemental irrigation at monthly intervals increased root size and total root yield (Anonymous 1970, Sajjapongse & Roan 1982). Unless hilling procedures are practiced, the lateral expansion of the roots will force the roots toward the soil surface, making them more accessible for weevil attack.

The water table depth influences root dry matter and vine growth (Ghuman & Lal 1983, Pardales & Escalante 1979). Increasing the water table depth from 30 cm to 91 cm increased vine growth and marketable storage root production (Pardales & Escalante 1979). Temporary flooding and high water table are both detrimental to root quality (Anonymous 1970, Collins & Wilson 1988). These conditions are also unfavorable for weevils. Collins and Wilson (1988) demonstrated that variability for tolerance to temporary flooding is present among sweet potato clones. It may be possible, therefore, to manipulate soil moisture, and thereby minimize weevil attack to the storage roots.

Cultivation

Soil aeration increases the activity of cell division and expansion (Wilson 1982) and increases storage root formation (Chua & Kays 1980). Soil compaction reduced leaf production and vine growth and affected root size (Sajjapongse & Roan 1982). Soils of medium compaction (bulk density) produced the largest roots because excessive vine growth induced from loose soil conditions was minimized. Thus, soil cultivation during early growth stages may have a positive effect on plant growth, by encouraging deeper root growth, and covering exposed enlarged roots at the soil surface, where weevils are most likely to gain access to storage roots.

Mulching with black plastic increased early leaf and vine growth and resulted in a higher total marketable yield by increasing soil temperature and possibly by reducing soil compaction (Hochmuth & Howell 1983). Few other studies have been reported on the effects of mulching on sweet potato growth and interactions with weevil. Nevertheless, it has been suggested that in regions of high insect population pressure, plastic mulch is not an effective means of control (Jansson *et al.* 1987).

Harvesting

As previously mentioned, the time to harvest can vary. One of two harvesting strategies is typically utilized once the sweet potato roots have reached a harvestable size. The first is a destructive, once-over harvest when storage root enlargement has met the local requirements for size and demand is present. During this procedure, the entire plant is uprooted. The second harvesting strategy is a non-destructive extended harvest, where plants are allowed to remain in the field for prolonged periods (Rose 1979). During non-destructive harvest, the soil is removed where roots have enlarged and the largest roots are removed from the plant. Following root removal, the soil is replaced to allow the smaller roots to reach an adequate size. Non-destructive harvests are practiced because sweet potato is a trailing vine, producing roots at locations where the vine comes in contact with the soil surface and total yields for the planting are increased. Once the plantings are unmanageable or when production declines, the plants are uprooted. The latter practice presents problems in weevil-infested areas, by providing a continuum for the weevils.

Although it is desirable to harvest the crop between 90 and 120 DAP to reduce the time the crop is exposed to weevil attack, availability of labor and equipment and/or low demand for storage roots may alter harvest schedules (Chen 1984, Chen & Yang, 1980). Some researchers have recommend that harvest be delayed until 175 to 200 DAP to realize maximum yields (Badillo & Lugo 1977). During this period, storage roots continue to enlarge. This practice is acceptable if demands for storage roots and land are low. However, in weevil-infested regions, the chances for attack are more likely.

CULTIVAR EFFECTS

Cultivars vary in their times to harvest due to differences in efficiencies of storage root enlargement over time (O'Hair *et al.* 1986) and some cultivars have greater variation in their growth and yield patterns than others (Lowe & Wilson 1975). Thus, when it is essential to minimize variation in crop growth and storage root production from one experimental period to another, care must be taken in selecting sweet potato cultivars. 'Georgia Jet' is often listed as being very efficient (Bhagsari 1980), producing harvestable roots earlier than 90 DAP (Harmon 1984). Thus, there may be opportunities to avoid severe weevil attack, by adopting cultivars with shorter cropping periods. Environment x genotype interactions occur in sweet potato as demonstrated by variation among cultivars in their growth habits during wet, dry, or cool periods of the year (Collins *et al.* 1987, Hahn & Leuschner 1982, Hochmuth & Howell 1983, King 1985) and among locations within a relatively small region (O'Hair *et al.* 1986). Selection for improved storage root growth during adverse conditions could result in fewer problems with weevil attack and cultivar recommendations should be based on locally conducted yield trials.

Genetic variation in percentage storage root dry matter and protein content has been observed in sweet potato (Collins *et al.* 1987). Weevils avoid types high in dry matter (Hahn & Leuschner 1981). Also, protein quality of roots may affect weevil population growth. For these reasons, the effects of these variables sweet potato growth and concomitant attack by these weevils merit additional attention (Hahn & Leuschner 1982).

Jansson *et al.* (1990) observed that the population dynamics of *C. formicarius* was influenced by cultivar, indicating that it is possible to reduce weevil populations by choosing a cultivar on which weevil populations grow less rapidly.

ENVIRONMENTAL EFFECTS

Sweet potato growth and weevil population dynamics are also influenced by environmental conditions (Bacusmo *et al.* 1988, Hahn & Leuschner 1982). Fortunately, conditions favoring the proliferation of weevils, such as dry conditions, may often be avoided for sweet potato production because such conditions result in suboptimal sweet potato growth. For example, dry conditions tend to favor weevil population growth, primarily because of soil cracking. Although these conditions are avoided at the time of planting, they may be encountered near the time for harvest. Therefore, field studies that evaluate the efficacy of weevil control programs may be affected by the soil moisture conditions during the latter plant growth stages.

Seasonal effects of environmental conditions on sweet potato growth have been observed in most production areas (Badillo 1976). Unfortunately, there are several interrelated seasonal factors which can affect plant growth and subsequent attack by weevils. In many trials, the individual factors including daylength, temperature, and rainfall are not separable and the relative importance of each by itself is unknown.

Daylength

When compared to plants grown under natural daylengths of 11.5 to 12.5 h, long days of 18 h reduced storage root yield and increased leaf and stem growth (McDavid & Alamu 1980). Short days of 8 h reduced stem and root growth when compared to plants grown under natural daylengths. Thus, there appears to be an unknown triggering mechanism which encourages storage root enlargement, which is associated with daylength. This triggering mechanism was suggested to be hormonal in nature. Since storage root initiation is a relatively easy event to document, it leaves room to speculate that many other changes in the plants physio-chemical status may occur during growth and be affected by the environment and simultaneously influence weevil performance.

Temperature

Air temperature can have an effect on early growth of sweet potato (Agata & Takeda 1982). High temperatures during the first 60 DAP reduced the number of storage roots produced and increased top growth, whereas cool temperatures reduced root size (Sajjapongse et al. 1988). A mean daily temperature of 22 to 23°C was found to be optimal for storage root production. Nevertheless, Agata (1982) reported that mean solar radiation was more important ($r = 0.94$) in determining changes in root growth rate than mean air temperature ($r = 0.52$).

Rainfall

Lack of moisture reduces plant growth and encourages soil cracking, thereby providing easy access to storage roots for weevils (Hahn & Leuschner 1982, Sajjapongse & Roan 1982). Rainfall, which is correlated with lower solar radiation, has been associated with increased vine growth and reduced storage root yield and number per plant (Agata 1982, Gollifer 1980, King 1985). Thus, very wet periods, which are not favorable for weevil attack, are also not conducive to root enlargement.

Seasonal variations in rainfall can also cause considerable variation in sweet potato growth, with the greatest variations occurring for storage root number and size during the wet season (Haynes & Wholey 1971). As a consequence of this variation, accessibility of roots for weevil attack at or near the soil surface may vary more during the wet season than during the dry season, due to variation in the rate and amount of lateral enlargement of storage roots.

SUMMARY

Growth and development of the sweet potato can be quite variable, depending on cultural practices, genotype, and environmental conditions. All have a significant effect on weevil population dynamics. Most growers have control over the cultivars and cultural practices, which can be adjusted to minimize the degree of damage caused by weevils. In the case of plant growth and insect damage, modifications of cultural practices may reduce losses due to weevil attack. These include use of healthy, weevil-free propagules, planting during periods of low weevil activity, judicious use of N or organic fertilizers to avoid excess amounts and late applications, hilling of soil around the plants during their growth, irrigation during extended dry periods, use of early maturing cultivars, and harvest of storage roots in a timely process.

ACKNOWLEDGMENTS

This is Florida Agricultural Experiment Station Journal Series no. R-00876.

REFERENCES

Acock, A.C. & J.O. Garner, Jr. 1987. Activity of cell-wall-bound invertase during storage root initiation in the sweet potato. HortScience 22:586-588.

Agata, W. 1982. The characteristics of dry matter and yield production in sweet potato under field conditions, pp. 119-127. *In* R.L. Villareal & T.D. Griggs (eds.). Sweet potato. Proceedings of the first international symposium. Asian Vegetable Research and Development Center Publ. No. 82-172, Tainan, Taiwan.

Agata, W. & T. Takeda. 1982. Studies on matter production in sweet potato plants. 1. The characteristics of dry matter and yield production. J. Fac. Agric. Kyushu Univ. 27:65-73.

Anonymous. 1970. Thirty years of cooperative sweet potato research. Southern Coop. Ser. Bull. No. 159, Louisiana State Univ., Baton Rouge.

Bacusmo, J.L., W.W. Collins & A. Jones. 1988. Effects of fertilization on stability of yield and yield components of sweet potato clones. J. Am. Soc. Hort. Sci. 113:261-264.

Badillo, J.F. 1976. Effect of planting season on yield of sweetpotato cultivars. J. Agric. Univ. Puerto Rico 60:163-171.

Badillo, J.F. & M.A. Lugo. 1976. Effect of four levels of N, P, K, and micronutrients on sweetpotato yields in an oxisol. J. Agric. Univ. Puerto Rico 60:597-605.

Badillo, J.F. & M.A. Lugo. 1977. Sweet potato production in oxisols under a high level of technology. Univ. Puerto Rico Agric. Exp. Stn. Bull. No. 256.

Bartolini, P.U. 1982. Timing and frequency of topping sweet potato at varying levels of nitrogen, pp. 209-214. *In* R.L. Villareal & T.D. Griggs (eds.). Sweet potato. Proceedings of the first international symposium. Asian Vegetable Research and Development Center Publ. No. 82-172, Tainan, Taiwan.

Bautista, A.T. & R.M. Santiago. 1981. Growth and yield of sweet potato as influenced by different potassium levels in three soil types. Ann. Trop. Res. 3:177-186.

Bhagsari, A.S. 1980. Growth analysis of 20 sweet potato genotypes. HortScience 15:279.

Bourke, R.M. 1977. A long term rotation trial in New Britain, Papua New Guinea. Proc. Intern. Soc. Trop. Root Crops. 3:382-388.

Bourke, R.M. 1985. Influence of nitrogen and potassium fertilizer on growth of sweet potato (*Ipomoea batatas*) in Papua New Guinea. Field Crops Res. 12:363-375.

Bouwkamp, J.C. & L.E. Scott. 1980. Effect of plant density on yield and yield components of sweet potato. Ann. Trop. Res. 2:1-11.

Bryan, J.E. 1983. The importance of planting material in root and tuber crop production, pp. 3-6. *In* Global workshop on root and tuber crop production. International Center for Tropical Agriculture, Cali, Colombia.

Burdeos, A.T. & D.P. Gapasin. 1980. Effect of soil depth on the degree of sweet potato weevil infestation. Ann. Trop. Res. 2:224-231.

Chen, L.H. 1984. Sweet potato production systems modeling. Trans. Am. Soc. Agric. Eng. 27:683-687.

Chen, L.H. & C.C. Yang. 1980. Optimum starting date for the harvest of sweet potatoes. Trans. Am. Soc. Agric. Eng. 23:284-287.

Chen, L.H., T.S. Younis & M. Allison. 1982. Horizontal transplanting of sweet potatoes. Trans. Am. Soc. Agric. Eng. 25:1524-1528.

Chua, L.K. & S.J. Kays. 1980. Effect of soil oxygen concentration on sweet potato storage root induction and/or development. HortScience 15:280.

Cockerham, K.L., O.T. Deen, M.B. Christian & L.D. Newsom. 1954. The biology of the sweet potato weevil. La. Agric. Exp. Stn. Tech. Bull. No. 483.

Collins, W.W. & L.G. Wilson. 1988. Reaction of sweet potatoes to flooding. HortScience 23:1079.

Collins, W.W., L.G. Wilson, S. Arrendell & L.F. Dicken 1987. Genotype x environment interactions in sweet potato yield and quality factors. J. Am. Soc. Hort. Sci. 112:579-583.

De Vries, C.A., J.D. Ferwerda & M. Flach. 1967. Choice of food crops in relation to actual and potential production in the tropics. Neth. J. Agric. Sci. 15:241-248

Ghuman, B.S. & R. Lal. 1983. Growth and plant-water relations of sweet potato (*Ipomoea batata*) as affected by soil moisture regimes. Plant Soil 70:95-106.

Gollifer, D. E. 1980. A time of planting trial with sweet potatoes. Trop. Agric. 57:363-367

Hahn, S.K. 1984. Utilization, production constraints and improvement potential of tropical root crops, pp. 91-97. *In* D.L. Hawksworth (ed.). Advancing agriculture production in Africa. CAB Farnham Royal, Slough, U.K.

Hahn, S.K. & Y. Hozyo. 1984. Sweet potato, pp. 551-567. *In* P.R. Goldsworthy & N.M. Fischer (eds.). The physiology of tropical field crops. J. Wiley, London.

Hahn, S.K. & K. Leuschner. 1981. Resistance of sweet potato cultivars to African sweet potato weevil. Crop Sci. 21:499-503.

Hahn, S.K. & K. Leuschner. 1982. Breeding sweet potato for weevil resistance, pp. 331-336. *In* R.L. Villareal & T.D. Griggs (eds.). Sweet potato. Proceedings of the first international symposium. Asian Vegetable Research and Development Center Publ. No. 82-172, Tainan, Taiwan.

Hall, M.R. 1986. Length, nodes underground, and orientation of transplants in relation to yields of sweet potato. HortScience 21:88-89.

Hall, M.R. 1987. Shoot apex and nodes exposed above-ground influenced growth of sweet potato vine cuttings. HortScience 22:230-232.

Harmon, S.A. 1984. 'Georgia Jet' an early, high yielding high quality sweet potato cultivar for Georgia. Res. Rpt. No. 193. Univ. of Georgia, Athens.

Haynes, P.H. & D.W. Wholey. 1971. Variability in commercial sweet potatoes (*Ipomoea batatas* - L. Lam) in Trinidad. Exp. Agric. 7:27-32.

Hochmuth, G.H. & J.C. Howell, Jr. 1983. Effect of black plastic mulch and raised beds on sweet potato growth and root yield in a northern region. HortScience 18:467-468.

Hozyo, Y. 1982. Photosynthetic activity and carbon dioxide diffusion resistance as factors in plant production in sweet potato plants, pp. 129-133. *In* R.L. Villareal & T.D. Griggs (eds.). Sweet potato. Proceedings of the first international symposium. Asian Vegetable Research and Development Center Publ. No. 82-172, Tainan, Taiwan.

Hsi, D.C.H. 1980. Evaluating sweet potatoes for root piece propagation in New Mexico. HortScience 15:280.

Huett, D.O. 1982. Evaluation of sources of propagating-material for sweet-potato production. Scient. Hort. 16:1-7.

Indira, P. & T. Kurian. 1977. A study on the comparative anatomical changes undergoing tuberization in the roots of cassava and sweet potato. J. Root Crops 3:29-32.

Jansson, R.K. & A.G.B. Hunsberger. 1991. Diel and ontogenetic patterns of oviposition in sweetpotato weevil, *Cylas formicarius elegantulus* (Summers) (Coleoptera: Curculionidae). Environ. Entomol. (In press).

Jansson, R.K., H.H. Bryan & K.A. Sorensen. 1987. Within-vine distribution and damage of sweetpotato weevil, *Cylas formicarius elegantulus* (Coleoptera: Curculionidae), on four cultivars of sweet potato in southern Florida. Fla. Entomol. 70:523-526.

Jansson, R.K., A.G.B. Hunsberger, S.H. Lecrone & S.K. O'Hair. 1990. Seasonal abundance, population growth, and within-plant distribution of sweetpotato weevil (Coleoptera: Curculionidae) on sweet potato in southern Florida. Environ. Entomol. 19:313-321.

Kim, Y.C.P. & K. Woo. 1972. Studies on propagation method using tuberous cutting in the sweet potato. 1. Influence of the various parts of tuberous cutting on sprout producing, tuber forming and vine growth in the sweet potato. J. Korean Soc. Hort. Sci. 11:29-33.

Kim, Y.C., K.W. Park, Y.S. Kim & S.J. Choi. 1983. Studies on the sweet potato (*Ipomoea batatas*) propagation with tuber pieces. IV. Effects of irrigation method on the growth, yield and several quality criteria of sweet potato. Thesis Collection Agric. Forest., Korea University 23:91-96.

King, G.A. 1985. The effect of time of planting on yield of six varieties of sweet potato (*Ipomoea batatas* (L.) Lam.) in the southern coastal lowlands of Papua New Guinea. Trop. Agric. 62:225-228.

Leuschner, K. 1979. Screening for sweet potato weevil resisrtance. Proc. Intern. Soc. Trop. Root Crops. 5:

Lowe, S.B. & L.A. Wilson. 1974. Comparative analysis of tuber development in six sweet potato (*Ipomoea batatas* [L.] Lam) cultivars. I. Tuber initiation, tuber growth and partition of assimilate. Ann. Bot. 38:307-317.

Martin, F.W. 1984. Effect of age of planting on yields of sweet potato from cuttings. Trop. Root Tuber Crops Newsl. 15:22-25.

McDavid, C.R. & S. Alamu. 1980. Effect of daylength on the growth and development of whole plants and rooted leaves of sweet potato (*Ipomoea batatas*). Trop. Agric. 57:113-119.

Nawale, R.N. 1981. A study on the stage of harvest in sweet potato variety, H-268 under Konkan (Maharashtra) conditions. J. Root Crops 7:29-31.

Nottingham, S.F., K.C. Son, D.D. Wilson, R.F. Severson & S.J. Kays. 1987. Feeding and oviposition preferences of the sweet potato weevil, *Cylas formicarius elegantulus*, on the outer periderm and exposed inner core of storage roots of selected sweet potato cultivars. Entomol. Exp. Appl. 45:271-275.

Nottingham, S.F., K.C. Son, D.D. Wilson, R.F. Severson & S.J. Kays. 1988. Feeding by adult sweet potato weevils, *Cylas formicarius elegantulus*, on sweet potato leaves. Entomol. Exp. Appl. 48:157-163.

Nzima, M.D.S. & D.A. Del Rosario. 1982. Physiological basis of low tuber yield from basal sweet potato (*Ipomoea batatas* (L.) Lam.) cuttings. Phil. Agric. 65:119-129.

O'Hair, S.K., J.M. Dangler, P. Everett, R.B. Forbes, S.J. Locascio, S.M. Olson, J.R. Shumaker & J.M. White. 1986. Cruciferous and root crops for year-round biomass production, pp. 173-184. *In* W.H. Smith (ed.). Biomass energy development. Plenum, New York.

Pardales, J.R., Jr. & M.C. Escalante. 1979. The effect of various water table depths on the growth and yield of sweet potato. Philip. J. Crop Sci. 4:58-59.

Pole, F.S. 1988. Vine thickness in sweet potato (*Ipomoea batatas*): its inheritance and relationship to weevil damage. M.A. Thesis, University of the South Pacific, Western Samoa.

Pushkaran, K., P.S. Nair & K. Gopakumar. 1975. Analysis of yield and its components in sweet potato. Agric. Res. J. Kerala. 14:153-159.

Ramanujam, T. & P. Indira. 1979. Maturity studies in sweet potato (*Ipomoea batatas* (L.) Lam.). J. Root Crops 5:43-45.

Ray, P.K., S. Mishra & S.S. Mishra. 1983. Sweet potato productivity as affected by recurrent use of vines as planting-material. Scient. Hort. 20:319-322.

Roberts-Nkrumah, L.B., T.U. Ferguson & L.A. Wilson. 1986a. Responses of four sweet potato cultivars to levels of shade: 1. Dry matter production, shoot morphology and leaf anatomy. Trop. Agric. 63:258-264.

Roberts-Nkrumah, L.B., L.A. Wilson & T.U. Ferguson. 1986b. Responses of four sweet potato cultivars to levels of shade: 2. Tuberization. Trop. Agric. 63:265-270.

Rose, C.J. 1979. Comparison of single and progressive harvesting of sweet potato (*Ipomoea batatas* (L.) Lam.). Papua New Guinea Agric. J. 30:61-64.

Sajjapongse, A. & Y.C. Roan. 1982. Physical factors affecting root yield of sweet potato (*Ipomoea batatas* (L.) Lam., pp. 203-208. *In* R.L. Villareal & T.D. Griggs (eds.). Sweet potato. Proceedings of the first international symposium. Asian Vegetable Research and Development Center Publ. No. 82-172, Tainan, Taiwan.

Sajjapongse, A., M.H. Wu & Y.C. Roan. 1988. Effect of planting date on growth and yield of sweet potatoes. HortScience 23:698-699.

Singh, B., S.S. Yazdani, R. Singh & S.F. Hameed. 1984. Effect of intercropping on the incidence of sweet potato weevil, *Cylas formicarius* Fabr., in sweet potato (*Ipomoea batatas* Lam.). J. Entomol. Res. 8:193-195.

Sutherland, J.A. 1986. Damage by *Cylas formicarius* Fab. to sweet potato vines and tubers, and the effect of infestations on total yield in Papua New Guinea. Trop. Pest Manage. 32:316-323.

Talleyrand, H. & M.A. Lugo. 1976. Effect of five levels and three sources of N on sweetpotato yields in an Ultisol. J. Agric. Univ. Puerto Rico 60:9-14.

Vasquez, E.A. & D.P. Gapasin. 1980. Stems and tubers for rearing sweet potato weevil (*Cylas formicarius elegantulus*). Ann. Trop. Res. 2:80-88.

Wilson. L.A. 1973. Effect of different levels of nitrate-nitrogen supply on early tuber growth of two sweet potato cultivars. Trop. Agric. 50:53-54

Wilson, L.A. 1977a. Components of tuber yield in sweet potato. Proc. Intern. Soc. Trop. Root Crops. 3:83-86.

Wilson, L.A. 1977b. Root crops, pp. 187-236. *In* P. de T. Alvim & T.T. Kozlowski (eds.). Ecophysiology of tropical crops, Academic Press, New York.

Wilson, L.A. 1982. Tuberization in sweet potato (*Ipomoea batatas* (L.) Lam.), pp. 79-94. *In* R.L. Villareal & T.D. Griggs (eds.). Sweet potato. Proceedings of the first international symposium. Asian Vegetable Research and Development Center Publ. No. 82-172, Tainan, Taiwan.

Wilson. L.A., & S.B. Lowe. 1973. Quantitative morphogenesis of root types in the sweet potato (*Ipomoea batatas* (L.) Lam.) root systems during growth from stem cuttings. Trop. Agric. 50:343-345.

5. Sex Pheromone of *Cylas formicarius*: History and Implications of Chemistry in Weevil Management

Robert R. Heath and James A. Coffelt
U.S. Department of Agriculture, Agricultural Research Service
Insect Attractants, Behavior, and Basic Biology Research Laboratory
Gainesville, Florida 32604 U.S.A.

Fredrick I. Proshold
U.S. Department of Agriculture, Agricultural Research Service
Methods Development
Otis Air Force Base, Otis, Massachusetts 20542 U.S.A.

Richard K. Jansson
Tropical Research and Education Center
University of Florida, I.F.A.S.
Homestead, Florida 33031 U.S.A.

Phillip E. Sonnet
U.S. Department of Agriculture, Agricultural Research Service
Eastern Regional Research Center
Philadelphia, Pennsylvania 19118 U.S.A.

Sweet potato, *Ipomoea batatas* (L.) Lam., is considered the seventh most important human food crop in the world (FAO 1984) and of all root and tuber crops, it is surpassed in importance by only the potato, *Solanum tuberosum* L., (Martin 1983). The sweetpotato weevil, *Cylas formicarius* (Fabricius) (= *C. f. elegantulus* [Summers] and *C. f. formicarius* [F.]), are the most devastating pests of sweet potato worldwide (Chalfant *et al.* 1990, Edmond 1971, Schalk & Jones 1985, Sutherland 1986). Low level pre-and postharvest infestations reduce both quality and marketable yield and can render sweet potato storage roots unfit for consumption (Proshold 1983) because of toxic sesquiterpenes that are produced by the roots in response to weevil feeding (Akazawa *et. al.* 1960, Uritani *et. al.* 1975). These chemicals impart a bitter taste to the storage roots. Losses due to sweetpotato weevil and plant diseases that often follow weevil attack have been estimated at 35 to 95% (Anonymous 1978).

The cryptic feeding habits of the sweetpotato weevil larvae and the nocturnal activity of the adults make it difficult to detect sweetpotato weevil infestations. Additionally, these factors limit the effectiveness of chemical insecticides applied for weevil management. Until recently, no suitable methods were available for detecting low level infestations of this weevil. Thus, a sensitive detection tool was needed to help growers monitor and control infestations of this pest. Traps baited with attractants, such as sex pheromones and food lures, have great potential for monitoring many insect populations. Early research indicated that adult female weevils produce a pheromone which is only attractive to adult males (AVRDC 1976, Louton 1975, Russo 1973). Research was undertaken in 1975 by scientists in the U.S. Department of Agriculture, Agricultural Research Service, to isolate and bioassay a sex pheromone of *C. formicarius* (Coffelt *et al.* 1978). Further studies resulted in the isolation, identification, and synthesis of the active female-produced pheromone, (Z)-3-dodecen-1-ol (E)-2-butenoate (Heath *et al.* 1986). This chapter will highlight the important aspects of the pheromone project and provide an overview of the current research regarding the use of the pheromone with particular emphasis on the chemical aspects of the pheromone. For a more complete description of the techniques used for the isolating and identifying pheromones, the reader is referred to Heath and Tumlinson (1984).

PHEROMONE IDENTIFICATION

Laboratory and Field Bioassays

All materials (combined, active, and inactive fractions) beginning with the unpurified volatiles (crude extract) and continuing through the purification process, were bioassayed in serial dilution in the laboratory to ensure that no loss in activity had occurred. Laboratory bioassays have been described by Coffelt *et al.* (1978). The activation bioassay consisted of exposing the weevil, which was contained in a Teflon [R] coated vial (1.6 x 5 cm), to a small amount of the test extract. A positive response was defined as antennal elevation and locomotion. A replicate consisted of exposing the test sample and a solvent control sample to ten individual adult male weevils. All fractions were bioassayed using serial dilutions of 10^{-1}, 10^{-2}, 10^{-3}, and 10^{-4} of pheromone collected from one female equivalent day (FED), i.e., the amount of pheromone collected from one adult female over a 24 hour period.

Field tests of crude and purified materials were conducted in 0.2-ha plots of sweet potato at the U.S. Department of Agriculture, Federal Experiment Station, St. Croix, U.S. Virgin Islands. Dosages of 0.0, 0.2, 2.0 and 20 FED were prepared in 50 μl of solvent and applied to 22 cm^2 glass cover slips. Cover slips were then placed in a Teflon^R-lined Petri dish which was supported by a wooden platform at a height equal to that of the plant canopy. The numbers of adult male *C. formicarius* attracted to each dosage were counted and compared to controls after a 15 minute exposure period. This bioassay, which compared captures (numbers of attracted male weevils) for the crude material with various isolates obtained during the purification of the pheromone, indicated that a single active fraction and fractions that contained the active component were as active as the crude material and no loss in activity had occurred during pheromone purification (Heath *et al.* 1986).

Collection and Purification of the Pheromone

Insect volatiles, which contained the pheromone, were collected continuously on an adsorbent (Chromosorb 102) over a 30-day period from virgin *C. formicarius* females cultured on sweet potato storage roots. The volatiles were extracted from the adsorbent and purified on silica using a gravity flow column. The active fractions from the gravity flow column were then concentrated and chromatographed by high performance liquid chromatography (HPLC). The active fractions eluted from the HPLC column were combined, concentrated, and further purified by gas chromatography (GC) on an apolar methyl silicone column. The material required further purification by GC and this was effected using a polar polyethylene glycol column.

The biologically active material that was obtained from the HPLC column was further purified by GC on a packed methyl silicone (OV-101) column and was contained within a 2-min fraction with a Kovat's Index (KI) of 1780-1800 (Kovats 1965). No increase or decrease in biological activity was observed when the active fraction was recombined with other GC fractions and subsequently bioassayed over a range of 1×10^{-4} to 1×10^{-1} FEDs in the laboratory.

Further purification on a polar (Carbowax 20M) column yielded a single peak with KI=2200. This single compound was determined by regression analysis to be as attractive as the original isolate when compared in laboratory and field bioassays (Heath *et al.* 1986). Quantitative analysis of this material indicated that a FED was equivalent to about 4 picograms (1 pg = 1 x 10^{-12} g). The active peak from the Carbowax 20M column was found to be >99.8% pure when analyzed on the cyanopropyl methyl silicone, methyl silicone, polyethylene glycol, and cholesteryl-p-chlorocinnamate capillary columns (Heath *et al.* 1979).

Identification of the Pheromone

Complete details of the identification of the sex pheromone of *C. formicarius* have been reported (Heath *et al.* 1986); thus, only a brief description of the methods are provided. Identification of (Z)-3-dodecen-1-ol (E)-2-butenoate was difficult because no precedent for this structure existed at the time of the identification. Also only limited amounts of the isolated compound were available (approximately 4 x 10^{-6} g obtained after collecting 850,000 virgin female equivalent hours), thereby limiting investigations to use of gas chromatography-mass spectrometry (GC-MS) and proton nuclear magnetic resonance (PMR) spectroscopy which used micro-analytical techniques developed in our laboratory (Heath & Tumlinson 1984).

Spectroscopic data obtained on the isolated active peak and used for the identification of the pheromone is shown in Figures 5.1 and 5.2. The following structural information was elucidated from the mass spectral data. The isobutane chemical ionization mass spectrum (CIMS) (Figure 5.1A) established that the molecular weight of the sex attractant was 252 with diagnostic peaks at m/e 251 (M-1), 253 (M+1), 254 (M+2), and 291 (M+39). The fragment ion at m/e 167 (M+1-86) suggested the loss of butenoic acid ($C_4H_6O_2$) from the parent molecule. The methane CIMS (Figure 5.1B), in addition to showing a peak at (M+1-86), provided a base peak at m/e 87 [$C_4H_7O_2^+$] consistent with protonated butenoic acid. Based on the mass spectral data, the structure proposed was a butenoate of a 12-carbon alcohol which contained one degree of unsaturation. Ozonolysis of the sex attractant (approximately 50 ng/ozonolysis) produced two major products that were identified as nonanal and decanal by comparison of their retention times on a 50 m CW-20M capillary column and their mass spectra with synthetic standards.

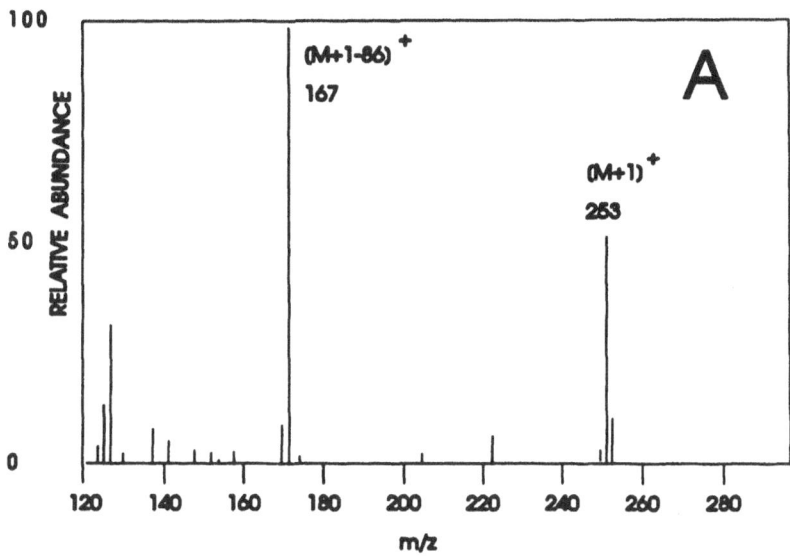

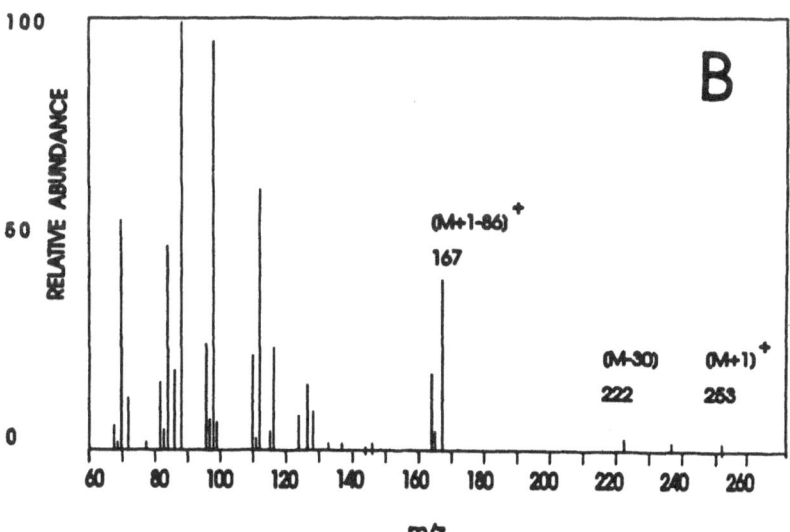

Fig. 5.1. Mass spectra of (Z)-3-dodecen-1-ol (E)-2-butenoate. Spectra A obtained using isobutane as the reagent gas and spectra B obtained using methane as the reagent gas.

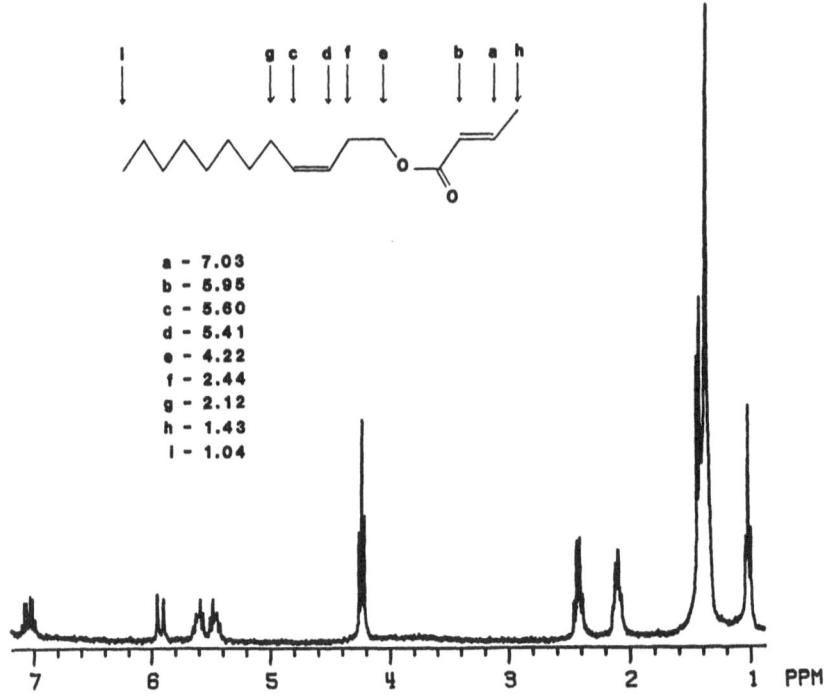

a – 7.03
b – 5.95
c – 5.60
d – 5.41
e – 4.22
f – 2.44
g – 2.12
h – 1.43
I – 1.04

Fig. 5.2. Proton magnetic resonance spectra of (Z)-3-dodecen-1-ol (E)-2-butenoate obtained at 300 MHz (Offset relative to tetramethylsilane and sample run in C_6D_6).

High-field 300 MHz PMR spectroscopy with decoupling experiments provided the following structural information. The spectrum (Figure 5.2) indicated four olefinic protons. The proton giving rise to the signal at $\delta = 7.03$ (1H sextet) was coupled to the olefinic proton at $\delta = 5.95$ (1H, d) and the methyl protons of $\delta = 1.43$ (3H, d). Thus, the crotonate moiety, O=C-C$\underline{H}$=CH-CH$_3$, was established. The olefinic proton $\delta = 5.41$ (1H, m) was coupled to the methylene protons at $\delta = 2.44$ (2H, q), which also were coupled to the methylene protons at $\delta = 4.22$ (2H, t), CH=C$\underline{H}$-C$\underline{H}_2$-O. Thus, the position of the olefinic bond in the alcohol chain was unequivocally established as being in the three-position. Examination of PMR spectra of synthetic samples of (Z)- and (E)-3-dodecen-1-ol and the methyl esters of (Z)-and (E)-2-butenoate indicated that the geometry of the olefinic bonds were (Z)- 3-dodecen-1-ol (E)-2-butenoate.

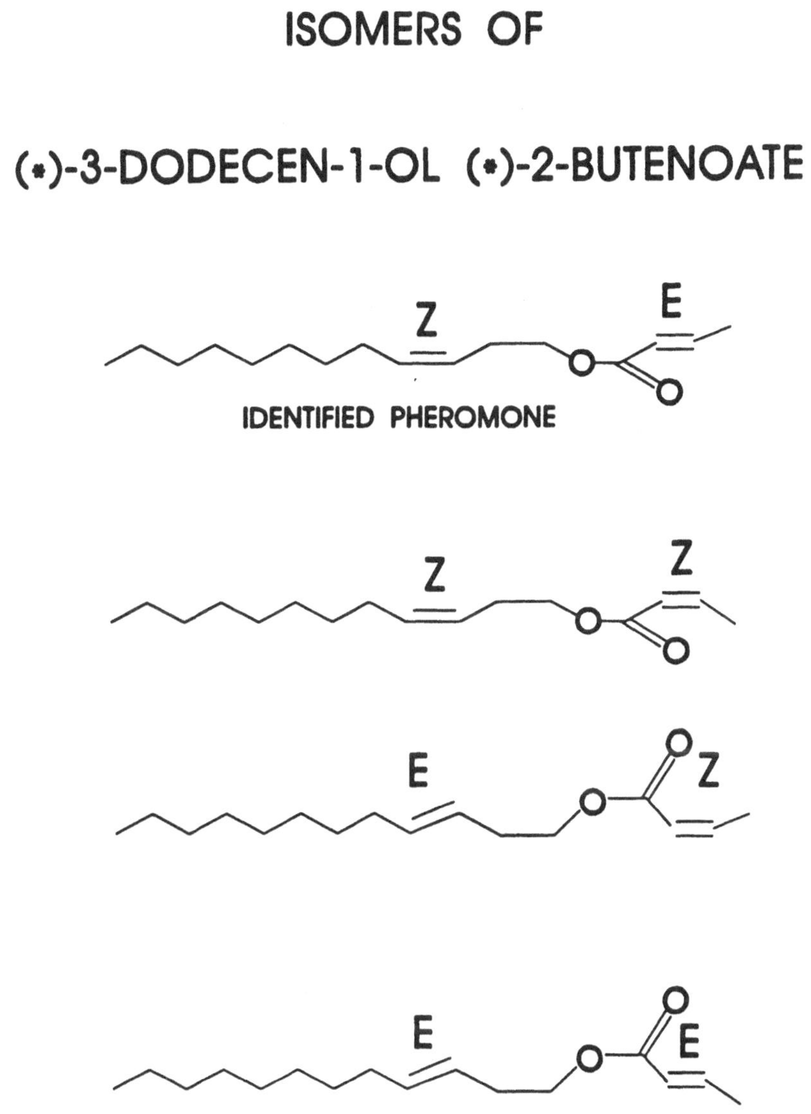

ISOMERS OF

(∗)-3-DODECEN-1-OL (∗)-2-BUTENOATE

IDENTIFIED PHEROMONE

Fig. 5.3. Structure of the identified sex pheromone of *C. formicarius* and the three possible geometric isomers.

In addition to (Z)-3-dodecen-1-ol (E)-2-butenoate, the other three possible geometric isomers were synthesized. The four isomeric forms shown in Figure 5.3 were separated to baseline on a cyanopropyl methyl silicone capillary column (Heath *et al.* 1980). The isolated natural sex attractant had a retention time identical to that of the synthetic (Z)-3-dodecen-1-ol (E)-2-butenoate. The synthetic (Z)-3-dodecen-1-ol (E)-2-butenoate was purified by Ag$^+$ ion paired chromatography (Heath & Sonnet 1980) to > 99%. The synthetic pheromone was identical to the purified natural material by all analytical procedures described. PMR and capillary GC data established that only one compound identical to the synthetic (Z)-3-dodecen-1-ol (E)-2-butenoate had been isolated as the natural sex attractant. Ozonolysis of the synthetic sex attractant produced the same two aldehyde products (nonanal and decanal) that were produced by ozonolysis of the natural sex attractant material.

It appears that the sex pheromone of *C. formicarius* is the first insect sex pheromone identified which uses a crotonate ester as part of the pheromone molecule. The synthetic material is biologically active and attracts *C. formicarius* males, and its attractiveness is comparable or greater than that produced by *C. formicarius* females and comparable to that of the purified natural pheromone.

RESEARCH SUBSEQUENT TO THE IDENTIFICATION OF THE PHEROMONE

The availability of synthetic pheromone provides numerous opportunities for both basic and applied research world-wide. Considerable research is required to realize the full potential of the pheromone, some of these studies are reported elsewhere (see Chapter 6). Earlier studies conducted after the pheromone was identified focused on the effectiveness of the pheromone at catching male weevils, the development of an effective trap, and behavioral responses of males to synthetic analogs of the pheromone.

Effectiveness of the Pheromone

Studies were conducted to compare the numbers of male weevils caught in a light trap, and in pheromone traps baited with one virgin female, three virgin females, or rubber septa loaded with 100 ng or 10 μg of the synthetic sex pheromone. Virgin females were held in a fiberglass tetrahedral cage, which was placed within the pheromone traps. Synthetic pheromone was dissolved in hexane and applied to rubber septa; septa were allowed to equilibrate 24 hours prior to use. In a 24-hour test, PFT-1 traps (*vide infra*) (Lingren *et al.* 1980, see below) baited with 100 ng or 10 μg of synthetic compound caught 75 and 275 males, respectively, compared with only two males (and one female) caught in the light trap, and two and five males caught in traps baited with one or three virgin females, respectively. These data indicated that the pheromone would provide a viable tool for detecting and monitoring sweetpotato weevils.

Development of a Trap

Studies that developed a trap for monitoring *C. formicarius* with sex pheromone have been reported (Proshold *et al.* 1986). The effectiveness of nine different trap types baited with 10 μg of synthetic pheromone formulated on methylene chloride-extracted rubber septa was evaluated under field conditions in St. Croix, U.S. Virgin Islands. Trap types tested were: light trap; sticky trap; water trap; PFT-1 (a plastic funnel live trap used for monitoring the pink bollworm [Lingren *et al.* 1980]), set within a truncated cone made from standard hardware cloth; PFT-2 (trap similar to PFT-1 but with larger entrance holes); black PFT-1 trap (truncated cone was painted black); yellow PFT-1 trap (truncated cone was painted yellow); orange PFT-1 trap (truncated cone was painted orange); and AFT (aluminum funnel trap similar to PFT-1 but made with an aluminum funnel). Results indicated that the AFT trap was most effective at capturing weevils attracted to the pheromone (Figure 5.4). Subsequent studies indicated that the effectiveness of the trap was dependent on the trap height and that the optimum height of the openings to the funnel was at the height of the plant canopy (Proshold *et al.* 1986). No differences in trap catch were found when the funnel diameter varied from 5.5 to 25.5 cm in traps baited with either 100 ng or 10 μg of synthetic pheromone. Based on this information, a trap was developed (Proshold *et al.* 1986). An illustration of this trap is shown in Figure 5.5. The trap consists of a plastic funnel (19.7 cm outside diameter at top) supported on a wire-framed base. Four holes (5.6 cm diameter) were placed equidistant from each other through the side surface of the funnel such that the wire mesh base was level with the openings to the funnel. A plastic vial (5.2 cm outside diameter, 15 cm in length) was attached to the bottom of the funnel to collect weevils. A plastic top (25 x 25 cm) was glued to the top of the funnel to minimize exposure of the lure to rain and sun. A wire was attached to the top plate and the lure was pinned to a cork fastened to the wire. More recent studies using a modification of this trap are reported elsewhere (see Chapter 6). It was apparent, even in

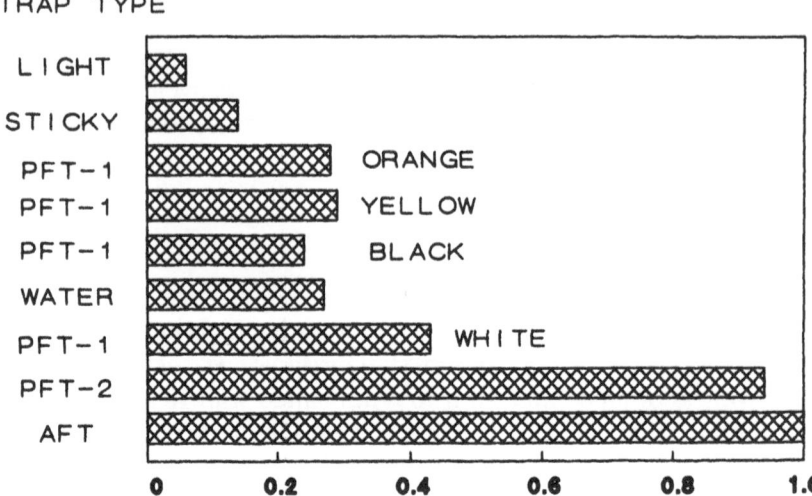

Fig. 5.4. Comparison of the effectiveness of different trap types at capturing male *C. formicarius* weevils attracted to sex pheromone. Data are based on back transformation of ´Y ± 0.05 transformed male counts. Traps are ranked relative to the AFT trap (= 1.00).

these early studies, that a ladder, such as that provided by the wire-framed base, was needed to facilitate weevil movement up to the pheromone trap.

Attractiveness of Pheromone Analogs

The attractiveness of several synthetic pheromone analogs were compared with the synthetic pheromone in the field. The chemicals tested were the pheromone, (Z)-3-dodecen-1-ol (E)-2-butenoate, and the analogs, (E)-3-dodecen-1-ol (E)-2-butenoate, (Z)-2-dodecen-1-ol (E)-2-butenoate, (Z)-4-dodecen-1-ol (E)-2-butenoate, and (Z)-3-dodecen-1-ol butanoate. Results showed that the identified pheromone was the most attractive chemical tested (Figure 5.6). Attractiveness of (E)-3-dodecen-1-ol (E)-2-butenoate and the (Z)-2-dodecen-1-ol (E)-2-butenoate to weevils was likely due to trace amounts of (Z)-3-dodecen-1-ol (E)-2-butenoate present in the synthetic material, and this was later confirmed by GC analysis of these synthetics. It should be noted that the configuration, position, and

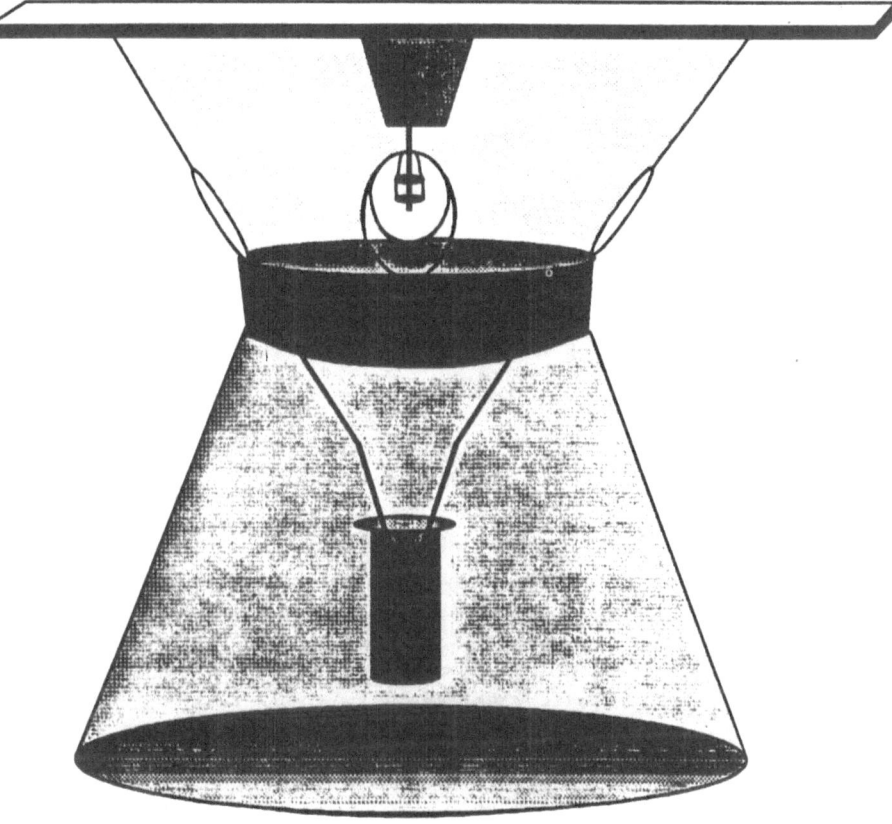

Fig. 5.5. Illustration of plastic funnel trap (developed by F.I.P.) used to capture male *C. formicarius* weevils attracted to pheromone.

presence of the double bonds in the pheromone molecule is important for the capturing of weevils, and the analogs examined did not appear to result in a para-pheromonal response.

CURRENT PHEROMONE RESEARCH - CHEMISTRY

More than 40 scientists world-wide are using the pheromone of *C. formicarius*. The pheromone has been synthesized in several countries (e.g., United States, Canada, Japan, India, Taiwan, and the Netherlands) and tested using different sources and amounts of

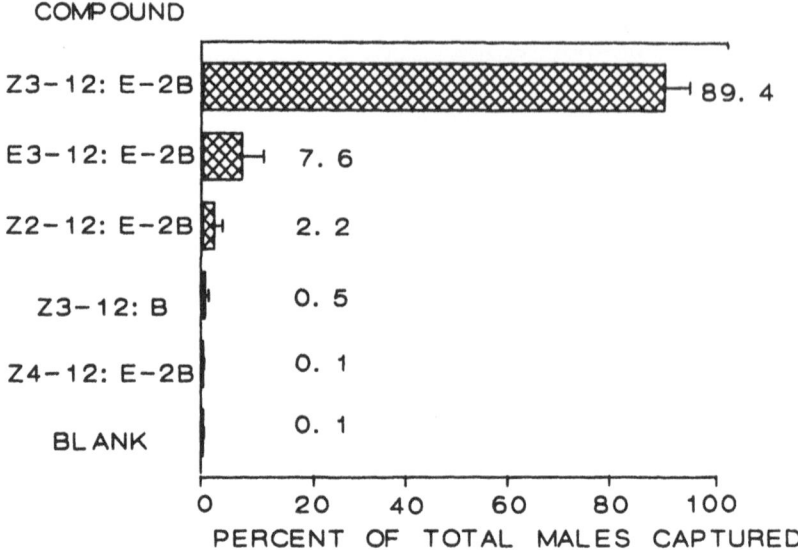

Fig. 5.6. Percentage of male *C. formicarius* caught in traps baited with different synthetic analogs of the pheromone and the pheromone. Z3-12:E-2B = (Z)-3-dodecen-1-ol (E)-2-butenoate; E3-12:E2B = (E)-3-dodecen-1-ol (E)-2-butenoate; Z2-12:E-2B = (Z)-2-dodecen-1-ol (E)-2-butenoate; Z3-12:B = (Z)-3-dodecen-1-ol butanoate; and Z4-12:E-2B = (Z)-4-dodecen-1-ol (E)-2-butenoate.

pheromone on a variety of substrates in various trap types. Also, the attractiveness of the pheromone to other closely related weevils, such as *C. brunneus* (Fabricius) and *C. puncticicollis* (Boheman), to synthetic pheromone has been tested (B. Parker, pers. comm.). Although these studies are important to each researcher within their local region, the significance of each of these studies is difficult to assess because of a lack of a standardized protocol for testing. For example, there is considerable variation in the release rate of pheromone off the different lures that have been used world-wide. Also, analytical data on the purity of the pheromone synthesized and used in tests in different geographic regions is lacking. For these reasons, direct comparisons among research groups is not possible; however, standardized protocols for laboratory and field testing are being developed.

Commercial Development of the Pheromone Lure

The development of protocols for use in a pheromone-trap monitoring system for weevil management is near completion (see Chapter 6). An integral part of such protocols is the establishment of an analytical protocol for the pheromone. Currently, laboratory protocols are being developed to provide the necessary information regarding pheromone purity, pheromone release rate, and lure (mechanism of release) design. In the United States, some of this research is being done through a cooperative effort with the company (AgriSense, Fresno, California) that obtained the license for the patent of the pheromone (Heath *et al.* 1988).

Spectroscopic and chromatographic analyses, such as those that led to the identification of the natural pheromone, have been used to verify the structure of synthetic pheromone. Additional spectroscopic methods that are used to verify the structure of synthetic pheromone include infrared spectroscopy and C^{13} nuclear magnetic resonance.

Currently, analytical methods for determining the emission rate of the pheromone from formulations are difficult to develop due to the extremely small amounts (about 100 pg/hour) of material released from lures. Although release rates from substrates, such as methylene chloride-extracted rubber septa, have been described (Heath *et al.* 1986) at low (ng/hour) and high (pg/hour) levels, these methods can not be used if the background contaminants from the formulation obscure the detection of the pheromone in the analysis. Similarly, impurities from the formulation substrate, which may occur in large amounts may rapidly degrade the performance and destroy the analytical capillary columns used in GC analysis.

The current analytical protocol for determining the release rate of the *C. formicarius* pheromone off of a formulation consists of collecting volatiles emitted from the formulation for 4 to 8 hours, extracting the pheromone from the adsorbent, and then analyzing by GC. The system used to collect pheromone emitted from a lure shown in Figure 5.7. Compressed air, purified by passing through two charcoal filters, enters a glass T which contains a stainless steel frit near the upwind end of the chamber and provides laminar air flow over the lure. The air flow (about 1 l/minute) passes through a collector trap (Super-Q), which has been prepared by packing the adsorbent (approximately 60 mg) in glass tubes (4 cm long x 4.0 mm inside diameter) resulting in a bed length of 15 mm. Two stainless steel frits contain the adsorbent. The Super-Q traps are cleaned by soxhlet extraction with methylene chloride for 24 hours prior to use. Volatiles collected on the Super-Q

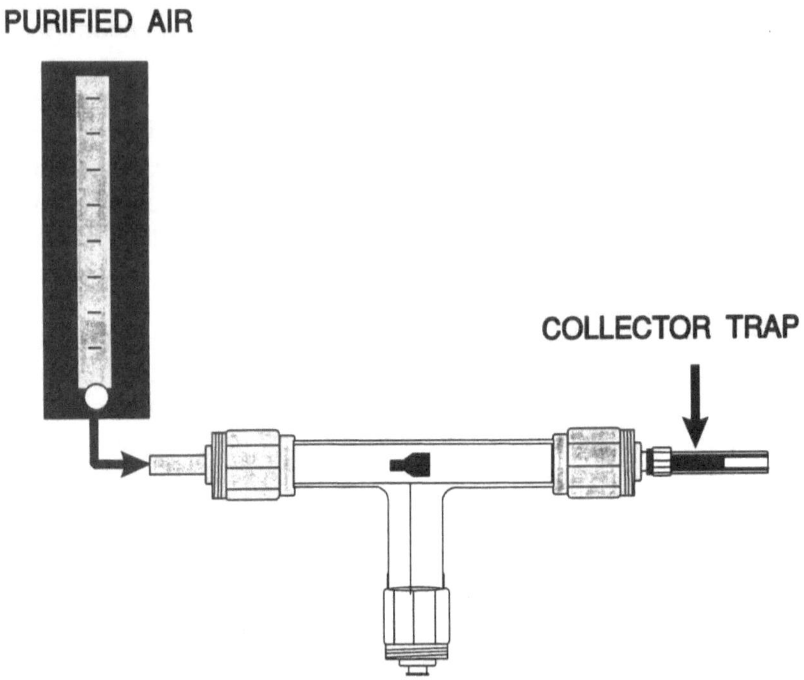

PURIFIED AIR

COLLECTOR TRAP

Fig. 5.7. Illustration of system used to collect pheromone emitted from rubber septa or polymeric substrate.

traces are eluted with methylene chloride (50 μl) and tetradecane (2 ng) is added as an internal standard.

Gas chromatographic analyses is conducted using a capillary gas chromatograph, equipped with splitless capillary and cool-on column injectors and flame ionization detectors. Helium is used as the carrier gas at a linear flow velocity of 18 cm/second. Columns used are either a apolar capillary column or a polar capillary column (typically 50 m x 0.25 mm inside diameter). Columns are operated at 60°C for 2 minutes, then temperature programmed at 30°C/minute to 160°C. The identity of the peak is confirmed by GC-MS analyses using one of the capillary columns described above, coupled to a mass spectrometer using chemical ionization with either methane or isobutane.

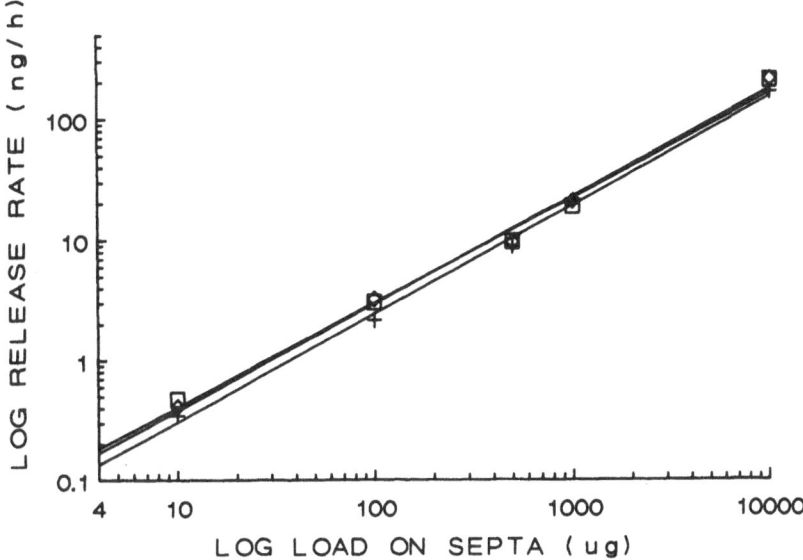

Fig. 5.8. Release rate of pheromone from septa obtained from extracted A. H. Thomas septa = squares, (Y release rate [ng/hr] = 0.02 x septa load - 0.36, r^2 = 0.99), extracted Wheaton septa = triangles, (Y release rate [ng/hr] = 0.16 x septa load + 1.28, r^2 = 0.99) and nonextracted Wheaton septa = plus (Y release rate [ng/hr] = 0.02 x septa load - 0.02, r^2 = 0.99).

Measurement of Pheromone Emission

Based on the above methods we have investigated the release rate of the pheromone off rubber septa obtained from two different commercial sources: A.H. Thomas Scientific (no. 8753-d22), Swedesboro, New Jersey and Wheaton (no. 224091), Millville, New Jersey. Septa from A.H. Thomas have been used extensively in our research. These septa have been soxhlet extracted for 24 hours with methylene chloride. These septa were compared with methylene chloride extracted and nonextracted septa obtained from Wheaton. Septa were loaded with 10, 100, 500, 1,000, and 10,000 µg of synthetic pheromone and release rates were measured 72 to 96 hours after septa were equilibrated in a laboratory hood with an air flow of approximately 1,320 m/hour. Results are shown in Figure 5.8. Release rate off septa increased with an increase in pheromone dosage. More importantly, release rates of pheromone did not differ among the three septa types. Field bioassays demonstrated that these lures caught comparable percentages of male weevils indicating that soxhlet extraction did not increase lure attractiveness (see Chapter 6). Septa from Wheaton, which are less expensive (currently $0.07 each compared with $0.14 each for septa from Thomas Scientific), do not require extraction with methylene chloride prior to use. Large scale production of lures which require soxhlet extraction is expensive, requires specialized laboratories, and presents safety and health considerations.

Several commercial formulations that incorporate the pheromone into a polymeric substrates have been developed by AgriSense. However, attempts to determine release rates of the pheromone off these substrates were unsuccessful using the analytical methods described above because of emission of compounds that co-elute with the pheromone. This type of formulation method does provide the potential for preparation of large number of lures at a significantly reduced cost compared to using rubber septa. For this reason, considerably more research is warranted in the development of polymer-based lures with reduced background emission and/or development of analytical techniques that can be used for analyses of the pheromone in the presence of volatile contaminants.

REFERENCES

Akazawa, T., I. Uritani & H. Kubota. 1960. Isolation of ipomoeamarone and two coumarin derivatives from sweet potato roots injured by the weevil, *Cylas formicarius elegantulus*. Arch. Biochem. Biosphys. 88:150-156.

Anonymous. 1978. Postharvest food losses in developing countries. FAO 206-208.

AVRDC. 1976. AVRDC progress report summaries, 1975. Asian Vegetable Research and Development Center, Shanhua, Tainan, Taiwan, Republic of China.

Chalfant, R.B., R.K. Jansson, D.R. Seal & J.M. Schalk. 1990. Ecology and management of sweet potato insects. Annu. Rev. Entomol. 35:157-180.

Coffelt, J.A., K.W. Vick, L.L. Sower & W.T. McClellan. 1978. Sex pheromone of the sweetpotato weevil, *Cylas formicarius elegantulus*: laboratory bioassay and evidence for a multiple component system. Environ. Entomol. 7:756-758.

Edmond, J.B. 1971. Sweet potato pests. Part 4: Destructive insects, pp. 196-207. *In* J.B. Edmond & G.R. Ammerman (eds.). Sweet potatoes: production, processing and marketing. Avi, Westport, Connecticut.

FAO. 1984. Food production yearbook. Food and Agriculture Organization, Rome.

Heath, R.R. & P.E. Sonnet. 1980. Technique for *in situ* coating of Ag$^+$ on to silica gel in HPLC columns for the separation of geometrical isomers. J. Liquid Chromatogr. 3:1129-1135.

Heath, R.R. & J.H. Tumlinson. 1984. Techniques for purifying, analyzing, and identifying pheromones, pp. 287-322. *In* H.E. Hummel & T.A. Miller (eds.). Techniques in pheromone research. Springer-Verlag, New York.

Heath, R.R., G.E. Burnsed, J.H. Tumlinson & R.E. Doolittle. 1980. Separation of a series of positional and geometrical isomers of olefinic aliphatic primary alcohols and acetates by capillary gas chromatography. J. Chromatogr. 189:199-208.

Heath, R.R., J.R. Jordan, P.E. Sonnet & J.H. Tumlinson. 1979. Potential for the separation of insect pheromones by gas chromatography on columns coated with cholesteryl cinnamate, a liquid-crystal phase. J. HRC & CC 2:712-714.

Heath, R.R., P.E.A. Teal, J.H. Tumlinson & L.J. Mengelkoch. 1986a. Prediction of release ratios of multicomponent pheromones from rubber septa. J. Chem. Ecol. 12:2133-2143.

Heath, R.R., J.A. Coffelt, P.E. Sonnet, F.I. Proshold, B. Dueben & J.H. Tumlinson. 1986b. Identification of a sex pheromone produced by female sweetpotato weevil, *Cylas formicarius elegantulus* (Summers). J. Chem. Ecol. 12:1489-1503.

Heath, R.R., J.A. Coffelt, F.I. Proshold, P.E. Sonnet & J.H. Tumlinson. 1988. (Z)-3-dodecen-l-ol (E)-2-butenoate and its use in monitoring the sweetpotato weevil. U.S. Patent No. 4,732,756.

Kovats, E. 1965. Retention index system. Adv. Chromatogr. 1:229-234.

Louton, P.A. 1975. Localization of sex pheromone production in female *Cylas formicarius elegantulus* (Summers). M.S. Thesis, Louisiana State University, Baton Rouge.

Lingren, P.D., J. Barton, W. Shelton, & J.R. Raulston. 1980. Night vision goggles for design evaluation and comparative efficiency determination of a pheromone trap for capuring live adult male pink bollworms. J. Econ. Entomol. 73:622-630.

Martin, F.W. 1983. Goals in breeding the sweet potato for the Caribbean and Latin America. Proc. Am. Soc. Hort. Sci., Trop. Reg. 27(B):61-70.

Proshold, F.I. 1983. Mating activity and movement of *Cylas formicarius elegantulus* (Coleoptera: Curculionidae) on sweet potato. Proc. Am. Soc. Hort. Sci., Trop. Reg. 27(B):81-92.

Proshold, F.I., J.L. Gonzalez, C. Asencio & R.R. Heath. 1986. A trap for monitoring the sweetpotato weevil (Coleoptera: Curculionidae) using pheromone or live females as bait. J. Econ. Entomol. 79:641-647.

Russo, A. 1973. Studies directed toward the isolation and structure determination of sex and food attractants in *Cylas formicarius elegantulus* (Summers). Ph.D. Dissertation, Louisiana State University, Baton Rouge.

Schalk, J.M. & A. Jones. 1985. Major insect pests, pp. 59-78. *In* J.C. Bouwkamp (ed.). Sweet potato products: a natural resource for the tropics. CRC Press, Boca Raton, Florida.

Sutherland, J.A. 1986. Damage by *Cylas formicarius* (Fab.) to sweet potato vines and tubers, and the effect of infestations on total yield in Papua New Guinea. Trop. Pest Manage. 32:316-323.

Uritani, I., H. Saito, H. Honda & W.K. Kim. 1975. Induction of furanoterpenoids in sweet potato roots by the larval components of the sweet potato weevils. Agric. Biol. Chem. 39:1857-1862.

6. Use of Sex Pheromone for Monitoring and Managing *Cylas formicarius*

Richard K. Jansson and Linda J. Mason
Tropical Research and Education Center
University of Florida, I.F.A.S.
Homestead, Florida 33031 U.S.A.

Robert R. Heath
U.S. Department of Agriculture, Agricultural Research Service
Insect Attractants, Behavior, and Basic Biology Research Laboratory
Gainesville, Florida 32604 U.S.A.

SEX PHEROMONES

Pheromones are semiochemicals secreted by an organism to the outside that cause a specific reaction in a receiving organism of the same species (Nordlund 1981). These chemicals have been classified on the basis of the type of interaction mediated, such as sex pheromones, alarm pheromones, and epideictic pheromones (Nordlund 1981). Sex pheromones have become the most publicized semiochemical, especially in terms of their usefulness in integrated pest management (IPM) programs (Roelofs 1981). Sex pheromone communication systems are highly sensitive and specific. Specificity is achieved by the uniqueness of the chemical structure (e.g., variation of position or geometry of double bonds, chirality of structure, etc.), which may be an isolating mechanism in certain species, and by the specific blend of pheromone components (in multicomponent pheromones) (Bestmann & Vostrowsky 1988, Lanier & Burkholder 1974). However, specificity also allows for exploitation as demonstrated by the numbers of reviews and books on the use of sex pheromones for monitoring and managing insect populations (e.g., Birch & Haynes 1982, Jutsum & Gordon 1989, Kydonieus & Beroza 1982, Mayer & McLaughlin 1990, Mitchell 1981, Morgan & Mandava 1988, Ridgway *et al.* 1990, Roelofs 1979, 1981, Roelofs & Carde 1977).

Sex pheromones, like other nonconventional control tactics, have become increasingly more important in pest management due to a greater awareness of the deleterious effects of conventional chemical insecticides on agroecosystems (e.g., development of insecticide-resistant populations, effects on nontarget beneficial organisms, environmental contamination, and outbreaks of secondary pest problems).

This chapter examines the potential uses of a sex pheromone of the sweetpotato weevil, *Cylas formicarius* (Fabricius) (= *C. f. elegantulus* [Summers] and *C. f. formicarius* [Fabricius]; see Chapter 2), for managing this insect. This pest is the most important biotic factor limiting sweet potato production world-wide (Chalfant *et al.* 1990). The previous chapter reviewed some of the initial research conducted on the pheromone and discussed the importance of pheromone and formulation chemistry in pheromone trap monitoring programs for this weevil (see Chapter 5). Recently, considerable research has been conducted to develop a pheromone-trap monitoring system for this weevil in the southern United States and the Caribbean basin. This chapter reviews the development of this monitoring system and the potential of the pheromone for reducing weevil populations by mass trapping and mating disuption. Use of the pheromone as a regulatory tool to detect low level weevil infestations in quarantined areas is discussed later (see Chapter 13). The future outlook of sex pheromone in weevil management programs in developed and developing countries is discussed.

THE SWEETPOTATO WEEVIL SYSTEM

Sweetpotato weevils are cryptic by nature and develop within the vines and roots of plants. Approximately 80-90% of the weevil population within vines and roots is distributed below the soil surface (Jansson *et al.* 1990a). For this reason, sampling for weevils is physically demanding and time consuming. Until recently, there were no published sampling plans available for this weevil. Jansson and McSorley (1990) recently developed a sampling plan for this weevil and found that an excessive number of samples were needed to reliably estimate weevil densities (see Chapter 8). Thus, an alternative method is needed to monitor this insect pest.

Historically, the lack of reliable and cost-effective methods to detect the presence of insect pests at low population densities has been a major constraint to optimum utilization of control methods (Knipling 1982). Sex pheromones, however, have often been effective at detecting insect populations, even at low population densities. For this reason, studies were initiated to assess the potential of the female-produced sex pheromone of *C. formicarius* as a monitoring tool for this pest. Previous researchers demonstrated that the female-produced pheromone attracted only males (Coffelt *et al.* 1978). Subsequent studies isolated, identified, and synthesized the pheromone (Heath *et al.* 1986). Complete details of the early work on this pheromone were reported previously (see Chapter 5).

SEX PHEROMONE AS A MONITORING TOOL

The development of a pheromone-trap monitoring program for *C. formicarius* has relied on the use of systems analysis. There are four principal components of the pheromone-trap system for *C. formicarius* (Figure 6.1). These include (1) the pheromone (e.g., dosage, purity, longevity, source, formulation/lure type, and release rate off lures), (2) the trap (design, height), (3) the weevil (behavioral, chemical, physiological, and population ecology), and (4) the environment (climatic conditions, such as wind speed, wind direction, rainfall, and temperature, which affect plume structural dynamics, release rates, and weevil behavior, and field [plant] age, which affects plume structure and weevil behavior). The effects of most of these components on weevil trap captures have been assessed and are presented below.

Pheromone

Of central importance to the development of a pheromone-trap monitoring system are the effects of pheromone dosage, purity, longevity, production source, longevity, and lure (formulation) type on insect captures in traps. Several studies have been conducted to assess the response of male *C. formicarius* to changes in these factors (Jansson *et al.* 1989, 1990c, 1992a,b,c).

Pheromone Dosage. The effects of pheromone dosage on weevil trap captures have been examined in Puerto Rico, Florida, Texas, North Carolina, and Louisiana (Hammond *et al.* 1989, Jansson *et al.* 1990c, 1992c, Mason *et al.* 1990). In general, weevil counts in traps increased with an increase in pheromone dosage (e.g., see Figure 6.2). The regression relationship between total weevil counts and dosage did not differ ($P > 0.05$) among trials despite large differences in weevil densities (Jansson *et al.* 1990c). They noted, however, that

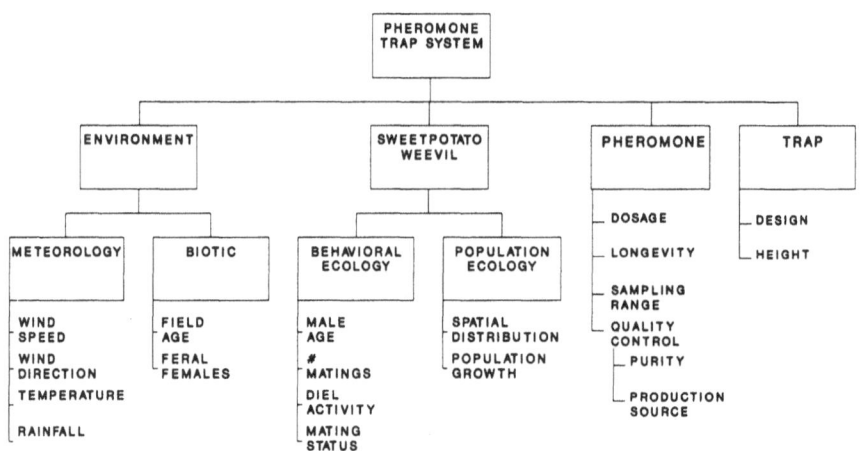

Fig. 6.1. Diagram of the components of the pheromone-trap monitoring system that affect numbers of male *C. formicarius* caught in pheromone-baited traps.

weevil trap counts were not consistently positively correlated with pheromone dosage on each night of each trial, especially during periods of low weevil density. Most weevils were caught on the first or one of the first night(s), which was also observed by Proshold *et al.* (1986).

In a more recent study, the effect of pheromone dosage on weevil counts in pheromone traps was assessed in four regions of the southern United States (Jansson *et al.* 1992c). They found that the local population density of *C. formicarius* present within each region may affect the relationship between weevil trap counts and dosage. For example, in southern Florida, where weevil populations were very dense, and southeastern Texas, where populations were moderately dense, trap counts were consistently positively correlated with dosage, and the relationship between these two variables was similar between these regions (Figure 6.3). Conversely, in Louisiana, where weevil

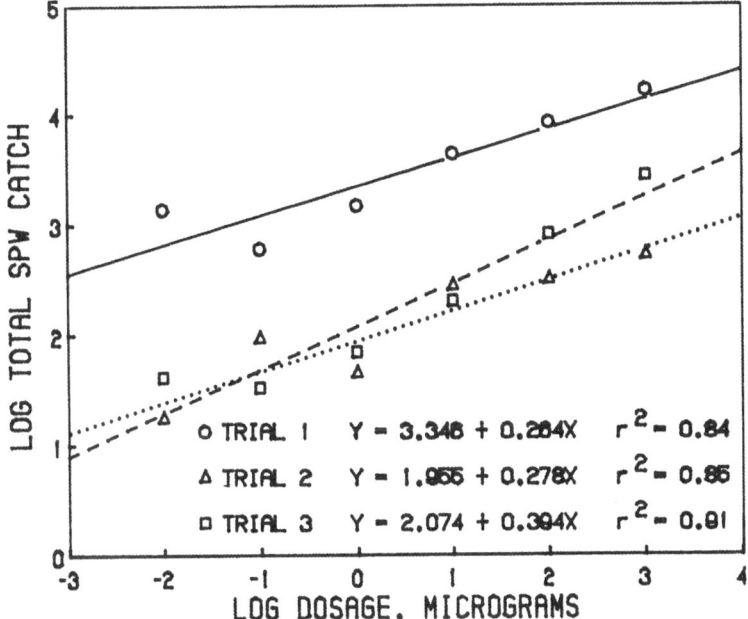

Fig. 6.2. Log$_{10}$-transformed total numbers of *C. formicarius* males caught in plastic funnel traps baited with different dosages (μg) of synthetic sex pheromone in southern Florida (from Jansson *et al.* 1990c; reprinted with permission from Trop. Pest Manage. [Taylor and Francis, Ltd., London]).

populations were sparse to moderately dense, and in North Carolina, where weevil populations were sparse, trap counts were not consistently positively correlated with dosage, especially during periods of low density. During periods of moderate weevil density, however, such as found during one trial (3) in Louisiana, the relationship between weevil counts and dosage resembled those found in Florida and Texas. Thus, in areas with moderate to high weevil densities, dosage significantly affected trap catch, whereas, in areas with low densities, dosage did not significantly affect trap catch. For this reason, a low dosage, such as 10 μg, is probably adequate for monitoring this insect world-wide. Use of the pheromone for mating disruption or mass trapping, however, may require higher dosages, such as 100 μg or 1 mg.

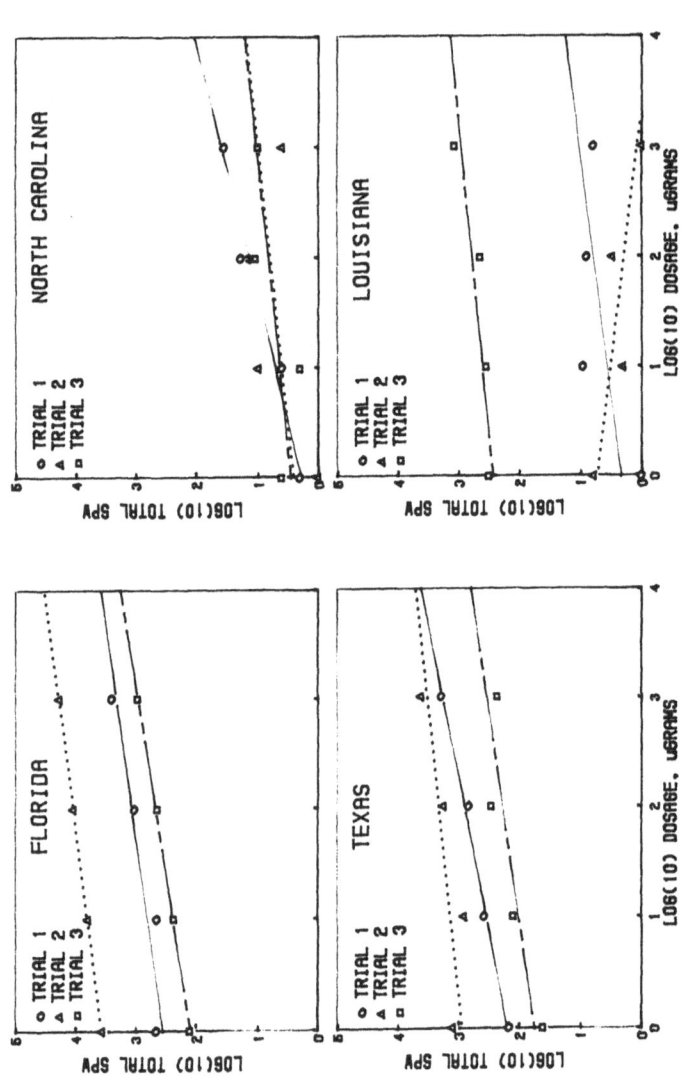

Fig. 6.3. Log$_{10}$-transformed total *C. formicarius* males caught in traps baited with different dosages (μg) of synthetic sex pheromone formulated on methylene chloride-extracted rubber septa in four regions of the southern United States (from Jansson *et al.* 1992c).

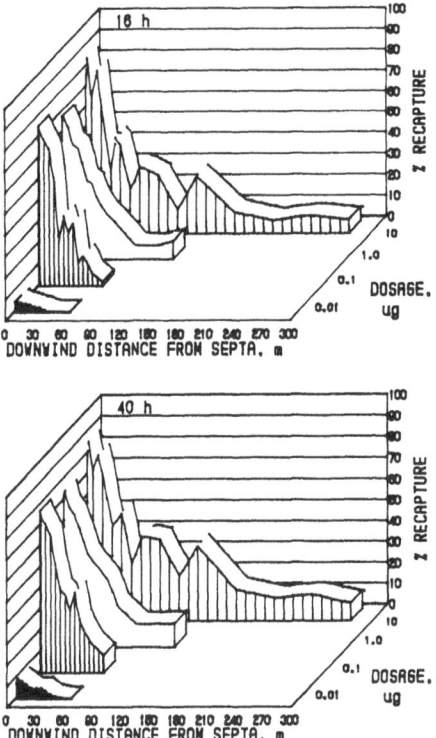

Fig. 6.4. Percentages of marked *C. formicarius* males recaptured at 16 and 40 hours in plastic funnel traps baited with different dosages of synthetic sex pheromone when released at different distances downwind of the pheromone source (from Mason *et al.* 1990; reprinted with permission from J. Chem. Ecol. [Plenum Publishing Corp., New York]).

In another study, Mason *et al.* (1990) determined the sampling range of plastic funnel traps baited with different dosages of synthetic sex pheromone. Male *C. formicarius* were marked with enamel paint and released from various distances downwind from the pheromone source. Traps counts were recorded 16 and 40 hours after release. They found that percentages of males caught in traps decreased with a increase in release distance, and with a decrease in pheromone dosage at each release distance (Figure 6.4). For example, when traps were baited with 10 μg, over 80% of males were recaptured when released 5 m downwind of the traps compared with about 10% recaptured when released 280 m downwind of the traps. Corresponding levels of males (i.e., 10%) were recaptured at 120, 70, and 5 m when traps were baited with 1 μg, 100 ng, and 10 ng, respectively. These data showed that

Table 6.1. Parameter estimates for arcsine-transformed percentages of the total *C. formicarius* catch per replicate per night regressed on lure age in southern Florida (Jansson *et al.* 1990c).

Trial	Intercept[a]	Slope[a]	P[b]	r^2
1	0.059(0.002)a	-0.0007(0.0001)a	****	0.34
2	0.057(0.001)a	-0.0005(0.0001)ab	****	0.33
3	0.052(0.003)a	-0.0004(0.0001)b	***	0.08
4	0.056(0.002)a	-0.0005(0.0001)ab	****	0.20

[a] Means within a column followed by the same letter are not significantly different ($P > 0.05$) by the general linear test.

[b] Significance level of regression: ***, $P < 0.001$; ****, $P < 0.0001$.

pheromone dosage affects the distance that weevils can be attracted to pheromone-baited traps in a given time period.

Pheromone Longevity. The length of time that lures remain attractive to male weevils has been assessed in Florida and the U.S. Virgin Islands (Jansson *et al.* 1990c, Mason *et al.* 1990). Trap captures were compared among methylene chloride-extracted rubber septa loaded with 10 μg of synthetic sex pheromone and aged outdoors for different durations (0 to 30 days). Trap counts were significantly negatively correlated with lure age (Table 6.1). Parameter estimates did not differ among the four trials indicating that the effects of aging on lure attractiveness to weevils were consistent. Although trap catch was negatively correlated with lure age, slopes of the regression analyses were shallow and coefficients of determination (r^2) were low. From a practical perspective, differences in attractive- ness of 0- to 30-day-old lures to weevils were minimal. These data indicate that lures are attractive to male weevils for at least 30 days under field conditions. In an earlier study conducted in St. Croix, U.S. Virgin Islands, septa aged outdoors in the sun or shade were as attractive to male weevils as fresh septa for up to 7 weeks (49 days) (Jansson *et al.* 1990c). In a more recent study, Mason *et al.* (1990) found that septa aged outdoors for 64 days were as attractive as fresh lures. Collectively, these data indicate that the sex pheromone of *C. formicarius* is relatively stable on methylene chloride-extracted rubber septa and remains attractive to male weevils for at least 30 to 64 days.

Pheromone Purity. The purity of *C. formicarius* pheromone, (Z)-3-dodecen-1-ol (E)-2-butenoate, may vary depending on the synthetic method used. Isomerization of the two olefinic bonds can result in production of three additional geometric isomers. Also, several analogs of the pheromone (see Chapter 5) can be produced during synthesis. These intermediates and byproducts may significantly reduce the attractiveness of a pheromone. In addition, even small amounts of unwanted compounds may completely inhibit the attractiveness of a pheromone. Such changes in pheromone purity are especially important in quality control programs. For these reasons, the effect of pheromone purity on weevil trap captures was determined. Trials were conducted in four regions of the southern United States. Purities of 75, 85, 92, and 99% were tested. Purity was varied by altering the reaction time and temperature used in the coupling of the crotonyl chloride to (Z)-3-dodecen-1-ol. The pheromone designated 99% was further purified using $AgNO_3$ high performance liquid chromatography as previously described (Heath & Sonnet 1980). Pheromone load (10 μg) on methylene chloride-extracted rubber septa was held constant.

The relationship between percentages of weevils caught in pheromone traps and pheromone purity varied among the four regions of the southern United States with different weevil densities (Figure 6.5) (Jansson *et al.* 1992b). In regions with high densities, such as Florida, slope estimates were shallow (0.003-0.004) and regressions were not significant in two of three trials, suggesting that pheromone purity did not appreciably affect weevil captures. Similar results were found in Texas, where intermediate densities were present. Conversely, in regions with low weevil densities, such as Louisiana, percentages of weevils caught were significantly positively correlated with pheromone purity (slope = 0.013; $P < 0.001$). Inconsistent results were found in North Carolina, where weevil densities were the lowest of all test sites. Percentages of weevils caught were positively correlated (slope = 0.011-0.016) with purity in two of three trials, although regressions were not consistently significant. These data suggest that pheromone purity may affect trap captures in regions with low weevil densities, but is less important in regions with intermediate and high densities. Considering the importance that this pheromone may have as a regulatory tool in regions with low weevil densities, it is essential that very pure pheromone be used in monitoring programs. Thus, syntheses and purification procedures used to produce very pure (99%) pheromone product are suggested for best results.

Pheromone Source. Sex pheromone of *C. formicarius* has been synthesized in the United States, Canada, Japan, India, and Taiwan. The method used to synthesize pheromone at each of these locations is not known. In order to develop a standardized pheromone-trap

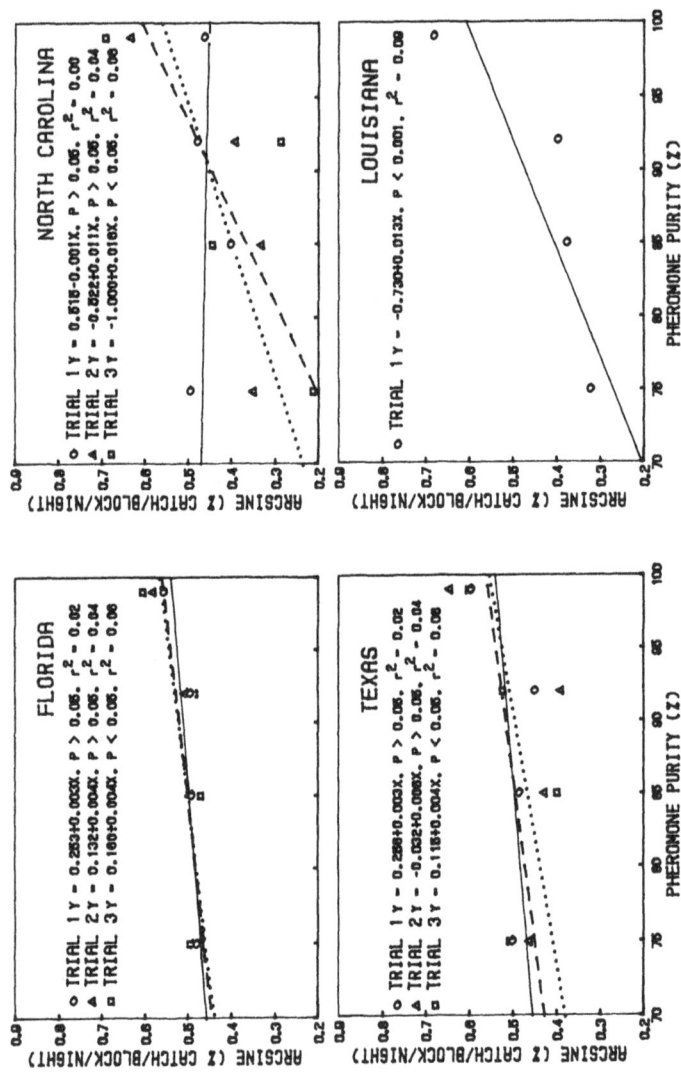

Fig. 6.5. Relationship between arcsine-transformed percentages of *C. formicarius* males caught per night in traps baited with 10 μg of pheromone with different levels of purity (75, 85, 92, and 99%) in four regions of the southern United States.

monitoring system for use world-wide, the attractant used in traps must be chemically identical to that used in studies that developed the system. Quality control of pheromone is central to the development of a reliable pheromone-trap monitoring system. For this reason, studies were conducted to compare the attractiveness of pheromone supplied by two sources: U.S. Department of Agriculture, Agricultural Research Service, Insect Attractants, Behavior, and Basic Biology Research Laboratory, Gainesville, Florida (by R.R. Heath); and AgriSense (Fresno, California; the company to whom the patent for the pheromone [Heath *et al.* 1988] was licensed). Chemical analyses showed that the commercially-synthesized pheromone was comparable to that produced by the U.S.D.A. (Jansson *et al.* 1992b). Weevil counts were compared in traps baited with two dosages (10 and 100 μg) of each of the the two pheromone sources. Experiments were conducted in four regions of the southern United States.

Results showed that the source of pheromone production did not affect trap captures. Percentages of weevils caught did not consistently differ between traps baited with U.S.D.A. and AgriSense pheromone at dosages of 10 and 100 μg (Figure 6.6) (Jansson *et al.* 1992b). Although pheromone produced by these two sources did not differ in their attractiveness to weevils, it is imperative that other producers of the pheromone in other countries, such as India, Taiwan, etc., compare their synthetic pheromone with a known standard for attractiveness to weevils.

Lure Type. Previous studies used pheromone loaded on rubber septa that were soxhlet extracted for 24 hours with methylene chloride. Such lures are not practical for commercial development, nor for use in developing countries. Extraction procedures are costly, labor intensive, and involve the use of methylene chloride, a costly and hazardous chemical. For this reason, two studies were conducted to compare different lure/formulation types for attractiveness to weevils in the field. One experiment compared the attractiveness of methylene chloride-extracted rubber septa loaded with three dosages of AgriSense pheromone (1, 10, and 100 μg) and one dosage (10 μg) of U.S.D.A. pheromone with commercially-formulated selibate lures loaded with two dosages of pheromone (10 and 100 μg). The second experiment compared the attractiveness of pheromone (10 μg) loaded on extracted and nonextracted rubber septa obtained from two sources, A.H. Thomas Scientific (Swedesboro, New Jersey; 5 x 9 mm rubber stoppers) and Wheaton Co. (Millville, New Jersey; 5 x 11 mm natural sleeve rubber stoppers).

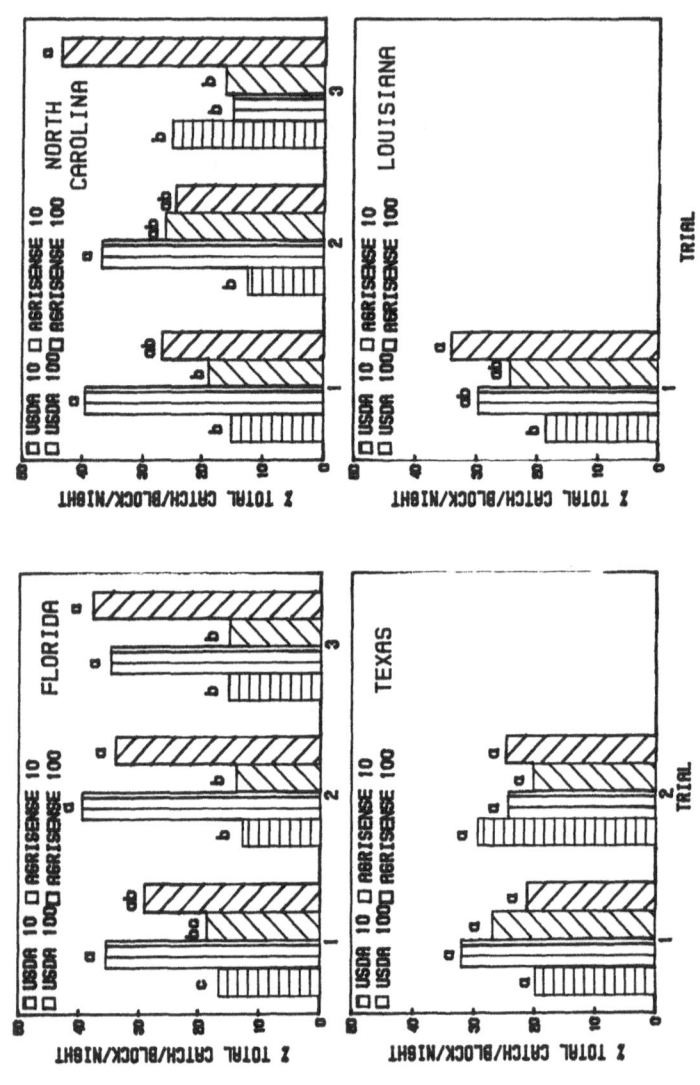

Fig. 6.6. Percentages of *C. formicarius* males caught per trap per night in traps baited with 10 and 100 μg of synthetic sex pheromone produced by two sources (U.S.D.A. and AgriSense) in each of four regions of the southern United States. Bar graphs with different letters within a trial indicate significant differences among treatments by the Waller-Duncan *K*-ratio *t*-test (from Jansson *et al.* 1992b).

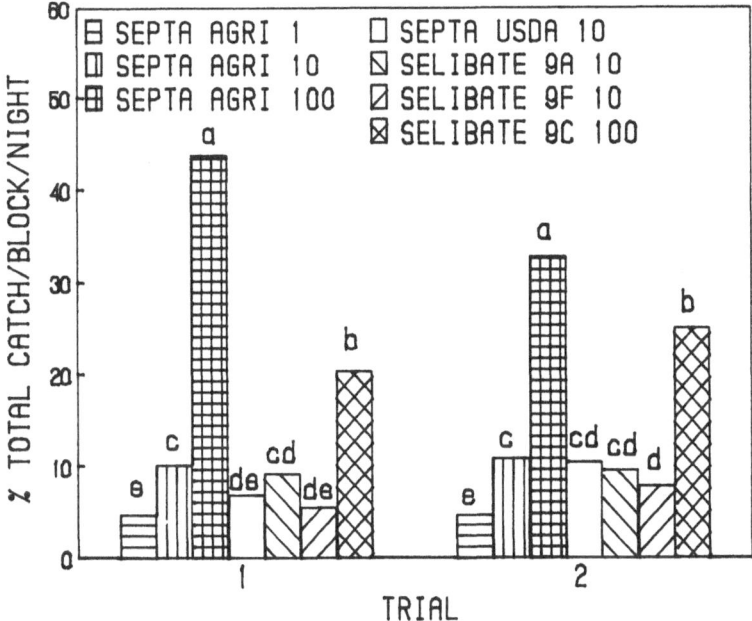

Fig. 6.7. Percentage of *C. formicarius* male trap catch per replicate in traps baited with one of seven different lures: methylene chloride extracted rubber septa loaded with 1, 10, and 100 μg of AgriSense pheromone or 10 μg of U.S.D.A. pheromone, two selibate (polymer-based) lure formulations loaded with 10 μg of AgriSense pheromone, and one selibate formulation loaded with 100 μg of AgriSense pheromone (from Jansson *et al.* 1992a).

In the first experiment, weevil captures differed among the various lures tested (Figure 6.7) (Jansson *et al.* 1992a). Percentages of weevils caught did not consistently differ among lures loaded with 10 μg. Rubber septa loaded with 10 μg of pheromone (U.S.D.A. or AgriSense) were as attractive as commercially-prepared lures loaded with 10 μg, although traps baited with one selibate formulation (0.6 x 0.6 cm; 9F) tended to catch fewer weevils. Percentages of weevils caught differed between lures loaded with 100 μg of pheromone. Traps baited with pheromone formulated on rubber septa caught higher percentages of males than those baited with pheromone formulated on a selibate lure.

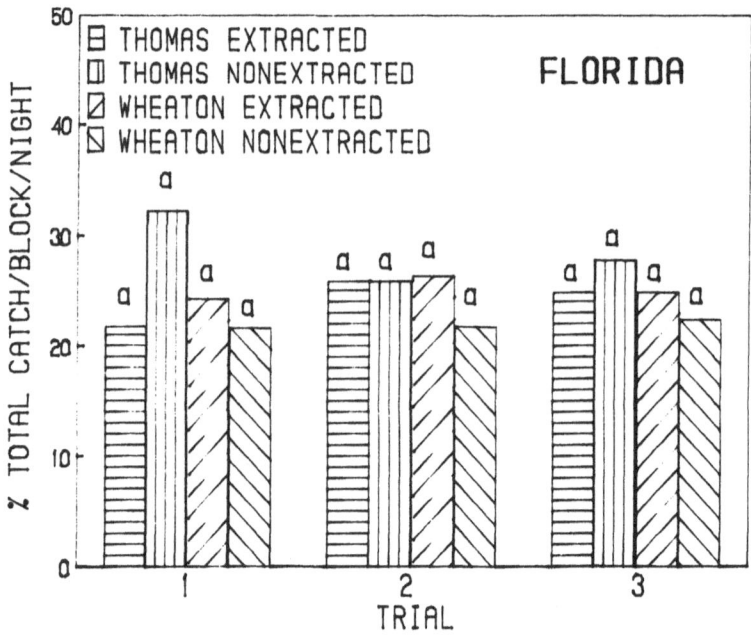

Fig. 6.8. Percentages of *C. formicarius* males caught per trap per night in traps baited with one of four different rubber septa: methylene chloride extracted and nonextracted septa obtained from two sources (A.H. Thomas Scientific, Swedesboro, New Jersey and Wheaton, Millville, New Jersey). Septa were loaded with 10 μg of U.S.D.A. pheromone (from Jansson *et al.* 1992a).

In the second experiment, percentages of weevils caught did not differ among the four lure types (Figure 6.8) (Jansson *et al.* 1992a). Extracted and nonextracted rubber septa obtained from the two sources were equally attractive to male weevils. For this reason, the use of nonextracted Wheaton rubber septa should be studied further for lure development. Soxhlet extraction of this lure was not necessary to increase its attractiveness to weevils. Elimination of this step will result in a cost savings and a safer method for mass producing lures. Additionally, the cost of septa from Wheaton is about one-third of that of septa from A.H. Thomas. Additional studies are needed to determine the effects of pheromone dosage, purity, and longevity on weevil captures in traps baited with nonextracted Wheaton septa to refine the current system.

Slow-release and fast-release dispensers composed of three-layer vinyl polymer laminates of different barrier thicknesses were compared for their attractiveness to *C. formicarius* in Louisiana (Hammond *et al.* 1989). They found that slow-release dispensers loaded with 25 μg of pheromone were more attractive than fast-release dispensers loaded with 25 μg of pheromone and methylene chloride-extracted rubber septa loaded with 10 μg of pheromone. More work is needed to develop alternative, commercially-produced lures.

Trap

A second important component in a pheromone-trap monitoring or management program is trap design. The behaviors of insects as they approach, enter, and try to escape from traps can be very different, and may affect selection of an appropriate design for a given insect (Birch & Haynes 1982). In addition to the chemical cue(s) of the pheromone, insects also use visual cues to locate traps. Thus, trap color may significantly affect trap selection, especially in insects that are active during daylight. Additionally, an insect's propensity to fly, and flight behaviors, may affect the selection of a proper trap design, as well as affect the selection of trap placement (location, height).

Trap Design and Height. Several studies have compared the effectiveness of different trap designs at collecting male *C. formicarius*. Proshold *et al.* (1986) found that a plastic funnel trap was the most effective trap design at collecting weevils. A complete description this trap and a modification of this trap are presented elsewhere (Proshold *et al.* 1986, Jansson *et al.* 1989, also see Chapter 5). They also determined the effects of various trap heights (9 to 85 cm) on weevil captures, and found that traps with a height equal to the height of the canopy were most effective at catching weevils. Observations of the behavior of *C. formicarius* indicated that male weevils move to the upper leaves at night to search for potential mates (Proshold 1983).

A recent study compared three types of commercial screen cone boll weevil traps for monitoring weevils in Louisiana (Hammond *et al.* 1989). A more recent study, however, found that the plastic funnel trap was more effective and more efficient at collecting male weevils than screen cone boll weevil traps (Jansson & Heath 1990, Jansson *et al.* 1989). Subsequent studies compared the efficacy of three trap types (plastic funnel, screen cone boll weevil, and Universal moth trap) in four regions of the southern United States (Figure 6.9). Screen cone boll weevil traps and Universal moth traps were suspended into the

Fig. 6.9. Three pheromone trap types tested in four regions of the southern United States: screen cone boll weevil trap (left), plastic funnel trap (center), and Universal moth trap (right).

plant canopy with the entrance holes on these traps raised to a height slightly above the canopy to minimize plant interference. Plant foliage was arranged around the traps to facilitate weevil entry into these traps. Plastic funnel traps rested on a wire-framed base with the openings in the funnel positioned slightly above the plant canopy.

Results indicated that the resident weevil density affected the ability to determine differences among traps. In regions with moderate to high weevil densities, such as Florida and Texas, differences among traps were very apparent, whereas in regions with lower densities, such as Louisiana and North Carolina, differences among traps were less apparent, especially early in the growing season (Figure 6.10) (Jansson *et al.* 1992c). In regions with moderate to high densities, the plastic funnel trap was the most effective trap followed in decreasing order by the Universal moth trap and the screen cone boll weevil trap. In regions with low densities, trap catch did not consistently differ among trap types.

Plastic funnel and Universal moth traps also minimized weevil escape. Between 2-3% of marked male weevils escaped from these traps overnight compared with approximately 73% escape from the screen cone boll weevil trap (Jansson *et al.* 1992c).

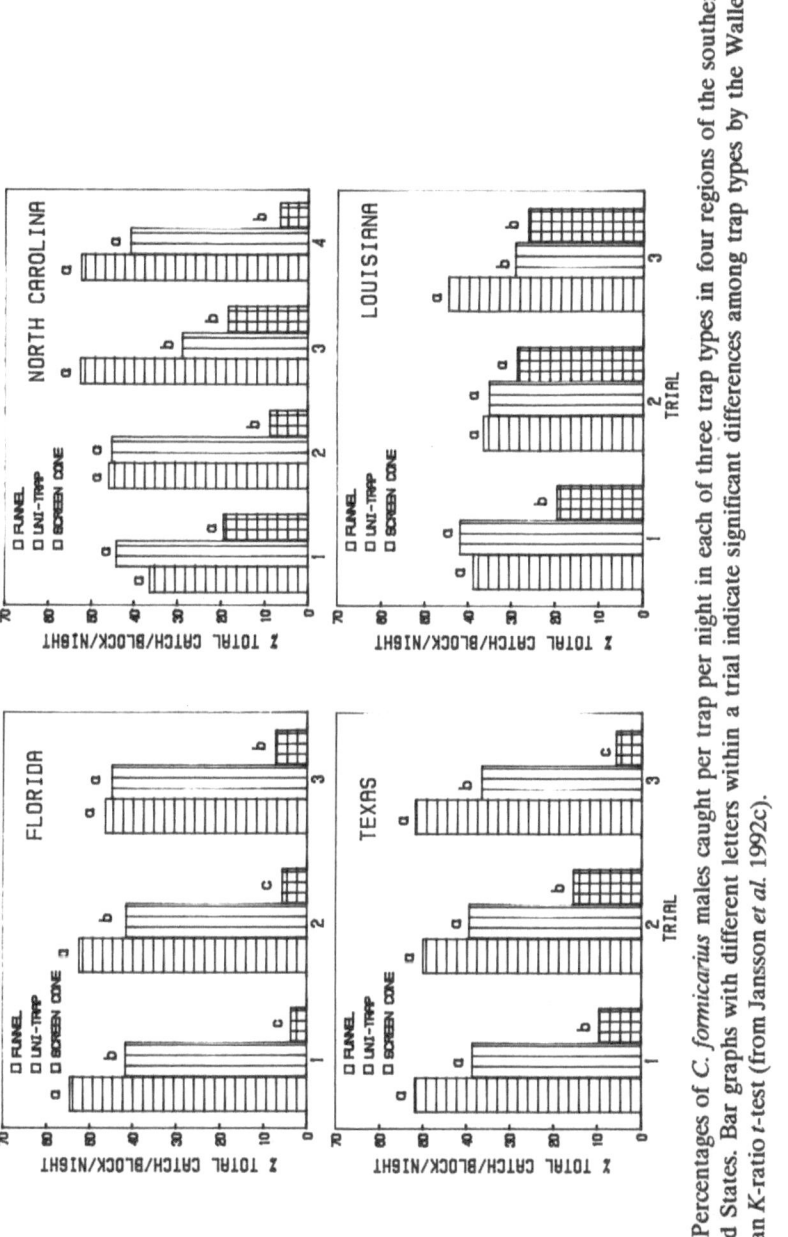

Fig. 6.10. Percentages of *C. formicarius* males caught per trap per night in each of three trap types in four regions of the southern United States. Bar graphs with different letters within a trial indicate significant differences among trap types by the Waller-Duncan *K*-ratio *t*-test (from Jansson *et al.* 1992c).

Further studies compared the effectiveness of modified Universal moth traps with the plastic funnel trap at collecting weevils. As mentioned previously, the wire-framed base of the plastic funnel trap serves as a ladder to facilitate weevil entry into this trap (see Chapter 5). Based on this belief, the following modifications of the Universal moth trap were tested: traps resting on a wire-framed base, traps with a wire-framed skirt located just below the entrance to the trap, and traps with a nylon mesh skirt attached immediately below the entrance to the trap. These modifications presumably aided in weevil entry into the trap. These traps were then compared to the conventional Universal moth trap (trap suspended into canopy with entrance holes raised slightly above the canopy) and the plastic funnel trap. Results showed that all trap types tested were comparable in their effectiveness at collecting weevil males (Figure 6.11) (Jansson *et al.* 1992c). Based on these data, the plastic funnel trap or the Universal moth trap appear to be appropriate for monitoring weevils.

Although the plastic funnel trap and Universal moth trap were most effective at collecting weevils, these traps may not be appropriate for use in developing countries. Universal moth traps are available commercially; however, they are costly. Conversely, the plastic funnel trap is not available commercially, but requires certain materials and labor, which may also make it too costly. For these reasons, studies are needed in developing countries to design practical, effective, and inexpensive traps that can be easily constructed from locally available materials. Talekar (1988) described an inexpensive water trough trap that was effective at collecting *C. formicarius* males in Taiwan. This trap has been tested by the International Potato Center in several developing countries (K.V. Raman, pers. comm.). An earthen pan water trap showed great promise at collecting weevils in Bangladesh (M.N. Islam, pers. comm.). Further studies are planned by the International Potato Center to identify a trap type that would be suitable for certain developing countries.

Physiological, Behavioral, and Population Ecology of *Cylas formicarius*

Responses of insects to pheromones may be influenced by the physiological, behavioral, and population ecology of insects. These factors may influence the effectiveness and reliability of a pheromone trapping program by increasing or decreasing trap catch, or by affecting interpretations of trap catch. The effects of several of these factors on trap catch of male *C. formicarius* have been examined,

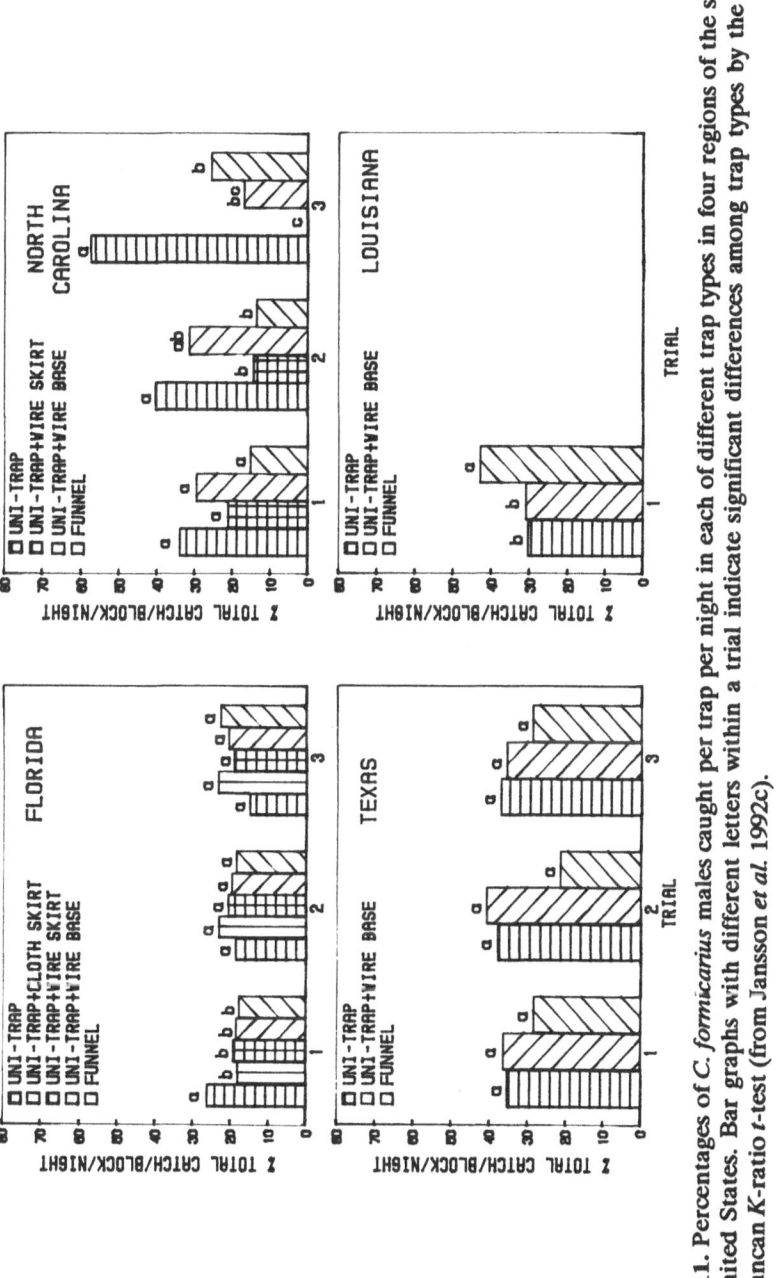

Fig. 6.11. Percentages of *C. formicarius* males caught per trap per night in each of different trap types in four regions of the southern United States. Bar graphs with different letters within a trial indicate significant differences among trap types by the Waller-Duncan *K*-ratio *t*-test (from Jansson *et al.* 1992c).

including previous exposure of males to the pheromone, mating status of males, age of males, and temporal and spatial patterns of trap catch. The importance of these factors in the pheromone-trap monitoring system for *C. formicarius* is presented below.

Previous Exposure to Pheromone. Previous exposure to the pheromone (Hawkins & Rust 1977) or a previous mating (Shorey & Gaston 1964) have been shown to influence an insect's response to sex pheromone. An insect which mates only once throughout its life may not respond to pheromone after a successful mating. However, the period between mating and subsequent receptivity to the pheromone could vary from minutes to hours, if an insect mates more than once (Shorey 1976). The refractory period might be due to a reduced sperm content from a recent mating, olfactory sensory adaptation, or pheromone habituation (Shorey 1976).

Laboratory-reared males (1 to 3 week old) with no previous exposure to pheromone were compared with pheromone-trap collected males to determine if previous exposure to pheromone influenced the percentages of marked *C. formicarius* males captured 16 and 40 h after release. Insects were released from various distances downwind from a 10 μg pheromone source. Percentages captured did not differ between laboratory-reared and pheromone-trap-collected males (Figure 6.12) (16 hours: slopes, $F_{1,52} = 0.62$, $P = 0.43$; intercepts, $F_{1,52} = 0.23$, $P = 0.63$; 40 hours: slopes, $F_{1,52} = 0.57$, $P = 0.45$; intercepts, $F_{1,52} = 0.20$, $P = 0.65$) (Mason *et al.* 1990). Thus, previous exposure to the pheromone did not influence percentages of males captured.

Mating Status. The influence of mating status of male *C. formicarius* on their response to sex pheromone was determined by comparing the response of virgin males to pheromone-baited traps with those of males that had mated when 1, 3, or 6 days old. All males were tested when they were 7 days old. Virgin males and males that had not mated since they were 1 day old exhibited the greatest positive response to the pheromone (Figure 6.13) (Mason & Jansson 1992b). Slopes did not differ ($P > 0.05$) among the four types of males; however, the intercept for virgin males was significantly different ($P < 0.05$) from those of males mated when 3 and 6 days old. Intercepts did not differ between virgin males and those that were mated when 1 day old, nor between males mated when 3 and 6 days old and males mated when 1 day old. These data indicate that the mating status of male weevils affects their response to synthetic sex pheromone. It is not known how often males mate in the field. More studies are needed to understand this more fully.

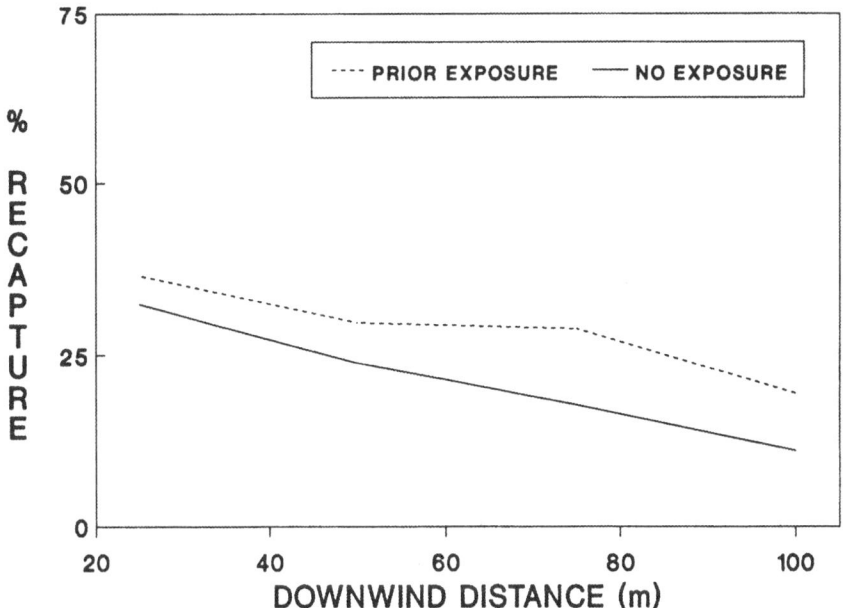

Fig. 6.12. Relationship between percentages of marked *C. formicarius* males caught in plastic funnel traps 16 and 40 hours after release and release distance downwind from the pheromone source (three traps baited with 10 μg of synthetic pheromone) for previously exposed (pheromone-trap collected) and laboratory-reared males (from Mason *et al.* 1990).

Male Age. Many insects synthesize, release, and respond to sex pheromones throughout their adult life. Others gradually attain an increased sexual responsiveness or become sexually responsive when they achieve a certain level of physiological maturity. Additionally, pheromone production and response behaviors may cease prior to natural death. The age at which insects first respond to pheromone and the percentage of individuals of various ages responding must be determined to estimate the proportion of the population caught in traps.

To determine the influence of male age on weevil response to synthetic sex pheromone, 1- to 43-day-old marked males were released 25 m downwind from a 10 μg pheromone source. The percentage of males recaptured was determined 16 hours later. Response to the pheromone increased with an increase in male age for males that were 1 to 14 days old (Figure 6.14) (Mason & Jansson 1992b). Percentages recaptured remained fairly constant for males between 15 and 43 days

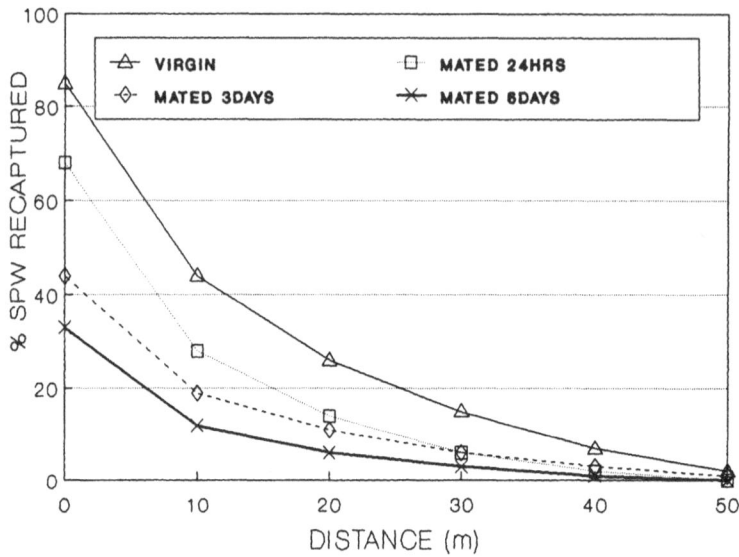

Fig. 6.13. Relationship between percentages of marked, 7-day-old *C. formicarius* males caught in plastic funnel traps 16 hours after release and release distance downwind from the pheromone source (three traps baited with 10 μg of synthetic pheromone) for virgin males and males that had mated when 1, 3, and 6 days old (from Mason and Jansson 1992b).

old. These data suggest that young males might be less physiologically prepared to mate, and less able to move long distances to a pheromone than older males. The longevity and age distribution of the male *C. formicarius* population in the field are not known. More research is needed to determine when and how these factors might influence counts in pheromone-baited traps.

Time of Day. Proshold *et al.* (1986) determined the trap catch patterns of adult male *C. formicarius* in the U.S. Virgin Islands. They found that peak capture occurred during the early evening hours, the time of greatest mating activity in the field. A recent study examined the relationship between time of day and weevil capture. The influence of time of day on trap counts was determined by collecting and counting males at 1 hour intervals from plastic funnel traps placed in commercial fields. Male weevils exhibited a distinct pattern in trap

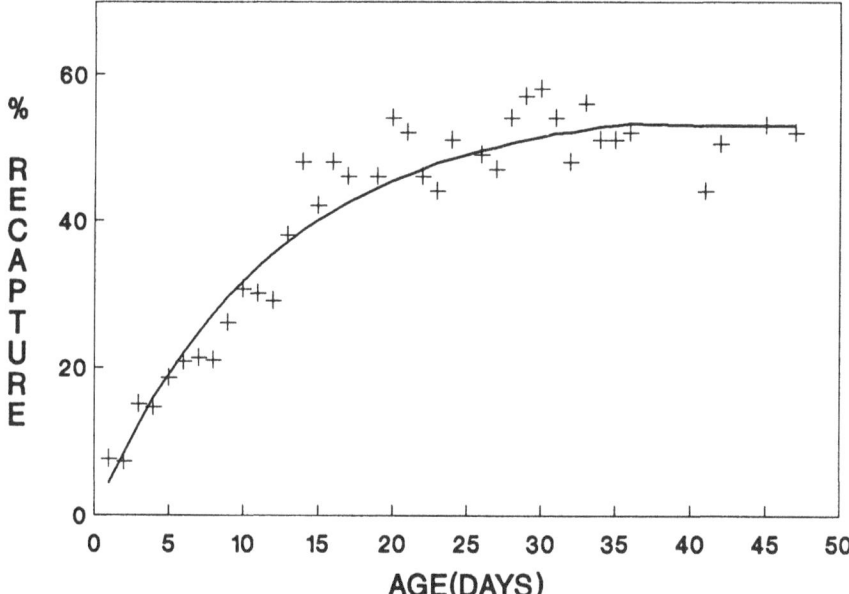

Fig. 6.14. Relationship between percentages of marked *C. formicarius* males caught 16 hours after release and male age (from Mason & Jansson 1992b).

catch; more weevils were caught a few hours after sunset (Figure 6.15) (Mason & Jansson 1992a). During daylight, few males were caught.

Temporal and Spatial Patterns of Trap Counts in Commercial Fields. Jansson *et al.* (1989) described temporal patterns of trap counts in several white-fleshed sweet potato fields in southern Florida. In this study, plastic funnel traps were baited with 10 μg of synthetic sex pheromone and placed overnight in commercial fields at approximately weekly intervals. They found that mean counts per trap were highest in each field during one of the first three sample dates; trap counts gradually declined over time (Figure 6.16). Trap counts peaked in one field on 22 September (5,785 males per trap). The maximum trap catch observed was 9,169 males in a single trap. A cyclical pattern of counts was observed at monthly intervals, and this pattern was probably related to monthly insecticide applications and the emergence of new weevil generations because of the time lag between peak counts approximated known developmental time requirements. They indicated that current insecticide-based management programs were not adequately managing this weevil in southern Florida. Also, the gradual decline in weevil counts over time

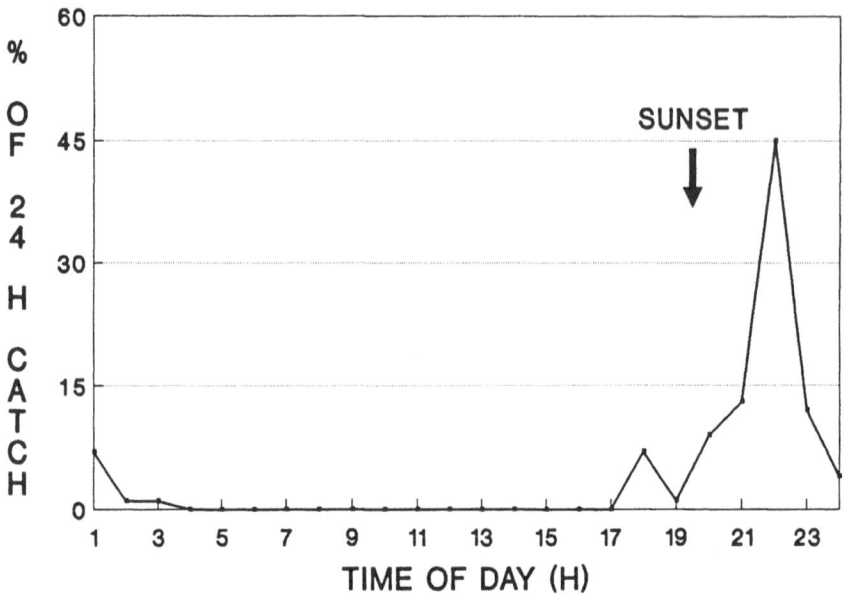

Fig. 6.15. Diel patterns of male *C. formicarius* trap counts in plastic funnel traps baited with 10 μg of synthetic sex pheromone in southern Florida (from Mason & Jansson 1992a).

was probably not due to the "trapping out" of male populations in these fields.

In the same study, Jansson *et al.* (1989) described spatial patterns of male trap counts in commercial fields. They showed that both Taylor's power law ($\log_{10}$ variance, s^2, regressed on $\log_{10}$ mean, x, of trap counts) and Iwao's patchiness regression (regression of mean crowding index, m, on x trap counts) provided a good fit to data collected from several commercial fields over several dates (e.g., see Figure 6.17). Slope estimates indicated that male populations were clumped in sweet potato fields. Despite the variability associated with field, field age, sample date, environmental conditions, insect behavior, and insecticide applications, the relationship between $\log_{10} s^2$ and $\log_{10} x$ of weevil counts was highly significant (Figure 6.17).

Based on these data, the optimal numbers of plastic funnel traps needed to reliably estimate *C. formicarius* populations in commercial fields (4-5 ha) was determined (Jansson *et al.* 1989). They showed that sample size requirements were not practical at high levels of precision (10 or 15%), but were more practical when less precision (25%) was acceptable (Figure 6.17). For example, when trap counts averaged

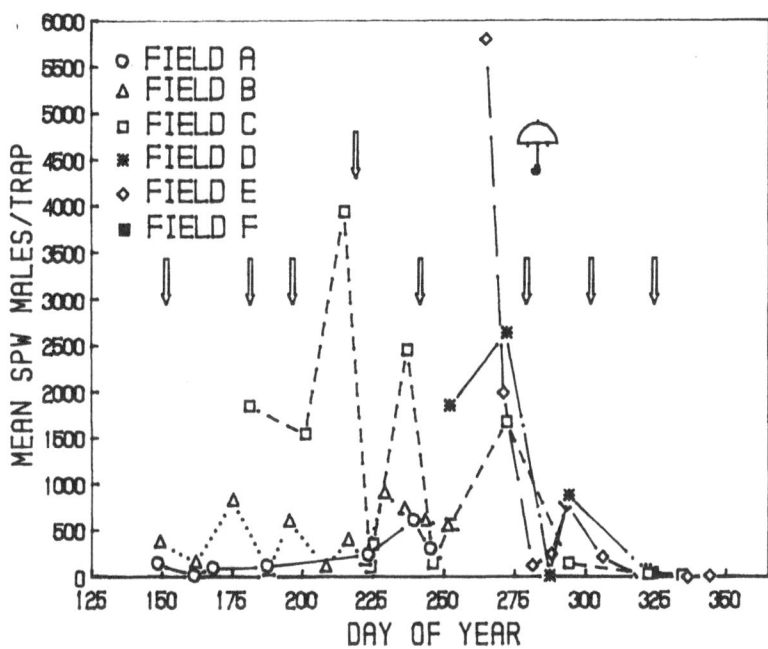

Fig. 6.16. Mean number of *C. formicarius* males caught per trap per night in six different commercial, white-fleshed sweet potato fields near Homestead. Arrows correspond to dates that chemical insecticides were applied. Umbrella indicates date of Hurricane Floyd (from Jansson *et al.* 1989; reprinted with permission from Environ. Entomol. [Entomological Society of America, Lanham, Maryland]).

1,000 male *C. formicarius* per trap per night, 104, 46, and 17 traps at a density of one trap per 0.5 ha were needed to reliably estimate weevil density at precision levels of 10, 15, and 25%, respectively. Conversely, when trap counts averaged 10 males per trap per night, 148, 66, and 24 traps were needed to reliably estimate weevil density at the three levels of precision, respectively. Interpretation of these data, however, was only valid for the pheromone-trap system used in this study. Changes in trap type, pheromone dosage, purity, source, lure age, etc. will probably affect spatial statistics and corresponding interpretations of reliability.

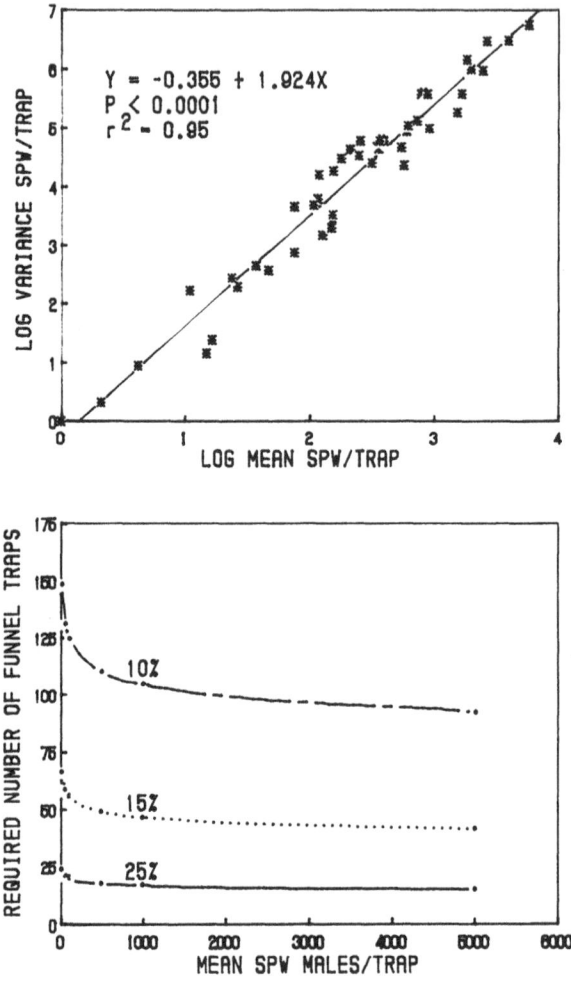

Fig. 6.17. Relationship between $\log_{10} s^2$ and $\log_{10} x$ of trap counts in plastic funnel traps baited with 10 μg of synthetic sex pheromone placed overnight in several commercial white-fleshed sweet potato fields in southern Florida (top). Optimal sample size requirements for funnel traps (1 trap/0.5 ha) in commercial white-fleshed sweet potato fields for 10, 15, and 25% levels of precision ($P < 0.05$) (bottom from Jansson *et al.* 1989; reprinted with permission from Environ. Entomol. [Entomological Society of America, Lanham, Maryland]).

Environmental Factors

Environmental factors that may influence behavioral responses of insects to pheromone include climatic conditions (i.e., wind direction, temperature, rainfall, etc.) and plant age, which may alter plume structure.

Field Age. *Cylas formicarius* males are confronted with a variety of distractions in cultivated sweet potato fields, including the presence of food plants and pheromone-releasing females. Additionally, pheromone plume characteristics may be affected by the size (i.e., height, distribution, etc.) of the plant canopy. A previous report (Mason *et al.* 1990) found that about 10% of marked males were recaptured in pheromone traps baited with 10 μg of pheromone 16 hours after release from as far away as 280 m downwind of the pheromone in fields devoid of plants. Further studies were conducted to determine the effects of commercial fields planted with sweet potato of different ages on levels of male recapture in pheromone-baited traps. Results showed that sampling range of pheromone traps was affected by the age of the planting (Figure 6.18) (Mason & Jansson 1992c). Approximately 10% of marked males were recaptured when released 80, 70, and 40 m downwind of pheromone traps in old fields (3 months old), young fields (1 month old) and very old fields (greater than 6 months old), respectively. The reasons for these differences may be related, in part, to differences in weevil densities between fields of different ages, pheromone adsorption onto the vegetation (Wall *et al.* 1981), and/or reduction in weevil movement by the plant canopy (Bell & Carde 1984).

Wind Direction. Wind speed and direction may alter pheromone plume shape and size, and create pheromone concentration gradients within the plume. Studies were conducted to determine the effects of wind direction on weevil movement to pheromone-baited traps. Prevailing winds in southern Florida are generally from the east-southeast. Marked males were released at the cardinal points (NESW) 25 m and 45 m away from a 10 μg pheromone source. Wind direction and speed were recorded at 10 minute intervals. The majority of insects recaptured were released downwind of the pheromone source (Figure 6.19A,B,C). However, a small percentage of males released upwind from the trap were also recaptured. Theoretically, males released upwind of the pheromone should not have come in contact with pheromone unless brief undetected changes in wind direction occurred, or males intercepted the plume while moving randomly in the direction of the plume. Thus, random movement by males might

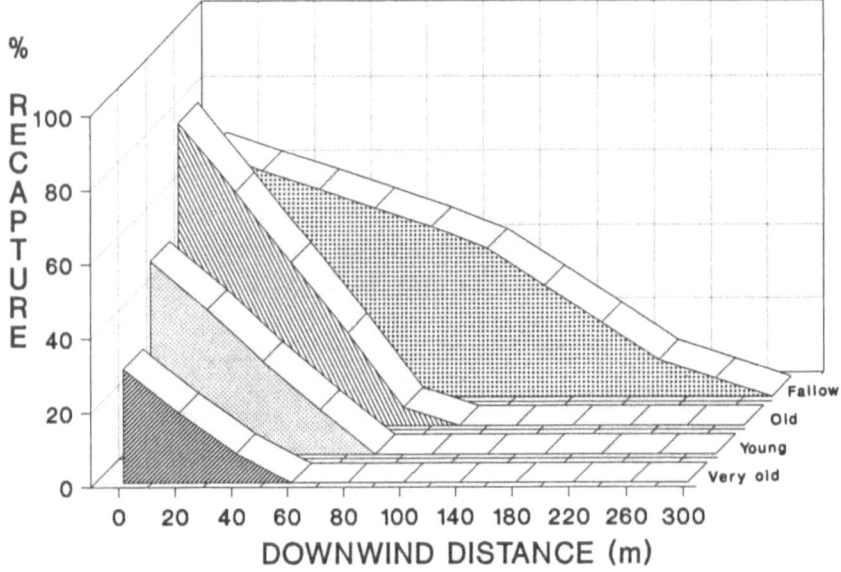

Fig. 6.18. Percentages of marked *C. formicarius* males captured after 16 hours from various release distances downwind of the pheromone source (plastic funnel traps baited with 10 μg of synthetic sex pheromone) in fields of different ages (from Mason & Jansson 1992c).

constitute a small percentage of the total pheromone trap catch. Also, nondirected plume gusts containing small amounts of pheromone might be sufficient to elicit an orientation response to pheromone. On nights when the wind direction varied considerably, males were recaptured from all release points with equal frequencies (Figure 6.19D)

SEX PHEROMONE AS A MANAGEMENT TOOL

Mating Disruption

Insect control achieved through modification of pheromone-mediated sexual communication is well known (Kydonieus & Beroza 1982). One approach that has great potential in insect management is a technique commonly called "male confusion" or mating disruption. The basic premise is that when the air is permeated with a sex

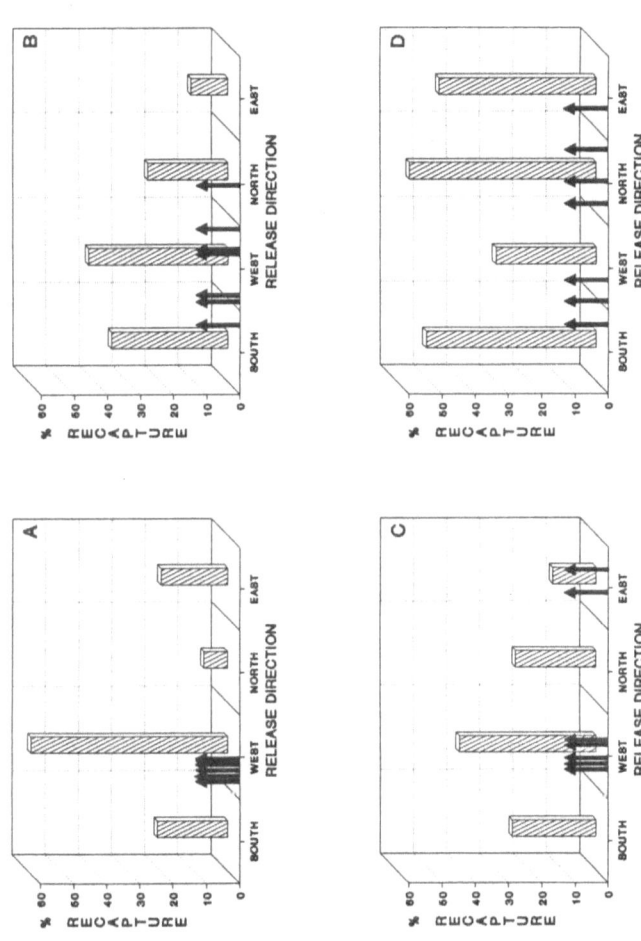

Fig. 6.19. Influence of wind direction on percentages of marked *C. formicarius* males captured when released at the cardinal compass points (NESW) around a plastic funnel trap baited with 10 μg of pheromone. Bars represent the percentages of males captured at each release point. Arrows represent the relative average direction of wind each hour (A-D represent four separate nights with differing wind directions).

attractant, the number of males successful at locating and mating with females will be reduced, thereby reducing the percentage of mated females in the population (Birch & Haynes 1982). Fewer mated females will then result in a lower pest population density in the next generation. It is believed that the decrease in mating success (i.e., poor male response to females) is due to habituation and/or sensory adaptation to the pheromone (Birch & Haynes 1982). This approach has been examined in several insect systems, such as the gypsy moth, *Lymantria dispar* (L.), (Beroza & Knipling 1972, Cameron *et al.* 1974, Plimmer 1982, Schwalbe *et al.* 1974, 1979, 1983, Schwalbe & Mastro 1988), cotton leafworm, *Spodoptera littoralis* (Boisd.) (Kehat *et al.* 1983, 1985, 1986), pink bollworm, *Pectinophora gossypiella* (Saunders) (Campion *et al.* 1987, Critchley *et al.* 1983, 1985, Henneberry *et al.* 1982, McLaughlin *et al.* 1972), sugar cane borers, *Chilo sacchariphagus* (Bojer) and *C. auricilius* (Nesbitt *et al.* 1980, 1986), cabbage looper, *Trichoplusia ni* (Hubner) (Shorey & Gaston 1964, Gaston *et al.* 1967, Shorey *et al.* 1967, 1972), codling moth, *Cydia pomonella* (L.) (Carde *et al.* 1977, Rothschild 1982, Moffitt & Westigard 1984), grape berry moth, *Endopiza viteana* Clemens (Taschenberg & Roelofs 1976), navel orangeworm, *Amyelois transitella* (Walker) (Curtis *et al.* 1985, Landolt *et al.* 1981, 1982), western pine shoot borer, *Eucosma sonomana* Kearfott (Daterman 1982), spruce budworm, *Choristoneura fumifurana* (Clemens) (Sanders & Seabrook 1982), Oriental fruit moth, *Grapholita* (= *Cydia*) *molesta* (Busck) (Gentry *et al.* 1982), peachtree borer, *Synanthedon exitiosa* (Say) (Yonce & Gentry 1982), Douglas-fir tussock moth, *Orgyia psuedotsugata* (McDunnough) (Sower & Daterman 1977, Sower 1982), and others (Ridgway *et al.* 1990, Roelofs 1979, Roelofs & Carde 1977). In general, some success at disrupting mating patterns was achieved in most of these systems, especially at low to moderate population densities. In certain cases, some mating still occurred despite considerable male confusion. In pink bollworm, mating disruptants reduced insecticide use by up to 64% (Birch & Haynes 1982). This resulted in both a savings of insecticide costs, and conservation of natural enemies, and clearly illustrates the ease with which this type of control strategy might be integrated into management programs for pests.

Failures in mating disruption have been due, in part, to a lack of knowledge on the mechanisms involved in the process (Birch & Haynes 1982). For example, little is known about pheromone emmission rates and concomitant quantities of pheromone needed to disorient an individual insect. Distribution dynamics of the pheromone in the air is also poorly understood (Rothschild 1982). Failures have also been due, in part, to immigration of mated females into pheromone-treated areas, increased success of populations at low levels, and complications associated with multicomponent pheromone blends (Rothschild 1982).

Recent studies showed that synthetic sex pheromone of *C. formicarius* can disorient adult males and perhaps disrupt mating patterns (Figure 6.20) (Mason & Jansson 1990, 1991, L.J. Mason *et al.*, unpubl. data, Talekar & Lee 1989). Male counts in traps baited with 1 ng of synthetic pheromone (to simulate feral, "calling" females) were considerably lower in fields in which the air was presumably permeated with synthetic sex pheromone (12 lures loaded with 100 μg of pheromone; lures placed 25 m apart) than in fields in which no pheromone was present. Similar results from a preliminary experiment were found in Taiwan (AVRDC 1988, Talekar & Lee 1989). However, it is not known if the mating disruptant decreased the percentage of mated females and reduced the population of *C. formicarius* and their damage to storage roots in pheromone-treated fields. Current studies in southern Florida are examining this approach more fully.

Mass Trapping

The success of pheromone-baited traps at catching insects suggested that insect populations might be managed by "trapping out" individuals in the population. Such an approach assumes that a reduction in the adult population will lead to a further population decrease in the next generation (Birch & Haynes 1982). Of central importance in mass trapping strategies is knowledge of the proportion of the population that is captured, and the relationship between it and the reduction in the population in the next generation. In certain cases, however, high captures may result in small population reductions due to: (1) increased reproductive rates in remaining individuals at lower densities; (2) remaining individuals may mate with more females; or (3) reduction in intraspecific competition (Birch & Haynes 1982). Conversely, if traps selectively remove those individuals that are most likely to mate, population reductions may be achieved.

Several studies have tested the potential of mass trapping strategies for pest control. This strategy was tested against at least twelve insect pests, such as spruce bark beetle, *Ips typographus* (L.) (Bakke 1982, Bakke & Riege 1982), gypsy moth (Webb 1982), Japanese beetle, *Popillia japonica* Newman (Ladd & Klein 1982), boll weevil,

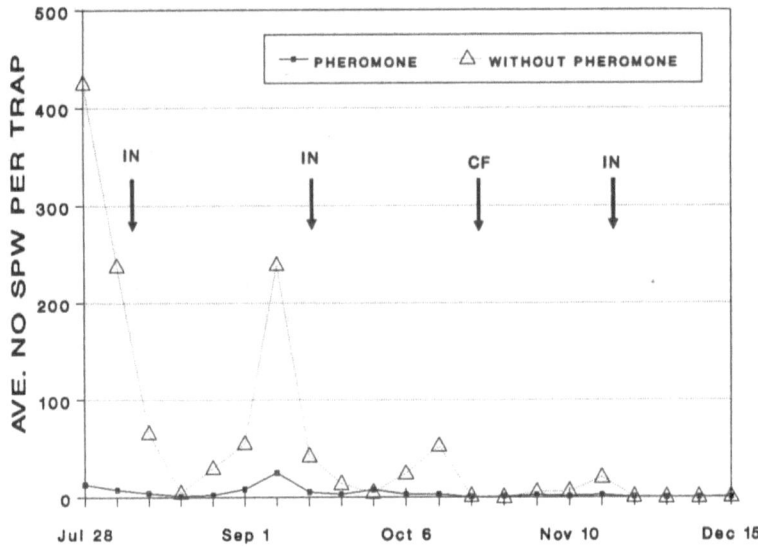

Fig. 6.20. An example of disorientation of male *C. formicarius* by high dosages (100 μg) of synthetic sex pheromone. Mean numbers of *C. formicarius* males caught in traps baited with 1 ng of synthetic sex pheromone and placed in pheromone-treated and nontreated plots (IN, insecticide application date; CF, cold front).

Anthonomus grandis grandis Boheman (Hardee 1982), codling moth, European elm bark beetle, *Scolytus multistriatus* (Marsham), and ambrosia beetle, *Gnathotrichus sulcatus* (LeConte) (Birch & Haynes 1982). Webb (1982) reviewed most studies that evaluated the potential of mass trapping as a management tool. Results with codling moth have been varied, and economic control has only been achieved at low densities. In contrast, economic control of redbanded leafroller, *Argyrotaenia velutinana* (Walker), has been more encouraging and was achieved when mass trapping was initiated early when population densities were low (Birch & Haynes 1982).

Male *C. formicarius* are extremely sensitive to the female-produced sex pheromone. Traps baited with low dosages of pheromone (10 μg) collect large numbers of adult *C. formicarius* males in a short period of time (16 hours) (e.g., see Figure 6.16). Maximum trap counts recorded in southern Florida were 9,818 and 10,277 males in individual traps in one night baited with 10 and 100 μg, respectively. A total of 211,332 *C. formicarius* males were caught during four consecutive nights in southern Florida in 20 plastic funnel traps baited with either 10 or 100 μg of pheromone. In Bangladesh, over 40,000 male *C. formicarius* were caught in one night in one trap baited with 1 mg of pheromone (M.N. Islam, pers. comm.). Because it is well known that trap captures increase with an increase in pheromone dosage, a mass trapping strategy would require at least 100 μg, but probably 1 mg, of pheromone per trap to increase the likelihood of success with such an approach. Talekar and Lee (1989) showed that weevil trap counts decreased over time in traps baited with 1 mg of pheromone. They suggested that the continuous trapping of males reduced weevil populations. The effect of male captures on storage root protection, however, was not assessed. Collectively, this information suggests that there may be potential for this sex pheromone to suppress *C. formicarius* populations by mass trapping of male populations, especially when one considers the large numbers of males that are caught in traps baited with high dosages of pheromone (e.g., 1 mg). The potential of this approach is currently being explored in southern Florida and in various developing countries in cooperation with the International Potato Center.

FUTURE OUTLOOK

The sex pheromone of *C. formicarius* has great potential for improving sweet potato production world-wide. It has great potential as a monitoring tool, and may help growers select and time management tactics to reduce weevil densities in seedbeds, in the field, and in storage. It may also help growers assess the success of current management programs for this weevil. This pheromone will also provide farmers and regulatory personnel with the ability to detect low weevil densities earlier in the growing season than was previously possible. For example, in Louisiana, where weevil densities are low to moderate, low level weevil infestations were detected in seedbeds within the first week that traps were placed in the seedbeds (Hammond *et al.* 1989).

Use of the pheromone in regulatory entomology will be especially important considering current domestic and international trade and movement restrictions of sweet potato. This chemical will benefit both

130

large scale commercial production, such as found in the United States, as well as small scale subsistence sweet potato systems, as found in developing countries. In many developing countries, this technology may ultimately help to make the transfer from subsistence farming to cash cropping more feasible.

Use of pheromone for mating disruption and/or mass trapping has potential for managing pests and can be easily integrated into IPM programs (Birch & Haynes 1982). This pheromone system is no exception. Determination of the management potential of this pheromone is central for integrating the pheromone with other management tactics, especially cultural controls and the use of entomopathogens for biological control of this weevil. Previous studies showed that entomopathogenic nematodes have great potential for managing *C. formicarius* populations and protecting fleshy roots from weevil damage (Jansson *et al.* 1990b, also see Chapter 9). Integration of these two approaches, however, should await completion of studies that assess each tactic individually.

ACKNOWLEDGMENTS

We thank A.M. Hammond (Department of Entomology, Louisiana State University, Baton Rouge), K.A. Sorensen (Department of Entomology, North Carolina State University, Raleigh), and J.V. Robinson (Texas Agricultural Experiment Station, Overton) for coordinating cooperative research projects in their respective regions of the southern United States. Additional thanks are extended to D.E. Forey and J.D. Knapp (AgriSense, Fresno, California) for their cooperative efforts. We also thank the many research associates and technicians who assisted with data collection at each of these locations. This research was supported by the U.S.D.A. under C.S.R.S. Special Grant Nos. 87-CSRS-2-3107 and 88-34135-3571 (to R.K.J.) managed by the Caribbean Basin Administrative Group (C.B.A.G.), and by C.S.R.S. Special Grant No. 88-34103-3262 (to R.K.J.) managed by the Southern Region IPM Program. This is Florida Agricultural Experiment Station Journal Series No. R-01189.

REFERENCES

AVRDC. 1988. Annual review 1987. Entomology. Asian Vegetable Research and Development Center, Shanhua, Tainan, Taiwan, Republic of China.

Bakke, A. 1982. Mass trapping of the spruce bark beetle *Ips typographus* in Norway as a part of an integrated control program, pp. 17-26. *In* A.F. Kydonieus & M. Beroza (eds.). Insect suppression with controlled release pheromone systems. Vol. 2. CRC Press, Boca Raton, Florida.

Bakke, A. & L. Riege. 1982. The pheromone of the spruce bark beetle *Ips typographus* and its potential use in the suppression of beetle populations, pp. 3-16. *In* A.F. Kydonieus & M. Beroza (eds.). Insect suppression with controlled release pheromone systems. Vol. 2. CRC Press, Boca Raton, Florida.

Bell, W.J. & R.T. Carde. 1984. Chemical ecology of insects. Sinauer Assoc., Inc., Sunderland, Massachusetts.

Beroza, M. & E.F. Knipling. 1972. Gypsy moth control with the sex attractant pheromone. Science 177:19-27.

Bestmann, H.J. & O. Vostrowsky. 1988. Pheromones of the Coleoptera, pp. 95-183. *In* E.D. Morgan & N.B. Mandava (eds.). CRC handbook of natural pesticides. Vol. IV. Pheromones. Part A. CRC Press, Boca Raton, Florida.

Birch, M.C. & K.F. Haynes. 1982. Insect pheromones. Edward Arnold, London.

Cameron, E.A., C.P. Schwalbe, M. Beroza & E.F. Knipling. 1974. Disruption of gypsy moth mating with microencapsulated disparlure. Science 183:972-973.

Campion, D.G., D.R. Hall & P.F. Prevett. 1987. Use of pheromones in crop and stored products pest management: control and monitoring. Insect Sci. Applic. 8:737-741.

Carde, R.T., T.C. Baker & P.J. Castrovillo. 1977. Disruption of sexual communication in *Laspeyresia pomonella* (codling moth), *Grapholitha molesta* (oriental fruit moth), and *G. prunivora* (lesser appleworm) with hollow fiber attractant sources. Entomol. Exp. Appl. 22:280-288.

Chalfant, R.B., R.K. Jansson, D.R. Seal & J.M. Schalk. 1990. Ecology and management of sweet potato insects. Annu. Rev. Entomol. 35:157-180.

Coffelt, J.A., K.W. Vick, L.L. Sower & W.T. McClellan. 1978. Sex pheromone of the sweetpotato weevil, *Cylas formicarius elegantulus*: laboratory bioassay and evidence for a multiple component system. Environ. Entomol. 7:756-758.

Critchley, B.R., D.G. Campion, L.J. McVeigh, E.M. McVeigh, G.G. Cavanagh, M.M. Hosny, A. Nasr El Sayed, A.A. Khidr & M. Naguib. 1985. Control of the pink bollworm, *Pectinophora gossypiella* (Saunders) (Lepidoptera: Gelechiidae), in Egypt by mating disruption using hollow fibre, laminate-flake and microencapsulated formulations of synthetic pheromone. Bull. Entomol. Res. 75:329-345.

Critchley, B.R., D.G. Campion, L.J. McVeigh, P. Hunter-Jones, D.R. Hall, A. Cork, B.F. Nesbitt, G.J. Marrs, A.R. Jutsum, M.M. Hosny & A. Nasr El Sayed. 1983. Control of pink bollworm, *Pectinophora gossypiella* (Saunders) (Lepidoptera: Gelechiidae) in Egypt by mating disruption using an aerially applied microencapsulated pheromone formulation. Bull. Entomol. Res. 73:289-299.

Curtis, C.E., P.J. Landolt & J.D. Clark. 1985. Disruption of navel orangeworm (Lepidoptera: Pyralidae) mating in large scale plots with synthetic sex pheromone. J. Econ. Entomol. 78:1425-1430.

Daterman, G.E. 1982. Control of western pine shoot borer damage by mating disruption - a reality, pp. 155-164. *In* A.F. Kydonieus & M. Beroza (eds.). Insect suppression with controlled release pheromone systems. Vol. 2. CRC Press, Boca Raton, Florida.

Gaston, L.K., H.H. Shorey & C.A. Saario. 1967. Insect population control by the use of sex pheromones to inhibit orientation between the sexes. Nature 213: 1155.

Gentry, C.R., C.E. Yonce & B.A. Bierl-Leonhardt. 1982. Oriental fruit moth: mating disruption trials with pheromone, pp. 107-116. *In* A.F. Kydonieus & M. Beroza (eds.). Insect suppression with controlled release pheromone systems. Vol. 2. CRC Press, Boca Raton, Florida.

Hammond, A.M., T.N. Hardy, S.J. Toth, Jr., L.H. Rolston, E.L. Freeman, L.A. Hampton, W.A. Hogan, Jr. & J.L. Bagent. 1989. Monitoring sweetpotato weevils with sex pheromone-baited traps. La. Agric. 32(3):12,13,18.

Hardee, D.D. 1982. Mass trapping and trap cropping of the boll weevil, *Anthonomus grandis* Boheman, pp. 65-71. *In* A.F. Kydonieus & M. Beroza (eds.). Insect suppression with controlled release pheromone systems. Vol. 2. CRC Press, Boca Raton, Florida.

Hawkins, W.A. & M.K. Rust. 1977. Factors influencing male sexual response in the American cockroach, *Periplaneta americana*. J. Chem. Ecol. 3:85-99.

Heath, R.R. & P.E. Sonnet. 1980. Techniques for *in situ* coating of Ag⁺ onto silica gel in HPLC columns for the separation of geometrical isomers. J. Liquid Chromatogr. 3:1129-1135.

Heath, R.R., J.A. Coffelt, F.I. Proshold, P.E. Sonnett & J.H. Tumlinson. 1988. (Z)-3-dodecen-1-ol (E)-2-butenoate and its use in monitoring the sweetpotato weevil. U.S. Patent No. 4,732,756.

Heath, R.R., J.A. Coffelt, P.E. Sonnett, F.I. Proshold, B. Dueben & J.H. Tumlinson. 1986. Identification of sex pheromone produced by female sweetpotato weevil, *Cylas formicarius elegantulus* (Summers). J. Chem. Ecol. 12:1489-1503.

Henneberry, T.J., J.M. Gillespie, L.A. Bariola, H.M. Flint, G.D. Butler, Jr., P.D. Lingren & A.F. Kydonieus. 1982. Mating disruption as a means of suppressing pink bollworm (Lepidoptera: Gelechiidae) and tobacco budworm (Lepidoptera: Noctuidae) populations on cotton, pp. 75-98. *In* A.F. Kydonieus & M. Beroza (eds.). Insect suppression with controlled release pheromone systems. Vol. 2. CRC Press, Boca Raton, Florida.

Jansson, R.K. & R.R. Heath. 1989. Development of a sex pheromone monitoring system for sweetpotato weevil management, pp. 543-552. *In* R.H. Howeler (ed.). Proceedings of the 8th International Symposium for Tropical Root Crops. International Center of Tropical Agriculture, Bangkok, Thailand.

Jansson, R.K. & R. McSorley. 1990. Sampling plan for the sweetpotato weevil (Coleoptera: Curculionidae) on sweet potato in southern Florida. J. Econ. Entomol. 83:1901-1906.

Jansson, R.K., R.R. Heath & J.A. Coffelt. 1989. Temporal and spatial patterns of sweetpotato weevil (Coleoptera: Curculionidae) counts in pheromone-baited traps in white-fleshed sweet potato fields in southern Florida. Environ. Entomol. 18:691-697.

Jansson, R.K., L.J. Mason & R.R. Heath. 1992a. Pheromone-trap monitoring system for sweetpotato weevil (Coleoptera: Apionidae) in the southern United States: effects of lure type, lure age, and storage duration. J. Econ. Entomol. (In review).

Jansson, R.K., A.G.B. Hunsberger, S.H. Lecrone & S.K. O'Hair. 1990a. Seasonal abundance, population growth, and within-plant distribution of sweetpotato weevil (Coleoptera: Curculionidae) on sweet potato in southern Florida. Environ. Entomol. 19:313-321.

Jansson, R.K., S.H. Lecrone, R.R. Gaugler & G.C. Smart, Jr. 1990b. Potential of entomopathogenic nematodes as biological control agents of sweetpotato weevil (Coleoptera: Curculionidae). J. Econ. Entomol. 83:1818-1826.

Jansson, R.K., F.I. Proshold, L.J. Mason, R.R. Heath & S.H. Lecrone. 1990c. Monitoring sweetpotato weevil (Coleoptera: Curculionidae) with sex pheromone: effects of dosage and age of septum. Trop. Pest Manage. 36:263-269.

Jansson, R.K., L.J. Mason, R.R. Heath, K.A. Sorensen, A.M. Hammond & J.V. Robinson. 1992b. Pheromone-trap monitoring system for sweetpotato weevil (Coleoptera: Apionidae) in the southern United States: effects of pheromone source and purity. Trop. Pest Manage. (In review).

Jansson, R.K., L.J. Mason, R.R. Heath, K.A. Sorensen, A.M. Hammond & J.V. Robinson. 1992c. Pheromone-trap monitoring system for sweetpotato weevil (Coleoptera: Apionidae) in the southern United States: effects of trap type and pheromone dosage. J. Econ. Entomol. (In review).

Jutsum, A.R. & R.F.S. Gordon (eds.). 1989. Insect pheromones in plant protection. J. Wiley, New York.

Kehat, M. E. Dunkelblum & S. Gothilf. 1983. Mating disruption of the cotton leafworm, *Spodoptera littoralis* (Lepidoptera: Noctuidae), by release of sex pheromone from widely separated hercon-laminated dispensers. Environ. Entomol. 12:1265-1269.

Kehat, M., S. Gothilf, E. Dunkelblum, N. Bar-Shavit & D. Gordon. 1985. Night observations on the cotton leafworm, *Spodoptera littoralis*: reliability of pheromone traps for population assessment and efficacy of widely separated pheromone dispensers for mating disruption. Phytoparasitica 13:215-220.

Kehat, M., E. Dunkelblum, S. Gothilf, N. Bar Shavit, D. Gordon & M. Harel. 1986. Mating disruption of the Egyptian cotton leafworm, *Spodoptera littoralis* (Lepidoptera: Noctuidae), in cotton with a polymeric aerosol formulation containing (Z,E)-9,11-tetradeca-dienyl acetate. J. Econ. Entomol. 79:1641-1644.

Knipling, E.F. 1982. Foreword. *In* A.F. Kydonieus & M. Beroza (eds.). Insect suppression with controlled release pheromone systems. Vol. I. CRC Press, Boca Raton, Florida.

Kydonieus, A.F. & M. Beroza (eds.). 1982. Insect suppression with controlled release pheromone systems. CRC Press, Boca Raton, Florida.

Ladd, T.L., Jr. & M.G. Klein. 1982. Trapping Japanese beetles with synthetic female sex pheromone and food-type lures, pp. 57-64. *In* A.F. Kydonieus & M. Beroza (eds.). Insect suppression with controlled release pheromone systems. Vol. 2. CRC Press, Boca Raton, Florida.

Landolt, P.J., C.E. Curtis, J.A. Coffelt, K.W. Vick & R.E. Doolittle. 1982. Field trials of potential navel orangeworm mating disruptants. J. Econ. Entomol. 75: 547-550.

Landolt, P.J., C.E. Curtis, J.A. Coffelt, K.W. Vick, P.E. Sonnett & R.E. Doolittle. 1981. Disruption of mating in the navel orangeworm with (Z,Z)-11-13-hexadecadienal. Environ. Entomol. 10: 745-750.

Lanier, G.M. & W.E. Burkholder. 1974. Pheromones in speciation of Coleoptera, p. 161. *In* M.C. Birch (ed). Pheromones. Elsevier, New York.

Leggett, J.E. & W.H. Cross. 1971. A new trap for capturing boll weevils. U.S.D.A. Coop. Econ. Insect Rept. 21:773-774.

Mason, L.J. & R.K. Jansson. 1990. Mass trapping and mating disruption: approaches for sweetpotato weevil management. Proc. Interamer. Soc. Trop. Hort. 34:89-95.

Mason, L.J. & R.K. Jansson. 1991. Potential of sex pheromone to control sweetpotato weevil (Coleoptera: Apionidae) populations by mating disruption. Fla. Entomol. (In review).

Mason, L.J. & R.K. Jansson. 1992a. Effects of environmental conditions on male sweetpotato weevil (Coleoptera: Apionidae) response to sex pheromone. J. Chem. Ecol. (In review).

Mason, L.J. & R.K. Jansson. 1992b. Effects of reproductive behavior and male age on response of sweetpotato weevil (Coleoptera: Apionidae) to sex pheromone. Environ. Entomol. (In review).

Mason, L.J. & R.K. Jansson. 1992c. Monitoring sweetpotato weevil (Coleoptera: Apionidae) with sex pheromone: effects of field age and trapping duration. J. Econ. Entomol. (In review).

Mason, L.J., R.K. Jansson & R.R. Heath. 1990. Sampling range of male sweetpotato weevils (*Cylas formicarius elegantulus*) (Summers) (Coleoptera: Curculionidae) to pheromone traps: influence of pheromone dosage and lure age. J. Chem. Ecol. 16:2493-2502.

Mayer, M.S. & J.R. McLaughlin. 1990. CRC handbook of insect pheromones and sex attractants. CRC Press, Boca Raton, Florida.

McLaughlin, J.R., H.H. Shorey, L.K. Gaston, R.S. Kaae & F.D. Stewart. 1972. Sex pheromones of Lepidoptera. XXXI. Disruption of sex pheromone communication in *Pectinophora gossypiella* with hexalure. Environ. Entomol. 1: 645-650.

136

Mitchell, E.R. (ed.). 1981. Management of insect pests with semiochemicals. Plenum, New York.

Moffitt, H.R. & P.H. Westigard. 1984. Suppression of codling moth (Lepidoptera: Tortricidae) populations on pears in southern Oregon through mating disruption with sex pheromone. J. Econ. Entomol. 77:1513-1519.

Morgan, E.D. & N.B. Mandava (eds.). 1988. Handbook of natural pesticides. Vol. IV. Pheromones. Parts A & B. CRC Press, Boca Raton, Florida.

Nesbitt, B.F., P.S. Beevor, D.R. Hall, R. Lester & J.R. Williams. 1980. Components of the sex pheromone of the female sugar cane borer, *Chilo sacchariphagus* (Bojer) (Lepidoptera: Pyralidae). Identification and field trials. J. Chem. Ecol. 6:385-394.

Nesbitt, B.F., P.S. Beevor, A. Cork, D.R. Hall, H. David & V. Nandagopal. 1986. The female sex pheromone of sugarcane stalk borer, *Chilo auricilius*. Identification of four components and field tests. J. Chem. Ecol. 12:1377-1388.

Nordlund, D.A. 1981. Semiochemicals: a review of the terminology, pp. 13-28. *In* D.A. Nordlund, R.L. Jones & W.J. Lewis (eds.). Semiochemicals: their role in pest control. J. Wiley, New York.

Plimmer, J.R. 1982. Disruption of mating in gypsy moth. pp. 135-154. *In* A.F. Kydonieus & M. Beroza (eds.). Insect suppression with controlled release pheromone systems. Vol. 2. CRC Press, Boca Raton, Florida.

Proshold, F.I. 1983. Mating activity and movement of *Cylas formicarius elegantulus* (Coleoptera: Curculionidae) on sweet potato. Proc. Amer. Soc. Hort. Sci. Trop. Region 27(B):81-92.

Proshold, F.I., J.L. Gonzalez, C. Asencio & R.R. Heath. 1986. A trap for monitoring the sweetpotato weevil (Coleoptera: Curculionidae) using pheromone or live females as bait. J. Econ. Entomol. 79:641-647.

Ridgway, R.L., R.M. Silverstein & M.N. Iscoe. 1990. Behavior-modifying chemicals for insect management: applications of pheromones and other attractants. Marcel-Dekker, Inc., New York.

Roelofs, W.L. (ed.). 1979. Establishing efficacy of sex attractants and disruptants for insect control. Entomol. Soc. Amer., Lanham, Maryland.

Roelofs, W.L. 1981. Attractive and aggregating pheromones, pp. 215-235. *In* D.A. Nordlund, R.L. Jones & W.J. Lewis (eds.). Semiochemicals: their role in pest control. J. Wiley, New York.

Roelofs, W.L. & R.T. Carde. 1977. Responses of Lepidoptera to synthetic sex pheromone chemicals and their analogs. Annu. Rev. Entomol. 22:377-405.

Rothschild, G.H.L. 1982. Suppression of mating in codling moths with synthetic sex pheromone and other compounds, pp. 117-134. *In* A.F. Kydonieus & M. Beroza (eds.). Insect suppression with controlled release pheromone systems. Vol. 2. CRC Press, Boca Raton, Florida.

Sanders, C.K. & W.D. Seabrook. 1982. Disruption of mating in the spruce budworm *Choristoneura fumifurana* (Clemens), pp. 175-186. *In* A.F. Kydonieus & M. Beroza (eds.). Insect suppression with controlled release pheromone systems. Vol. 2. CRC Press, Boca Raton, Florida.

Schwalbe, C.P. & V.C. Mastro. 1988. Gypsy moth mating disruption: dosage effects. J. Chem. Ecol. 14:581-588.

Schwalbe, C.P., E.C. Paszek, B.A. Bierl-Leonhardt & J.R. Plimmer. 1983. Disruption of gypsy moth (Lepidoptera: Lymantriidae) mating with disparlure. J. Econ. Entomol. 76:841-844.

Schwalbe, C.P., E.A. Cameron, D.J. Hall, J.V. Richerson, M. Beroza & L.J. Stevens. 1974. Field tests of microencapsulated disparlure for suppression of mating among wild and laboratory-reared gypsy moths. Environ. Entomol. 3: 589-592.

Schwalbe, C.P., E.C. Paszek, R.E. Webb, B.A. Bierl-Leonhardt, J.R. Plimmer, C.W. McComb & C.W. Dull. 1979. Field evaluation of controlled release formulations of disparlure for gypsy moth mating disruption. J. Econ. Entomol. 72: 322-326.

Shorey, H.H. 1976. Animal communication by pheromones. Academic, New York.

Shorey, H.H. & L.K. Gaston. 1964. Sex pheromones of noctuid moths. III. Inhibition of male responses to the sex pheromone of *Trichoplusia ni* (Lepidoptera: Noctuidae). Ann. Entomol. Soc. Amer. 57:775-779.

Shorey, H.H., L.K. Gaston & C.A. Saario. 1967. Sex pheromones of noctuid moths. XIV. Feasibility of behavioral control by disrupting pheromone communication in cabbage loopers. J. Econ. Entomol. 60:1541-1545.

Shorey, H.H., R.S. Kaae, L.K. Gaston & J.R. McLaughlin. 1972. Sex pheromones of Lepidoptera. XXX. Disruption of sex pheromone communication in *Trichoplusia ni* as a possible means of control. Environ. Entomol. 1:641-645.

Sower, L.L. 1982. Douglas-fir tussock moth disruption, pp. 165-174. *In* A.F. Kydonieus & M. Beroza (eds.). Insect suppression with controlled release pheromone systems. Vol. 2. CRC Press, Boca Raton, Florida.

Sower, L.L. & G.E. Daterman. 1977. Evaluation of synthetic sex pheromone as a control agent for Douglas-fir tussock moths. Environ. Entomol. 6:889-892.

Talekar, N.S. 1988. How to control sweetpotato weevil: a practical IPM approach. Asian Veg. Res. Devel. Cent. Bull. No. 88-292.

Talekar, N.S. & S.T. Lee. 1989. Studies on the utilization of a female sex pheromone for the management of sweetpotato weevil, *Cylas formicarius formicarius* (F.) (Coleoptera: Curculionidae). Bull. Inst. Zool. Acad. Sinica 23:245-252.

Talekar, N.S., R.M. Lai & K.W. Cheng. 1989. Integrated control of sweetpotato weevil at Penghu island. Plant Prot. Bull. (Taiwan) 31:185-192.

Taschenberg, E.F. & W.L. Roelofs. 1976. Pheromone communication disruption of the grape berry moth with microencapsulated and hollow fiber systems. Environ. Entomol. 5:688-691.

Wall, C., D.M. Sturgeon, A.R. Greenway & J.N. Perry. 1981. Contamination of vegetation with synthetic sex attractant released from traps for the pea moth, *Cydia nigricana*. Entomol. Exp. Appl. 30:111-115.

Webb, R.E. 1982. Mass trapping of the gypsy moth, pp. 27-56. *In* A.F. Kydonieus & M. Beroza (eds.). Insect suppression with controlled release pheromone systems. Vol. 2. CRC Press, Boca Raton, Florida.

Yonce, C.E. & C.R. Gentry. 1982. Disruption of mating of peachtree borer, pp. 99-106. *In* A.F. Kydonieus & M. Beroza (eds.). Insect suppression with controlled release pheromone systems. Vol. 2. CRC Press, Boca Raton, Florida.

7. Integrated Control of *Cylas formicarius*

Narayan S. Talekar
Asian Vegetable Research and Development Center
P.O. Box 42
Shanhua, Tainan, Taiwan, Republic of China

Sweetpotato weevil, *Cylas formicarius* (Fabricius), is the most destructive and widely distributed pest of sweet potato, *Ipomoea batatas* (L.) Lam., in tropical and subtropical areas of the world. Larvae of this pest tunnel through storage roots which results in major damage and economic yield loss (Chalfant *et al.* 1990). Weevil damage imparts a characteristic terpene odor to the roots, which renders even slightly damaged roots unfit for human consumption (Uritani *et al.* 1975). Larvae also feed inside the vines. The effect of weevil densities in vines on root yield has produced variable and often contradictory results (Cockerham *et al.* 1954, Mullen 1984, Sutherland 1986b, Talekar 1982). Adults make small superficial feeding holes on storage roots. Such damage, however, is considered minor compared with that caused by larval feeding in the roots. Adults also feed on sweet potato foliage (Franssen 1934, Nottingham *et al.* 1988), but the effect of foliar damage on root yield has not been demonstrated.

This insect can be difficult to control with applications of conventional chemical insecticides because larvae and pupae are cryptic. However, populations can be reduced by cultural practices and good sanitation because *C. formicarius* has a limited flight activity (Cockerham *et al.* 1954, Sherman & Tamashiro 1954), it has a limited host range (see Chapter 3), and it has a characteristic mode of entry into the host plant. Integration of these practices has potential for managing this insect on a sustainable basis.

CULTURAL CONTROL

Cultural pest control involves changing or modifying cultivation practices which directly or indirectly reduce the pest population. Cultural practices, such as crop rotation, intercropping, mulching, sanitation, etc., were the earliest control measures advocated for reducing damage by sweet potato weevils.

Crop Rotation

Rotation of crops, such as growing sweet potato in a field only once every five years (Anonymous 1954), avoiding planting sweet potato in the same area for two successive years (Ballou 1915, Chittenden 1919, Edward 1930, Holdaway 1941, Wedell 1951), or planting rice between two sweet potato crops (Franssen 1934) has long been suggested.

The usefulness of crop rotation in controlling the weevil was investigated in two experiments, each lasting 17 to 18 months in Taiwan (Talekar 1983). In the first experiment, planting rice after the harvest of the weevil-infested crop did not reduce weevil damage in the following sweet potato crop. Weevil densities and root damage were greater in sweet potatoes planted after rice than in those planted after sweet potato (Table 7.1). This was probably due, in part, to the weevil infestations in adjacent plots that were planted to sweet potato after harvesting the weevil-infested crop. Larger roots which were produced in plants grown after rice rotation may have also helped to increase weevil densities. Wild alternate hosts, *Ipomoea* sp., growing on the field borders, may also have helped to increase weevil infestations (Talekar 1983).

In the second experiment, weevil control was acceptable in a field in which rice rotation was integrated with field isolation in which sweet potato was planted away from a source of weevils. Conversely, in another field located near a weevil source, rotation with rice did not adequately improve weevil control (Figure 7.1). Thus, crop rotation, a useful agronomic practice to maintain soil fertility, will help to control sweetpotato weevil if integrated with other management approaches.

Intercropping

Intercropping may reduce the accessibility of host plants to pests because plants of the same crop species are more randomly distributed

Table 7.1. Effect of rice rotation and other management practices on *C. formicarius* population density, storage root damage and yield of sweet potato.

Management practice	Damaged roots, % SSS[a]	Damaged roots, % SRS[a]	No. of weevils/kg of root SSS	No. of weevils/kg of root SRS	No. of weevils/ crown SSS	No. of weevils/ crown SRS	Yield, t/ha SSS	Yield, t/ha SRS
'I123'[b]	19.9a	24.9a	9.0a	11.8a	4.2a	3.3b	5.1b	7.4ab
'AIS35-2'[c]	22.8a	27.9a	5.8a	11.0a	2.1a	5.5ab	1.6a	5.4ab
Sugarcane intercropping[d]	20.7a	24.2a	4.6a	11.7a	3.8a	3.8b	1.7a	4.6b
Corn intercropping[d]	17.9ab	25.7a	3.4a	11.0a	1.5a	5.2ab	1.1a	6.5ab
Methomyl[e]	13.7ab	13.6a	4.6a	7.7a	3.8a	6.9a	4.5b	9.1ab
Triazophos[e]	6.3b	14.2a	5.3a	6.0a	2.7a	4.2ab	8.4c	11.6a

Means in each vertical column followed by the same letter are not significantly different at 5% level by Duncan's multiple range test.

[a] SSS, Sweet potato-sweet potato-sweet potato; SRS, sweet potato-rice-sweet potato.

[b] Sweet potato cultivar moderately resistant to weevil.

[c] Weevil-susceptible check cultivar.

[d] 'AIS35-2' intercropped between sugarcane or corn.

[e] 'AIS35-2', methomyl, or triazophos applied (0.5 kg AI/ha) once every two weeks from two weeks after transplanting until two weeks before harvest.

(Litsinger & Moody 1976). Certain crops may also provide physical barriers for the movement of the pest insect. Little research has examined this approach for the management of sweetpotato weevil. In one experiment in Taiwan, sweet potato was planted between two rows of each of 68 crop species and weevil infestations of the roots were monitored. Intercropping with chickpea (*Cicer arietinum* L.), coriander (*Coriandrum sativum* L.), pumpkin (*Cucurbita moschata* Duch.), radish (*Raphanus sativus* L.), fennel (*Foeniculum vulgare* Mill.), blackgram (*Vigna mungo* L.) and yardlong bean (*Vigna unguiculata* L. [Wap.] *ssp. sesquipedalis*) reduced weevil infestations considerably. However, intercropping with blackgram, fennel, pumpkin, and yardlong bean also reduced sweet potato yields (AVRDC 1988). Similarly, Singh *et al.* (1984) observed reduced weevil damage when sweet potato was intercropped with a millet, proso (*Panicum miliaceum* L.) and sesamum (*Sesamum indicum* [L.]). These intercroppings also reduced sweet potato yields considerably. It is uncertain if the reduced yield (smaller or fewer roots) contributed to

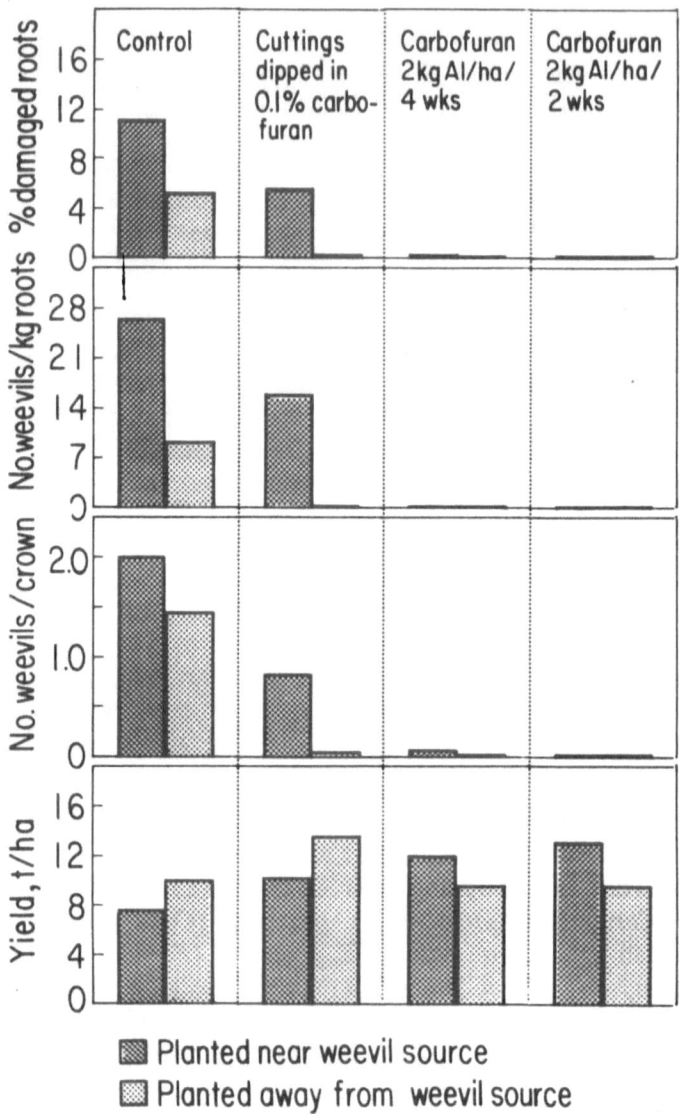

Fig. 7.1. Effects of various carbofuran treatments applied to sweet potato crops planted near and away from sources of weevil infestations on abundance and damage of sweetpotato weevil.

the lower weevil infestations. More research on the effects of intercropping on weevil damage and root yield is needed.

Mulching

Soil cracks are the major route of weevil access to roots. The enlargement of roots, especially in cultivars which set roots near the soil surface, and soil moisture stress can produce cracks and increase exposure of roots to weevil. Absence of cracks denies the weevil access to the roots. For example, in Taiwan, lower damage by *C. formicarius* occurred during the rainy season when soil cracks were minimized (AVRDC, unpubl. data). Similarly, the African sweetpotato weevil, *C. puncticollis* Boheman, which causes damage similar to that by the sweetpotato weevil, *C. formicarius*, is less damaging during the wet season than during the dry season in Nigeria (Hahn & Leuschner 1982). Others have reported similar findings (Leuschner 1982, Rajamma 1983, Sutherland 1986a). Prevention of soil cracking by hilling the area around the plant or irrigating frequently may also help to reduce weevil damage (Franssen 1934, Holdaway 1941, Sherman & Tamashiro 1954). Two experiments were conducted in Taiwan to study the potential of mulch for reducing sweetpotato weevil infestations (Talekar 1987b). Mulching materials, plastic film or rice straw, were spread over a sweet potato planting located near a weevil source shortly after planting. Plastic film and rice straw mulch reduced weevil infestations as compared with nonmulched plots (Figure 7.2). Mulches presumably conserved soil moisture and minimized soil cracking. The physical barrier made by mulching materials further reduced access of roots to the weevil (Talekar 1987b). Conversely, in the United States, Jansson *et al.* (1987) found that weevil damage to storage roots of four cultivars of sweet potato did not differ between plants grown in plots with a plastic mulch barrier and those grown in plots without a plastic mulch barrier.

Sanitation

Sanitation practices or clean cultivation, especially for the control of an insect that has a limited flying activity, may help protect the crop from insect infestations. These practices played an important role in pest control until the introduction and widespread use of chemical insecticides. A variety of sanitation methods have been recommended

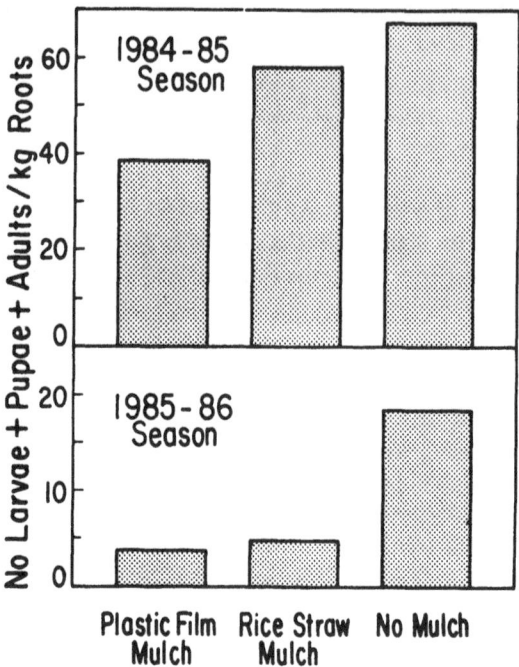

Fig. 7.2. Effects of plastic and rice straw mulch on the infestation of sweetpotato weevil during the 1984-85 and 1985-86 growing seasons at AVRDC, Taiwan.

for weevil control, and in some locations they are even legally enforced.

Destruction of Crop Residue. Destruction of the crop residue left in the field after harvest is important because weevils survive in roots and stems and infest succeeding or neighboring sweet potato plantings (Chittenden 1919, Eddy *et al.* 1942, Franssen 1934, Jansson *et al.* 1989, Wedell 1951). Crop rotation, in most cases, serves this purpose. However, in areas where sweet potato is a staple food and is planted year round, rotation is not always possible.

Flooding of infested fields was tested in Taiwan to induce rotting of the leftover plant materials and thereby reduce weevil densities from one planting to the next (Talekar 1987b). Standing water was maintained over recently harvested weevil-infested plots for 1 to 4 weeks. Two or more weeks of flooding considerably reduced the emergence of volunter sweet potato plants. Few plants emerged from flooded fields and these plants harbored few weevils. Conversely, a

Table 7.2. Emergence of volunteer sweet potato plants and corresponding weevil infestations after flooding of infested fields for varying lengths of time. [a]

Flooding duration, weeks	No. plants per plot	No. insects [b] per plant
0	15.6	2.2
1	4.2	0.2
2	0.3	0.0
3	3.3	0.0
4	0.4	0.1

[a] Flooding initiated on 11 November 1981; plants were sampled on 5 February 1982; plot sizes were 10 m long by 8 m wide; data are averages for three experiments.
[b] Total larvae, pupae, and adults.

large number of volunteer plants grew in the nonflooded control plots, all of which were infested with weevils (Table 7.2). These data show that flooding of fields between two consecutive sweet potato crops may reduce the immediate source of weevils from the field. This approach should be considered in areas where rotation is not possible.

Clean Cuttings. Sweetpotato weevil lays eggs in the vines, especially older portions, in the absence of storage roots or when the roots are inaccessible (AVRDC, unpubl. data). Planting of infested vines may spread the weevil infestation. Therefore, the use of weevil-free sweet potato cuttings is often advised (Ballou 1915, Franssen 1934, Holdaway 1941, Tucker 1937). Weevil-free cuttings can be produced by dipping them in a suitable insecticide solution before planting. In Taiwan, Talekar (1983) obtained substantial reduction in weevil damage when the cuttings were dipped in carbofuran (0.05% AI), even though the crop was planted in close proximity to a weevil source. The same dip treatment provided complete control of the weevil when the crop was planted away from weevil sources.

Recent findings in Taiwan showed that cuttings (25-30 cm long) taken from the fresh terminal growth, even from an infested crop, were rarely infested with weevils, whereas older portions of the stem were infested with weevils. The probability of finding weevils inside the stems decreased in younger cuttings (AVRDC 1990). This was further confirmed in a related study where 1- to 8-week-old weevil-free plants were exposed to the weevil in the field. The numbers of weevils in vines increased with increase in vine age (r = 0.92) (AVRDC 1990).

These results indicate that carry-over of the weevil from an infested crop to a new planting can be reduced by carefully selecting fresh cuttings.

Removal of Alternate Hosts. Several species of *Ipomoea*, in addition to sweet potato, and a few related convolvulaceous plants are also alternate hosts of *C. formicarius* (see Chapter 3). Sutherland (1986a) listed 30 such species and four additional species were recently found to harbor the weevil in Taiwan (AVRDC 1989). A more complete and correct list of host plants of *C. formicarius* was presented by Austin *et al.* (1990) and in a previous chapter (see Chapter 3). Among the convolvulaceous hosts, the insect overwhelmingly prefers sweet potato (Cockerham 1943). Presence of alternate hosts, most of which are perennial, may influence infestations of sweetpotato weevil. Removal of these hosts growing in the vicinity of sweet potato plantings is recommended as a control measure (Butani & Varma 1976, Chittenden 1919, Cockerham 1940, 1943, Edward 1930, Franssen 1934, Gonzales 1925, Ho 1970, Jayaramaiah 1975, Reinhard 1923, Subramaniam 1959, Wood 1976). However, indiscriminate elimination of wild *Ipomoea* to remove weevil sources may also have undesirable ecological effects. Availability of a synthetic sex pheromone of *C. formicarius* may help to detect "weevil-positive" *Ipomoea*. Using such an approach, only weevil-infested *Ipomoea* are removed which may help to reduce labor costs and minimize ecological effects. An alternative approach for managing wild *Ipomoea* species involves removing all *Ipomoea* from an area for one crop season and then allowing these plants to grow in subsequent seasons once the area is free of weevils. This approach resulted in the eradication of the weevil in the Kagoshima prefecture of Japan (Anonymous 1969). In Taiwan, removal of alternate hosts and volunteer sweet potato plants reduced the weevil densities considerably (Talekar 1983).

Other Cultural Practices

Other cultural practices which may help reduce weevil damage and which are often advocated include planting cuttings deep in the soil (Holdaway 1941), use of deep-rooted cultivars (Franssen 1934), and harvesting the crop early when roots are of acceptable size (Edward 1930, Holdaway 1941, Nawale 1981, Sherman & Tamashiro 1954, Sutherland 1986a,b). Planting weevil-resistant sweet potato cultivars also represents a potential cultural control method, however, cultivars with reliable levels of resistance to the weevil are not yet available (Talekar 1987a).

CHEMICAL CONTROL

Numerous chemical insecticides have been tested for the control of *C. formicarius* despite the hidden mode of the insect's life cycle. Sutherland (1986a) listed 59 different insecticides, including botanicals of unknown chemical composition, that were tested against sweetpotato weevil. These chemicals, most of which were applied as post-planting foliar sprays, resulted in varying levels of control.

Pre-plant Applications

Pre-plant insecticide applications have been used to manage weevils in planting material (vine cuttings). Insecticides with adequate water solubility are presumably transported through the vine and kill the weevils within the vine. This type of treatment is usually more economical than post-plant insecticide applications, and if combined with proper sanitation and other measures to prevent immigration of weevils from infested plants, may result in satisfactory control of the weevil (Ingram 1957, Sherman 1951, Sherman & Mitchell 1953, Sherman & Tamashiro 1954, Wolcott & Perez 1955).

Post-plant Applications

Control of the weevil is difficult with conventional spraying, dusting, or fumigation with presently available insecticides once weevils are present within the crown or storage roots. Control achieved by post-plant applications appears to be due to mortality of weevil adults in search of feeding or oviposition sites. Movement of adult weevils may facilitate the contact between the toxicant and the insect, thereby resulting in insect mortality. This method of control, however, requires frequent applications in order to kill adults that might migrate from other areas. This view concurs with Sakae (1988) in Japan. Frequent spraying of insecticides, however, is not cost-effective due to the low market price for sweet potato in developing countries.

SEX PHEROMONE

The existence of a female-produced sex pheromone in *C. formicarius* is well known (AVRDC 1976, Coffelt *et al.* 1978, Louton 1975, Russo 1973). Heath *et al.* (1986) isolated, identified, and synthesized this chemical, (Z)-3-dodecen-1-ol(E)-2-butenoate, which has great potential for attracting male sweetpotato weevils (Jansson *et al.* 1989, Proshold *et al.* 1986) and reducing the weevil populations in the field (Talekar & Lee 1989) (see Chapters 5 and 6). Because of its potency and relatively long persistence, this chemical may be used in various ways to manage *C. formicarius* in an integrated program (see below, also see Chapter 6).

BIOLOGICAL CONTROL

There are several predators and parasitoids that attack sweetpotato weevil, however, these natural enemies seem to be ineffective at managing weevil populations (see Chapter 9). Entomopathogens, such as fungi and entomogenous nematodes, however, have good potential for managing this weevil and could be included in integrated control programs in the future (see Chapter 9). No further mention will be made of this management approach because no studies have integrated these approaches into management programs for *C. formicarius* to date.

INTEGRATED CONTROL

In most cases, none of the above described control measures, used singly, will provide adequate control of *C. formicarius* where sweet potato is grown throughout the year and the weevil is endemic. However, a combination of tactics can provide satisfactory control of this pest. With the exception of biological control in combination with post-plant applications of insecticides, all other control measures are fully compatible. Certain biological control agents, such as predators and parasitoids, would be quickly eliminated by use of chemical insecticides.

Two principal sources of weevils that affect infestations of new sweet potato plantings are carry-over of the insect in cuttings taken from old infested field and immigration of weevils from alternate hosts or weevil-infested crops. To control the weevil successfully, these two weevil sources must be managed properly. Integration of some of the above control approaches has potential for reducing and possibly preventing the crop from being infested by the weevil from these two sources.

PRACTICALITY OF IPM: A CASE STUDY IN TAIWAN

The use of integrated pest management (IPM), based on laboratory, greenhouse, and field experiments, was tested in fields on Penghu, an island in the Taiwan Straits where *C. formicarius* is endemic. The experiment was conducted during three years, 1986 to 1988, in Jungtwun village and during one season in 1988 in Husyi village (Fig. 7.3). Although a recent study (Talekar 1983) found that only two factors (use of weevil-free cuttings and removal of alternate hosts) were adequate to control the weevil on an experimental farm, an expanded IPM program was used in these villages because of their long history of weevil infestations and the presence of numerous species of *Ipomoea* growing on Penghu island. The IPM program included crop rotation, dipping cuttings in carbofuran solution (0.05% AI) before planting, elimination of wild *Ipomoea* species from the field borders, hilling of soil to fill in soil cracks, application of carbofuran (2 kg AI/ha) once per month during the first year of the IPM program in Jungtwun only, and continuous trapping of male weevils in pheromone-baited traps with 100 μg of sex pheromone from planting until harvest (1987 and 1988 only).

Before IPM was tested, a survey of the Jungtwun township found that 100% of all plants and 75% of the roots were damaged by *C. formicarius*. After the first growing season, when 29 farmers practiced IPM in 3.5 ha of sweet potato grown in 40 parcels of land, 50% of the plants and 30% of the roots were damaged by weevils. The unacceptable levels of weevil damage were due to non-compliance of some farmers at practicing all components of IPM (Talekar *et al.* 1989).

In 1987 and 1988, evaluations of weevil control procedures were modified. At harvest, roots (5 kg) were sampled from each field, excised, and the numbers of weevil larvae, pupae, and adults found were recorded. The damaged and healthy portions of roots in each sample were weighed, and the percentage of roots damaged by the weevil was recorded.

In 1987, 69,315 male weevils were caught in pheromone-baited traps in 62 parcels of land (5 ha) in Jungtwun. The numbers of insects

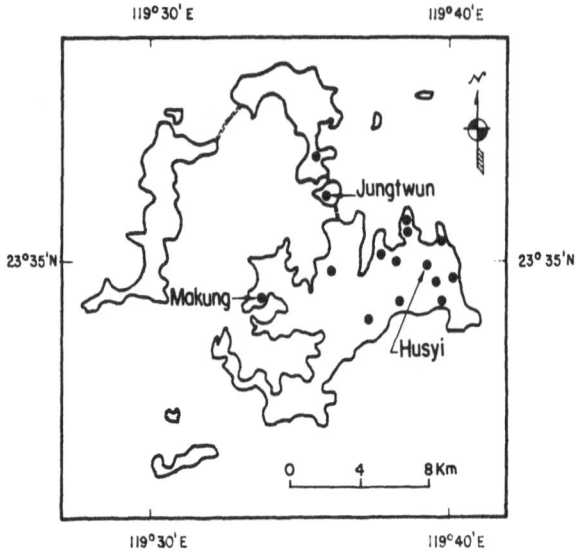

Fig. 7.3. Map of Penghu island showing the locations of experimental sites, Jungtwun and Husyi.

per kg of roots ranged from 0 to 98, with a mean of 14.5, and root damage ranged from 0 to 58% with a mean of 11.6% (Fig. 7.4) (Talekar *et al.* 1989). In most cases, root damage was less than 10% and in nine fields, no weevils were found. Farmers that integrated all components of IPM and moved the pheromone traps frequently within fields were successful at producing a weevil-free crop.

In 1988, 21,129 male weevils were trapped in 55 parcels of land (3 ha) in Jungtwun. The numbers of insects per kg of root ranged from 0 to 69.4 with a mean of 5.1, and root damage ranged from 0 to 46.4% with a mean of 3.6% (Fig. 7.4). In 85% of all farms, weevil damage to roots was less than 10%, and 31 farms were weevil-free. Thus, weevil damage was reduced considerably in 1988 compared with that found in 1987.

In Husyi, a total of 16,348 male weevils were caught in pheromone-baited traps. The numbers of weevils per kg of root ranged from 0 to 4 with a mean of 0.34, and root damage ranged from 0 to 4% with a mean of 0.4%. This level of damage was considerably less than the 40% root damage found in 1987 before IPM was implemented. In 1988, 81% of all sweet potato fields were free of weevils. Based on the above, it is possible to reduce sweetpotato weevil infestations in fields

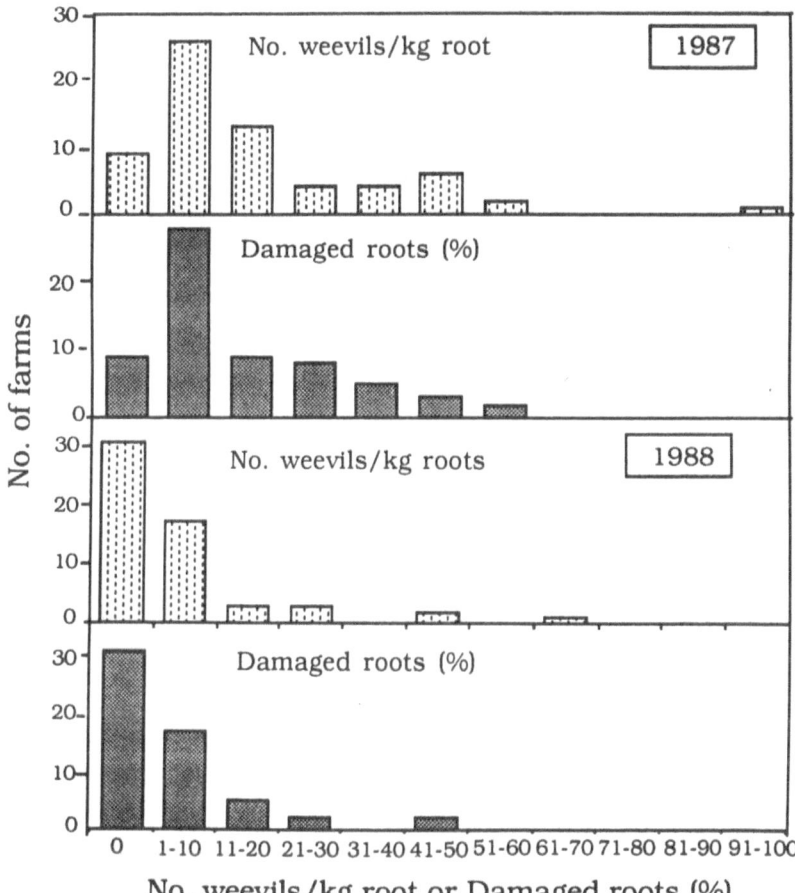

Fig. 7.4. Sweetpotato weevil infestation of sweet potato roots in Jungtwun in 1987 and 1988.

to a negligible level by adopting an IPM program. The contribution of sex pheromone in population reduction was difficult to ascertain; however, it was especially useful for indicating the status of control practices for farmers. It should be emphasized that the sex pheromone is not attractive to female weevils (see Chapter 5). For this reason, it is essential that the major sources of infestation, such as wild *Ipomoea*, be removed to reduce the possibility of mated females moving from wild host plants to sweet potato plantings. Also, it is essential that all growers in a community follow these control measures. If some growers do not adopt these practices, their fields may serve as sources of weevil infestations for the neighboring fields.

Other weevil species, such as *C. puncticollis*, *C. brunneus* (Fabricius), and *Euscepes postfasciatus* (Fairmaire), also damage sweet potato. The biology and the nature of damage caused by these species is similar to that of *C. formicarius*. Except for the use of sex pheromone, the integrated control approaches described in this paper for management of *C. formicarius* are applicable for managing other weevil species.

FUTURE OUTLOOK

The IPM program described is simple and readily adaptable for controlling sweet potato weevils under a wide range of sweet potato cultivation practices. It is based on a sound knowledge of the biology of the pest - the same knowledge that is used in developing and enforcing the legal ban on the movement of sweet potato plant parts to prevent potential immigration of the insect into certain areas in the United States and elsewhere. These techniques require few purchased inputs.

IPM, unlike other control practices, is sustainable over a long period and is well suited for developing countries of the tropics where sweet potato is grown on small farms and where family labor is readily available to maintain sanitation around the field. The utility of this technique has been successfully demonstrated on small farms in Taiwan.

The most important constraint of IPM is that it must be practiced by every grower in a community. This is not always possible, especially in developing countries where farmers lack education and extension services are poor or non-existent (see Chapter 1). In addition, over 90% of the world's sweet potato is grown in Asia where most countries are rapidly becoming self-sufficient and even experiencing surpluses in production of rice, a competing and preferred staple crop. As a result, sweet potato production, research, and development are declining. The decline in sweet potato production and better prices for competing commodities have led farmers to put less emphasis on the management of the sweet potato crop. On a global basis, this is a formidable obstacle and as long as this trend continues, IPM and other improved management practices will continue to be a low priority of sweet potato growers.

REFERENCES

Anonymous. 1954. Ways to control weevil (*Cylas formicarius elegantulus*) and diseases of sweet potatoes. Texas Agric. Coll. Ext. Leaflet No. 202.

Anonymous. 1969. Survey of occurrence of sweetpotato weevil on Tanegashima: No living weevil found. Kyushu Plant Prot. 301:3 (In Japanese).

Austin, D.F., R.K. Jansson & G.W. Wolfe. 1990. Convolvulaceae and Cylas: a proposed hypothesis on the origins of this plant insect relationship. Trop. Agric. (In press).

AVRDC. 1976. AVRDC sweet potato report 1975. Asian Vegetable Research and Development Center, Shanhua, Taiwan, Republic of China.

AVRDC. 1988. 1986 Progress Report. Asian Vegetable Research and Development Center, Shanhua, Taiwan, Republic of China.

AVRDC. 1989. AVRDC progress report summaries 1988. Asian Vegetable Research and Development Center, Shanhua, Taiwan, Republic of China.

AVRDC. 1990. AVRDC progress report summaries 1990. Asian Vegetable Research and Development Center, Shanhua, Taiwan, Republic of China (In press).

Ballou, H.A. 1915. The sweetpotato weevil. Agric. News 14:138.

Butani, D.K. & S. Varma. 1976. Pests of vegetables and their control: sweet potato. Pestic. 10(2):36-38.

Chalfant, R.B., R.K. Jansson, D.R. Seal & J.M. Schalk. 1990. Ecology and management of sweet potato insects. Annu. Rev. Entomol. 35:157-180.

Chittenden, F.H. 1919. The sweetpotato weevil and its control. U.S.D.A. Farm. Bull. No. 1020.

Cockerham, K.L. 1940. Wild hosts of the sweetpotato weevil. La. Agric. Exp. Stn. Bull. 323:38-39.

Cockerham, K.L. 1943. The host preference of sweetpotato weevil. J. Econ. Entomol. 36:471.

Cockerham, K.L., O.T. Dean, M.B. Christian & L.D. Newsom. 1954. The biology of the sweetpotato weevil. La. Agric. Exp. Stn. Tech. Bull. No. 483.

Coffelt, J.A., K.W. Vick, L.L. Sower & W.T. McClellan. 1978. Sex pheromone of the sweetpotato weevil *Cylas formicarius elegantulus*: laboratory bioassay and evidence for a multiple component system. Environ. Entomol. 7:756-758.

Eddy, C.O., E.H. Floyd & K.L. Cockerham. 1942. Sweetpotato weevil control, p. 42. *In* Annual report. La. Agric. Exp. Stn., Baton Rouge.

Edward, W.H. 1930. Insect pests of sweetpotato and cassava in Jamaica. Jamaica Entomol. Bull. No. 5.

Franssen, C.J.H. 1934. Insect pests of sweet potato crop in Java. Landbouw 10:205-225. (In Dutch with English summary).

Gonzales, S.S. 1925. The sweetpotato weevil (*Cylas formicarius* Fbr.). Philipp. Agric. 14:257-281.

Hahn, S.K. & K. Leuschner. 1982. Breeding of sweet potato for weevil resistance, pp. 331-336. *In* R.L. Villareal & T.D. Griggs (eds.). Sweet potato. Proceedings of the 1st international symposium. Asian Vegetable Research and Development Center, Shanhua, Taiwan, Republic of China.

Heath, R.R., J.A. Coffelt, P.E. Sonnett, F.I. Proshold, B. Dueben & J.H. Tumlinson. 1986. Identificaton of sex pheromone produced by female sweetpotato weevil, *Cylas formicarius elegantulus* (Summers). J. Chem. Ecol. 12:1489-1503.

Ho, T.H. 1970. Studies on some major pests of sweet potato and their control. Malay. Agric. J. 47:437-452.

Holdaway, F.G. 1941. Insects of sweet potato and their control. Ha. Agric. Exp. Stn. Progress Notes No. 26:1-7.

Ingram, W.R. 1967. Chemical control of sweetpotato weevil in Uganda. East Afr. Agric. Fores. J. 33:163-165.

Jansson, R.K., H.H. Bryan & K.A. Sorensen. 1987. Within-vine distribution and damage of the sweetpotato weevil, *Cylas formicarius elegantulus* (Coleoptera: Curculionidae), on four cultivars of sweet potato in southern Florida. Fla. Entomol. 70:523-526.

Jansson, R.K., R.R. Heath & J.A. Coffelt. 1989. Temporal and spatial patterns of sweetpotato weevil (Coleoptera: Curculionidae) counts in pheromone-baited traps in white-fleshed sweet potato fields in southern Florida. Environ. Entomol. 18:691-697.

Jayaramaiah, M. 1975. Reaction of sweetpotato varieties in the damage of the weevil *Cylas formicarius* (Fab.) (Coleoptera: Curculionidae) and the possibility of picking up infestation by weevil. Mysore J. Agric. Sci. 9:418-421.

Leuschner, K. 1982. Pest control for cassava and sweet potato, pp. 60-64. *In* Root crops in Eastern Africa: Proceedings of a workshop held at Kigali, Rwanda. International Development Research Centre, Ottawa, Canada.

Litsinger, J.A. & K. Moody. 1976. Integrated pest management in multiple cropping systems, pp. 293-316. *In* Multiple cropping. ASA Special Publication No. 17, American Society of Agronomy, Madison, Wisconsin.

Louton, P.A. 1975. Localization of sex pheromone production in female *Cylas formicarius elegantulus* (Summers). M.S. Thesis, Louisiana State University, Baton Rouge.

Mullen, M.A. 1984. Influence of sweetpotato weevil infestation on the yields of twelve sweet potato lines. J. Agric. Entomol. 1:227-230.

Nawale, R.N. 1981. A study on the stage of harvest of sweet potato variety H268 under Konkan (Maharashtra) conditions. J. Root Crops. 7:29-31.

Nottingham, S.F., K.C. Son, D.D. Wilson, R.F. Severson & S.J. Kays. 1988. Feeding by adult sweetpotato weevils, *Cylas formicarius elegantulus*, on sweet potato leaves. Entomol. Exp. Appl. 48:157-163.

Proshold, F.I., J.L. Gonzales, C. Asencio & R.R. Heath. 1986. A trap for monitoring the sweetpotato weevil (Coleoptera: Curculionidae) using pheromone or live females as biat. J. Econ. Entomol. 79:641-647.

Rajamma, P. 1983. Biology and bionomics of sweetpotato weevil, *Cylas formicarius* Fabr., pp. 87-92. *In* S.D. Goel (ed.). Proceedings of symposium on insect ecology and resource management. Sanatan Dharm College, Muzaffarnagar, India.

Reinhard, H.J. 1923. The sweetpotato weevil. Texas Agric. Exp. Stn. Bull. No. 308.

Russo, A. 1973. Studies directed towards the isolation and structure determination of sex and food attractants in *Cylas formicarius elegantulus* (Summers). Ph.D. Dissertation, Louisiana State University, Baton Rouge.

Sakae, M. 1988. Insect pests of Amani Islands. Gojo Insatsu, Kagoshima, Japan (In Japanese).

Sherman, M. 1951. Chemical control of sweet potato insects in Hawaii. J. Econ. Entomol. 44:652-656.

Sherman, M. & W.C. Mitchell. 1953. Control of sweet potato weevils and vine borer in Hawaii. J. Econ. Entomol. 46:389-393.

Sherman, M. & M. Tamashiro. 1954. The sweetpotato weevils of Hawaii, their biology and control. Ha. Agric. Exp. Stn. Tech. Bull. No. 23.

Singh, S.S., R. Yazdani & S.F. Hameed. 1984. Effect of intercropping on the incidence of sweetpotato weevil, *Cylas formicarius* Fabr., in sweet potato (*Ipomoea batatas* Lam.). J. Entomol. Res. 8:193-195.

Subramaniam, T.R. 1959. Observations on the biology of *Cylas formicarius* at Coimbatore. Madras Agric. J. 46:293-297.

Sutherland, J.A. 1986a. A review of the biology and control of the sweetpotato weevil *Cylas formicarius* (Fab). Trop. Pest Manage. 32:304-315.

Sutherland, J.A. 1986b. Damage by *Cylas formicarius* Fab. to sweet potato vines and tubers, and the effect of infestations on total yield in Papua New Guinea. Trop. Pest Manage. 32:316-323.

Talekar, N.S. 1982. Effects of a sweetpotato weevil (Coleoptera: Curculionidae) infestation on sweet potato root yields. J. Econ. Entomol. 75:1042-1044.

Talekar, N.S. 1983. Infestation of a sweetpotato weevil (Coleoptera: Curculionidae) as influenced by pest management techniques. J. Econ. Entomol. 76:342-344.

Talekar, N.S. 1987a. Feasibility of the use of resistant cultivar in sweetpotato weevil control. Insect Sci. Applic. 8:815-817.

Talekar, N.S. 1987b. Influence of cultural pest management techniques on the infestation of sweetpotato weevil. Insect Sci. Applic. 8: 809-814.

Talekar, N.S. & S.T. Lee. 1989. Studies on the utilization of a female sex pheromone for the management of sweetpotato weevil, *Cylas formicarius formicarius* (F.) (Coleoptera: Curculionidae). Bull. Inst. Zool. Acad. Sin. 21:281-288.

Talekar, N.S., R.M. Lai & K.W. Cheng. 1989. Integrated control of sweetpotato weevil at Penghu island. Plant Prot. Bull. (Taiwan) 31:185-191.

Tucker, R.W.E. 1937. The control of scarabee (*Euscepes batatas* Waterhouse) in Barbados. Barbados Dept. Sci. Agric., Agric. J. 6:133-154.

Uritani, I., T. Saito, H. Honda & W.K. Kim. 1975. Induction of furano-terpenoids in sweet potato roots by the larval components of the sweet potato weevils. Agric. Biol. Chem. 37:1857-1862.

Wedell, J.A. 1951. The sweetpotato weevil. Queensl. Agric. J. 73:25-26.

Wolcott, G.N. & M. Perez. 1955. Control of the sweetpotato weevil in Puerto Rico. J. Econ. Entomol. 48:486-487.

Wood, I.J.L. 1976. Sweet potato growing in Queensland. Queensl. Agric. J. 102:553-566.

8. Spatial Patterns of *Cylas formicarius* in Sweet Potato Fields and Development of a Sampling Plan

Robert McSorley
Department of Entomology and Nematology
University of Florida, I.F.A.S.
Gainesville, Florida 32611 U.S.A.

Richard K. Jansson
Tropical Research and Education Center
University of Florida, I.F.A.S.
Homestead, Florida 33031 U.S.A.

Sampling plans should be an integral part of management programs for most pests. They provide a basis for detecting infested locations and for monitoring a pest population once it is present. When a relationship exists between population density during the growing season and final yield, the regular monitoring of populations through a scouting program using appropriate sampling plans can provide the opportunity to make treatment decisions in an efficient and timely manner (Pohronezny *et al.* 1981, Waddill *et al.* 1981).

Before a sampling plan is developed, a few fundamental questions should first be considered. First, what is the size of the sample unit, i.e., the individual portion of the habitat that is to be examined? In what pattern should sample units be collected within the site? What is the spatial dispersion pattern of the pest? How many sample units should be collected from the site? For any pest-crop combination, it is likely that a number of possible sampling plans could be developed. Since these proposed plans can vary greatly in their levels of reliability, or precision, it is critical to select the plan that is reliable enough to meet the objectives of the study, yet not so intensive that it would exceed practical economic limitations.

To illustrate the steps involved in the development of a sampling plan, we consider the estimation of population density of the sweetpotato weevil, *Cylas formicarius* (Fabricius), in sweet potato (*Ipomoea batatas* [L.] Lam.) fields. The severe losses (up to 60-97%) due to this weevil and related species of *Cylas* (Ho 1970, Mullen 1984, Subramanian *et al.* 1977) provide incentive for growers to monitor this pest closely. In one study, densities as low as 0.8 weevils per 15 cm of vine were associated with 10-15% tuber damage at harvest (Sutherland 1986). To monitor weevil densities in the field, appropriate sampling plans would first have to be available. However, it is possible that such plans could be developed from available data (Jansson & McSorley 1990, Jansson *et al.* 1990) detailing the typical within-field distribution of *C. formicarius*.

SAMPLE UNITS

To obtain an estimate of *C. formicarius* density per plant, all larvae, pupae, and adults can be dissected from vines and roots. It is useful to compare two alternative sample units: the whole plant, and a zone extending from 15 cm below to 10 cm above the crown (i. e., the vine/soil surface interface). The sample unit from 15 cm below to 10 cm above the crown contains a high percentage of the total weevil population in vines and roots, ranging from 78.0 to 78.7% on 'Picadito' and 'Centennial' to 87.4 to 89.3% on 'Campeon' and 'Regal' (Jansson *et al.* 1990). In addition, the time for dissecting weevils from the zone 15 cm below to 10 cm above the crown is much less (12 minutes vs. 30 to 45 minutes) than the time required to dissect all weevils from the whole plant.

Total numbers of SPW in the whole plant (y) can be estimated from total numbers in the zone from 15 cm below to 10 cm above the crown (x) by the regression equation ($r^2 = 0.928$):

$$y = 0.148 + 1.137\,x \qquad (1)$$

This equation was derived from 3264 samples collected during the course of two experiments. Related equations, derived from data in each experiment, are of similar form, but Equation 1 is a general relationship applicable to both situations (Jansson & McSorley 1990). The high r^2 value indicates that numbers in the zone from 15 cm below to 10 cm above the crown are an excellent predictor of total numbers per plant.

SAMPLING PATTERN

The sample units obtained from a field or a plot could be collected in one of several different patterns (Elliott 1977, Southwood 1978). In random sampling, sample units are collected from random locations throughout the entire plot area. In stratified random sampling, the plot area is divided up into sections, or strata, and a sample unit is removed at random from each section. Another approach is systematic sampling, in which the location of the first sample unit within the entire plot area is determined at random, and subsequent sample units are removed at fixed intervals. No data are available which compare the utility of these three different patterns in sampling weevil populations. Available data on spatial dispersion of *C. formicarius* (Jansson & McSorley 1990) were obtained using a stratified random sampling pattern, by subdividing 0.2- or 0.4-ha plots into 33-m² sections, within which sample units were collected at random.

DISPERSION OF *Cylas formicarius* WITHIN FIELDS

Dispersion refers to the spatial arrangement of organisms in a defined space, such as a plot or a field. Distribution of *C. formicarius* within fields is not random, but shows the clumped or aggregated pattern typical of many insect populations (Southwood 1978). An

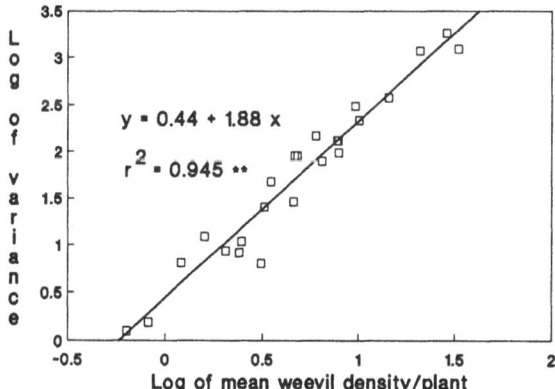

Fig. 8.1. Taylor's power law relationship between $\log_{10} s^2$ and $\log_{10} x$ for whole-plant samples from 24 plots. Each point is the mean and variance of 60 samples from a 0.2-ha plot (from Jansson & McSorley [1990]; reprinted with permission from J. Econ. Entomol. [Entomological Society of America, Lanham, Maryland]).

aggregated pattern of dispersion is indicated by a variance (s^2) greater than the mean (x), and can be described by certain types of contagious statistical distributions, of which the negative binomial distribution is often observed in entomology (Elliott 1977, Southwood 1978). However, since the underlying statistical distribution describing dispersion of a population may change with population density and other factors, distribution-free methods may be preferable for describing population dispersion more consistently (Taylor 1984).

Taylor's power law (Taylor 1961) is one such distribution-free approach and is based on the relationship between the variance and the mean over a range of densities. In this approach, the mean and variance are computed from all samples collected in a given plot (e.g., 60 sample units from a 0.2-ha plot). The logarithms (base 10) of the means and variances from a series of such plots can then be plotted (Figure 8.1), and the linear relationship between log s^2 and log x is given by Taylor's power law:

$$\log s^2 = \log a + b \log x \qquad (2)$$

The regression coefficients a and b are useful in developing sampling plans (Wilson & Room 1982). Taylor's parameter a is a sampling factor depending on the size of the sample unit (Grout 1985), and b is considered an index of dispersion by some authors (Elliott 1977).

In Table 8.1, coefficients of the best fits (highest r^2) from Taylor's power law regressions are summarized for several sets of data on *C. formicarius* dispersion within 0.2- or 0.4-ha plots (Table 8.1). The relatively high b values (1.47-1.91) reflect the fact that weevil populations are aggregated within fields, and show a significant departure from randomness, which would be suggested if $b = 1.0$ (Taylor 1984). For a given sampling scheme, similar values of a and b were obtained whether the sample unit was the whole plant or the zone from 15 cm below to 10 cm above the crown (Table 8.1). Among the three data sets examined, there were some differences in b values and considerable differences in a values (Table 8.1). An important reason for this may be in the size of plots sampled (0.2 vs 0.4 ha), but there are also differences in other factors as well, including cultivars, locations, and sampling times. Each relationship shown (Table 8.1) is fairly general, being derived for a series of fields, cultivars, and (or) sampling times. Thus, these relationships may be applicable under a wider range of conditions than those derived for more specific situations. Specific relationships have been obtained for the cultivars 'Centennial', 'Picadito' and 'Regal' grown in 0.2-ha plots (Jansson & McSorley 1990).

Table 8.1. Coefficients for Taylor's power law for selected data sets.

Data set used [a]	Sample unit [b]	Taylor's power law coefficients		
		r^2	a	b
8, 0.4-ha plots	Whole plant	0.962	8.13	1.55
various sampling times	-15 to +10 cm	0.952	6.92	1.47
12, 0.4-ha plots,	Whole plant	0.960	4.67	1.73
various fields and	-15 to +10 cm	0.956	4.78	1.74
sampling times				
24, 0.2-ha plots,	Whole plant	0.945	2.75	1.88
various cultivars, fields,	-15 to +10 cm	0.939	2.82	1.91
and sampling times				

[a] Data sets described by Jansson and McSorley (1990).

[b] Sample units included all weevils found in the whole plant, or in a zone 15 cm below to 10 cm above the crown.

NUMBER OF SAMPLE UNITS

The aggregated dispersion of the *C. formicarius* population within a plot poses a serious problem when sampling to obtain an estimate of average population density within that plot. A single sample unit removed at random could come from a location in which a high-density clump of weevils occurred or from a location with few weevils, in either case providing a poor estimate of mean weevil density in the plot. Removing and checking all vines and roots in the plot would provide an accurate estimate of mean density, but would be an impractical option. Therefore, it is necessary to achieve a balance between the number of sample units collected and the reliability of the population density estimate.

Reliability can be expressed in terms of an index of precision. In entomological sampling studies, D, the ratio between the half-width of the 95% confidence interval and the mean, is often used as an index of precision (Karandinos 1976). As a practical example, if D = 0.20 and x = 5.0, then the half width of the confidence interval is 1.0, and the full 95% confidence interval would extend from 4.0 to 6.0. If the level of precision were relaxed by increasing D to 0.50, then the 95% confidence interval would range from 2.5 to 7.5. Choice of a level of precision, D, is arbitrary and depends on how wide a confidence interval the pest manager can tolerate. Levels of D = 0.10 or 0.20 may be useful for detailed research studies involving life tables, while D = 0.50 may be satisfactory for damage assessment and control studies (Southwood 1978). Increasing the precision of sampling (lowering D) can be achieved by collecting more sample units, thereby increasing the cost of sampling.

If Taylor's power law holds for a given set of conditions, then the number of sample units (n) required per plot to reliably estimate population density at a specified level of precision is given by (Wilson & Room 1982):

$$n = (t/D)^2 a x^{(b-2)} \tag{3}$$

Here a and b are the coefficients of Taylor's power law, x is the mean density, t is a Student's t value approximating to 2 at the 95% probability level and n > 30 (Elliott 1977), and D is the half-width of the 95% confidence interval as a proportion of the mean.

Numbers of samples required to obtain selected levels of precision are illustrated for three data sets (Table 8.2). Sample numbers were calculated from Equation 3 using the appropriate a and b values from Table 8.1 for the sample unit 15 cm below to 10 cm above the crown. Mean densities of 1.0 and 5.0 weevils per whole plant correspond to densities of 0.75 and 4.27 weevils in the -15 cm to +10 cm sample unit (from Equation 1), and thus the latter densities were used in Equation 3 to obtain the numbers of samples shown (Table 8.2). For a given level of precision, sampling a higher mean density always required fewer samples than a lower density. At a given mean density, relaxing the precision requirements from D = 0.10 to D = 0.40 greatly decreased the number of samples required (Table 8.2). Obtaining a level of precision D = 0.10 is probably impractical, due to the excessive numbers of samples that would be required. At D = 0.40 and a density of 5.0 weevils per plant, 62 samples would be required from a 0.2-ha plot or 80 to 82 from a 0.4-ha plot to provide a 95% confidence interval of 3.0-7.0 weevils per whole plant. Similar numbers of samples would be required whether the sample unit was the whole plant or the zone from 15 cm below to 10 cm above the crown (Jansson & McSorley 1990), but the -15 cm to +10 cm zone requires less effort (approximately 1/3 the sampling time), and thus is preferable. In sampling a site in which density is not known, using the sampling plan for a low density will provide at least the desired level of precision for

Table 8.2. Numbers of samples required to achieve selected levels of reliability in sampling from three data sets.

Data set used [a]	Mean no. of *C. formicarius* per whole plant [b]	Minimum no. of samples needed for selected levels of reliability [c]		
		0.10	0.20	0.40
8, 0.4-ha plots,	1.0	3,224	806	202
various sampling times	5.0	1,282	321	80
12, 0.4-ha plots, various	1.0	2,060	515	129
fields and sampling times	5.0	1,311	328	82
24, 0.2-ha plots, various	1.0	1,158	289	72
cultivars, fields, and	5.0	990	247	62
sampling times				

[a] Data sets described by Jansson and McSorley (1990).

[b] The sample unit is a zone from 15 cm below to 10 cm above the crown. Mean densities of 1.0 and 5.0 weevils per whole plant correspond to densities of 0.75 and 4.27 weevils in the -15 to +10 cm zone, respectively, according to Equation 1.

[c] The index of reliability, D, is the ratio between the half-width of the 95% confidence interval and mean. Sample numbers are rounded up to the nearest whole number.

the density selected and any higher densities. Thus removing 72 sample units from a 0.2-ha site should provide a level of precision, D, less than 0.40 if the population is above 1.0 weevil per plant; however, if the mean density is less than 1.0 weevil per plant, then D will be greater than 0.40 and the confidence interval will be wider.

SEQUENTIAL SAMPLING PLAN

In sampling a 0.2-ha plot of unknown density, removing and checking 72 sample units for *C. formicarius* should be satisfactory in most cases, unless high precision is required or the threshold density targeted is much lower than 1.0 weevil per plant. However, in sampling a plot, it may soon become obvious that the population density is much higher or much lower than the designated threshold density. In that case, it may be possible to stop collecting samples before reaching the required number (e.g., 72), thereby saving time and labor. A sequential sampling plan (Ruesink & Kogan 1975, Southwood 1978) provides guidelines on how long to continue sampling for predetermined critical densities.

Data from the 24 0.2-ha plots were used to derive a sequential sampling plan for *C. formicarius* (Jansson & McSorley 1990). Predefined density classes are arbitrarily defined as "low" ($\leq$ 0.75 weevil per sample unit = 1.0 weevil per plant) or "high" ($\geq$ 4.27 weevil per sample unit = 5.0 weevils per plant). The sequential sampling chart (Figure 8.2) provides guidelines on cumulative numbers of weevils anticipated for the numbers of samples collected. For example, if 300 weevils were found in 20 samples, then the population density could clearly be designated as "high", and sampling could be terminated. However, if 20 samples yielded only 100 weevils, then the point would fall in a zone of uncertainty, and sampling would continue until the point moves into one of the defined zones, or until 72 samples were collected, when the constraints of Table 8.2 would apply.

Unfortunately, the zone of uncertainty at moderate densities may limit the practical use of sequential sampling plans in many instances. Often the need for sampling is greatest at those intermediate densities at which the sequential sampling plan is unresolved. Thus, predefined plans, such as those based on Taylor's power law, may have wider application in practice. In the example illustrated here (Figure 8.2), much additional work is obviously needed to better define density classes and reduce the size of the zone of uncertainty.

CONCLUSIONS AND FUTURE DIRECTIONS

In sampling to determine population densities of *C. formicarius*, dissection of all weevils in a sample unit extending from 15 cm below to 10 cm above the crown provides a reliable estimate of total weevil density per plant. Stratified random sampling plans are proposed for estimating weevil density per plant in 0.2- or 0.4-ha plots, based on field data collected in southern Florida. For example, examination of 72 sample units from 0.2 ha was sufficient to estimate densities of 1.0 weevil per plant or greater with a precision (D) of 40%. The proposed sampling plans were applicable across a range of fields, cultivars, and sampling dates in Florida, but should be tested in other geographic locations. The proposed sampling plans have not been applied in commercial fields, which may range in size from 2-40 ha. Dividing such large fields up into 0.2- or 0.4-ha units would require an excessive

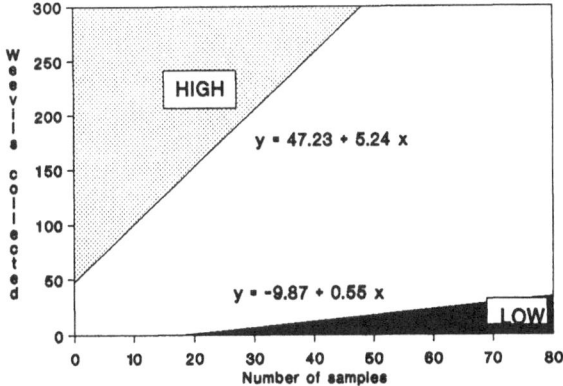

Fig. 8.2. Sequential sampling plan relating cumulative numbers of *C. formicarius* to number of samples for 0.2-ha plots. The sample unit is a zone 15 cm below to 10 cm above the crown, in which densities of 0.75 and 4.27 weevils per sample unit correspond to densities of 1.0 and 5.0 weevils per whole plant, respectively. The density classes are defined as low (≤ 1.0 weevil per whole plant) and high (≥ 5.0 weevils per plant). (from Jansson & McSorley [1990]; reprinted with permission from J. Econ. Entomol. [Entomological Society of America, Lanham, Maryland]).

number of samples. Possibly a few units could be selected to represent the entire field, but additional research would be needed to determine the number and arrangement of such units.

The plans outlined above can provide some initial guidelines in sampling for critical densities of *C. formicarius* However, additional research is needed to better define these critical densities. Sutherland (1986) found yield losses of 10-50% associated with densities as low as 0.8 weevils per 15 cm of vine at harvest. Thus, critical densities of *C. formicarius* per sample unit must be fairly low. Also, the strongest relationships between numbers of weevils in the -15 cm to +10 cm sample unit and numbers in the whole plant occur when both quantities are measured on the same date (Jansson & McSorley 1990). Relationships between total weevils per plant at harvest and numbers in the sample unit on some earlier date are not as well defined (Jansson & McSorley 1990), yet could provide some basis for predicting densities of weevils per plant at harvest and yield losses from samples collected early in the growing season. If critical densities are so low as to require excessive numbers of samples, then it may be necessary to incorporate pheromone traps (Jansson & Heath 1990, Jansson *et al.* 1989, also see Chapter 6) into the sampling program.

166

ACKNOWLEDGMENTS

This is Florida Agricultural Experiment Station Journal Series No. R-00172. We thank A.G.B. Hunsberger and S.H. Lecrone for assistance in data collection and L.T. Wilson (Department of Entomology, University of California, Davis) and C.S. Barfield (Department of Entomology and Nematology, University of Florida, Gainesville) for critically reviewing the manuscript. This research was supported by the U.S. Department of Agriculture under CSRS Special Grant Nos. 87-CRSR-2-3107 and 88-34135-3571 (to R.K.J.) managed by the Caribbean Basin Administrative Group (CBAG).

REFERENCES

Elliott, J.M. 1977. Some methods for the statistical analysis of samples of benthic invertebrates. Freshwater Biological Association Scientific Publication No. 25. The Ferry House, Ambleside, Cumbria, UK.

Grout, T.G. 1985. Binomial and sequential sampling of *Euseius tularensis* (Acari: Phytoseiidae), a predator of citrus red mite (Acari: Tetranychidae) and citrus thrips (Thysanoptera: Thripidae). J. Econ. Entomol. 78:567-570.

Ho, H.T. 1970. Studies on some major pests of sweet potatoes and their control. Malay. Agric. J. 47:437-452.

Jansson, R.K. & R.R. Heath. 1990. Development of a sex pheromone monitoring system for sweetpotato weevil management, pp. 543-552. *In* R.H. Howeler (ed.). Proceedings 8th symposium of the International Society for Tropical Root Crops, Centro Internacional Agriculture Tropical, Bangkok, Thailand.

Jansson, R.K. & R. McSorley. 1990. Sampling plans for the sweetpotato weevil (Coleoptera: Curculionidae) on sweet potato in southern Florida. J. Econ. Entomol. 18:1901-1906.

Jansson, R.K., R.R. Heath & J.A. Coffelt. 1989. Temporal and spatial patterns of sweetpotato weevil (Coleoptera: Curculionidae) counts in pheromone-baited traps in white-fleshed sweet potato fields in southern Florida. Environ. Entomol. 18:691-697.

Jansson, R.K., A.G.B. Hunsberger, S.H. Lecrone & S.K. O'Hair. 1990. Seasonal abundance, population growth, and within-plant distribution of sweetpotato weevil (Coleoptera: Curculionidae) on sweet potato in southern Florida. Environ. Entomol. 19:313-321.

Karandinos, M.G. 1976. Optimum sample size and comments on some published formulae. Bull. Entomol. Soc. Am. 22:417-421.

Mullen, M.A. 1984. Influence of sweetpotato weevil infestation on the yields of twelve sweet potato lines. J. Agric. Entomol. 1:227-230.

Pohronezny, K., R. McSorley & V.H. Waddill. 1981. Integrated management of pests of snap bean in Florida. Proc. Fla. State Hort. Soc. 94:137-140.

Ruesink, W.G. & M. Kogan. 1975. The quantitative basis of pest management: sampling and measuring, pp. 309-351. *In* R.L. Metcalf & W.H. Luckmann (eds.). Introduction to insect pest management. J. Wiley, New York.

Southwood, T.R.E. 1978. Ecological methods. Chapman & Hall, New York.

Subramanian, T.R., B.V. David, P. Thangavel & E.V. Abraham. 1977. Insect pest problems of tuber crops in Tamil, Nadu. J. Root Crops 3:43-50.

Sutherland, J.A. 1986. Damage by *Cylas formicarius* Fab. to sweet potato vines and tubers, and the effect of infestations on total yield in Papua New Guinea. Trop. Pest. Manage. 32:316-323.

Taylor, L.R. 1961. Aggregation, variance and the mean. Nature 189:732-735.

Taylor, L.R. 1984. Assessing and interpreting the spatial distributions of insect populations. Annu. Rev. Entomol. 29:321-357.

Waddill, V.H., R. McSorley & K. Pohronezny. 1981. Field monitoring: basis for integrated management of pests on snap beans. Trop. Agric. 58:157-169.

Wilson, L.T. & P.M. Room. 1982. The relative efficiency and reliability of three methods for sampling arthropods in Australian cotton fields. J. Austr. Entomol. Soc. 21:175-181.

9. Biological Control of *Cylas* spp.

Richard K. Jansson
Tropical Research and Education Center
University of Florida, I.F.A.S.
Homestead, Florida 33031 U.S.A.

Sweet potato weevils, *Cylas formicarius* (Fabricius) (= *C. f. elegantulus* [Summers], *C. f. formicarius* [Fabricius]; see Chapter 2), *C. puncticollis* (Boheman), and *C. brunneus* (Fabricius), are the most important biotic factors limiting sweet potato production world-wide (Chalfant *et al.* 1990). Their feeding on roots induces terpenoid production that may make even slightly damaged roots unpalatable for man and animals (Akazawa *et al.* 1960, Sato *et al.* 1981, Uritani *et al.* 1975). For this reason, even low weevil densities may cause devastating crop losses. For example, yield losses of between 60 to 100% are often observed in many developing countries.

Historically, management of these weevils has relied on cultural controls. These include the use of noninfested planting material (AVRDC 1988, Reinhard 1923, Rolston *et al.* 1983, Sherman & Tamashiro 1954, Talekar 1988), crop rotation (Smee 1965, Talekar 1983, 1988), removal of volunteer plants and crop debris (Boyden 1922, 1927, Jansson *et al.* 1989), prompt harvesting (Nawale 1981, Sherman & Tamashiro 1954, Sutherland 1986b), removal of alternate wild hosts (AVRDC 1988, Talekar 1983, 1988), planting away from weevil-infested fields (AVRDC 1988), filling in of soil cracks (banking) (AVRDC 1988, IITA 1986, Talekar 1988) or planting in light soils that do not crack (Reinhard 1923), irrigation to minimize soil cracking (Holdaway 1941, Sherman & Tamashiro 1954), hilling of plants (IITA 1975, Pardales & Cerna 1987), intercropping (Singh *et al.* 1984), and flooding (AVRDC 1983).

Chemical control has also been used for weevil management. Sutherland (1986a) listed over 40 reports on chemical control of sweet potato weevils. Recent studies indicated that chemical control may be effective at managing weevil populations (Chalfant *et al.* 1990,

Sutherland 1985, 1986c); however, the cryptic nature of these weevils, which develop within the vines and roots of sweet potato plants, may reduce the effectiveness of chemical insecticides at managing weevils. Unfortunately, in developing countries where over 95% of all sweet potatoes are grown (Horton 1988), the use of chemical insecticides for weevil management is too costly. In these countries, sweet potato is often a major staple and is produced in a low-input agricultural system. It is not uncommon for farmers to produce a crop with little to no monetary input (see Chapter 22). Additionally, sweet potato is often grown in remote areas, and transport of chemicals and equipment is not feasible. For these reasons, alternative, low-input, sustainable approaches are needed for weevil management world-wide.

Biological control is well suited for low-input cropping systems. It requires minimal physical input, is often cost-effective, and offers the potential for long-lasting control of the target pest. In addition, biological control is safe, specific for the target pest, and minimizes insecticide resistance, contamination of the environment, and human health problems associated with improper handling of pesticides. This paper reviews the current knowledge on biological control of these weevils using predators, parasitoids, and pathogens, and offers a promising outlook on the use of these approaches for managing weevils in the future. Also, although not discussed, the use of some of these biological control agents, except specific parasitoids, may be important for managing the West Indian sweet potato weevil, *Euscepes postfasciatus* (Fairmaire), since the biology of this weevil is similar to that of *Cylas* spp. (see Chapter 14).

PREDATORS AND PARASITOIDS

Several parasitoids and predators of *Cylas* spp. have been recorded (Table 9.1). Three predators have been reported, two of which are ants in the family Formicidae. These ants are generalist feeders. The argentine ant, *Iridomyrmex humilis* (Mayr), reportedly attacks *C. formicarius* in Louisiana (Cockerham *et al.* 1954), and may also infest homes and be a pest of poultry and certain crops (Arnett 1985). The big-headed ant, *Pheidole megacephala* (Fabricius), was reported to be an effective biological control agent of *C. formicarius* in Cuba (Castineiras *et al.* 1982). They showed that this ant was more effective than chemical insecticides at managing weevil populations. Yields in plots where the big-headed ant was used to control weevils were 21.5 mT/ha compared with only 7.8 mT/ha in plots that relied solely on

Table 9.1. Predators and parasitoids of *Cylas* spp.

Cylas host	Natural enemy species	Order: Family	Location	Reference
		PREDATORS		
formicarius	*Drapetis* s.s. *exilis*	Diptera: Empididae	India	Rajamma 1980
	Iridomyrmex humilis (Mayr)	Hymenoptera: Formicidae	U.S.A.	Cockerham *et al.* 1954
	Pheidole megacephala (Fabricius)	Hymenoptera: Formicidae	Cuba	Castineiras *et al.* 1982
		PARASITOIDS		
formicarius	*Agathis cylasovorus* Rohwer	Hymenoptera: Braconidae	Philippines	Anonymous 1985
	(as *Bassus cylasovorus* Rohwer)	Hymenoptera: Braconidae	Philippines	Gonzalez 1925
	Bracon sp.	Hymenoptera: Braconidae	India	Rajamma 1980
	(as *Microbracon* sp.)		U.S.A.	Cockerham 1944
	B. cylasovora Rohwer	Hymenoptera: Braconidae	Philippines	Anonymous 1985
	(as *M. cylasovorus* Rohwer)		Philippines	Gonzales 1925
	B. mellitor Say (as *M. mellitor* [Say])	Hymenoptera: Braconidae	U.S.A.	Cockerham 1944
	B. punctatus (Muesebeck) (as *M. punctatus* Muesebeck)	Hymenoptera: Braconidae	U.S.A.	Cockerham 1944
	Euderus purpureas Yoshimoto	Hymenoptera: Eulophidae	U.S.A.	Jansson & Lecrone 1991
	Metapelma spectabile Westwood (as *M. spectabilis* Westwood)	Hymenoptera: Braconidae	U.S.A.	Cockerham 1944

Table 9.1. (continued)

Cylas host	Natural enemy species	Order: Family	Location	Reference
		PARASITOIDS (continued)		
formicarius	Rhaconotus sp.	Hymenoptera: Braconidae	India	Rajamma 1980
	R. menippus Nixon	Hymenoptera: Braconidae	Philippines	Anonymous 1985
puncticollis	Bracon sp. (as Microbracon sp.)	Hymenoptera: Braconidae	Africa	Risbec 1947
	Macreupelmus sp.	Hymenoptera: Braconidae	Africa	Risbec 1947
	R. menippus var. africana Nixon	Hymenoptera: Braconidae	Africa	Risbec 1947
	Cheiloneurus [a] sp. (as Chiloneurus sp.)	Hymenoptera: Encyrtidae	Africa	Risbec 1947
	Eurytoma sp.	Hymenoptera: Eurytomidae	Africa	Risbec 1947
	Dinarmus sp. (as Bruchobias sp.)	Hymenoptera: Pteromalidae	Africa	Risbec 1947

[a] Many species of Cheiloneurus are hyperparasitoids (F. Bennett, pers. comm.).

chemical insecticides (Morales 1988). This ant is found in Florida, but it is not an important natural enemy of *C. formicarius* there. No other information is available on the use of this ant as a biological control agent. In pineapple, this ant displays a tending behavior for mealybugs, which increases the incidence of mealybug wilt disease (Beardsley *et al.* 1982, Reimer & Beardsley 1990).

Larvae of *C. formicarius* in India were attacked by predatory maggots, *Drapetis* s. s. *exilis* group (Diptera: Empididae) (Rajamma 1980). Maggots were observed to aggregate around weevil larvae hosts. Adult flies were also seen flying over weevil-infested fields and stored roots. The effects of this fly on weevil populations are not known.

Fifteen wasp parasitoids of *Cylas* spp. have been reported (Table 9.1). In general, most of these are not effective at suppressing weevil populations. Interestingly, parasitoids of *C. formicarius* have been reported from only three locations, India, Philippines, and the United States. Several other parasitoids of *C. puncticollis* have been reported in Africa; however, their impact on populations of *C. puncticollis* is not known. It is obvious from such limited information, that the search for parasitoids of *Cylas* spp. has not been extensive. Since the center of origin of *C. formicarius* is near India (see Chapter 2), extensive efforts should be undertaken in India to search more thoroughly for natural enemies of this weevil. Surveys should also be undertaken in Africa and Madagascar where the *Cylas* genus is most diverse (see Chapter 2). Surveys are also needed in South America, where *I. batatas* originated (see Chapter 3), before a final assessment can be made.

Although several parasitoids have been reported to attack *Cylas* weevils, no studies have examined the impact of these parasitoids on weevil populations. Studies are needed to identify the predominant parasitoids, investigate their biology as it relates to *Cylas* weevils, and determine the potential effectiveness of these parasitoids at suppressing weevil populations. These data will help to determine if some of these parasitoids may be important components of integrated management programs for these weevils.

ENTOMOPATHOGENIC FUNGI

Currently, over 750 species (85 genera) of entomopathogenic fungi are known, excluding 115 genera in the order Laboulbeniales (Anonymous 1979). Most entomopathogenic fungi are contained in Hyphomycetes, Zygomycetes, order Entomophthorales, and Ascomycetes. These fungi attack a wide variety of insect hosts that utilize most ecological niches (Carruthers & Soper 1987). For example, they attack aquatic insects, subterranean insects, phytophagous insects, entomogenous insects, and insects that attack man and animals.

These pathogens may enter insect hosts via several pathways. They may enter through natural body openings, such as the oral cavity, anus, and spiracles, or by direct penetration through the integument (Andreadis 1987). Several reports provide detailed information on the biology and potential of these fungi for managing insects (Allen *et al.*

1978, Burges 1981, Carruthers & Soper 1987, Ferron 1978, Fuxa & Tanada 1987, Ignoffo 1985).

All known fungal pathogens reported to attack *Cylas* spp. are listed in Table 9.2. The most predominant fungus isolated from *Cylas* spp. has been *Beauveria bassiana* (Bals.) Vuill., which is one of the most frequently recorded fungal pathogens attacking insects (Carruthers & Soper 1987). *Beauveria bassiana* has been isolated from *C. formicarius* in the continental United States (Cockerham *et al.* 1954, R.K. Jansson, unpubl. data), Hawaii (Sherman & Tamashiro 1954), Java (Kemner 1924), Cuba (Castineiras *et al.* 1984a, Diaz & Grillo 1986), and Taiwan (Su *et al.* 1988). It has also been isolated from *C. puncticollis* in Zimbabwe, Uganda, Kenya, and Rwanda, and from *C. brunneus* in Uganda and Rwanda (Allard 1990). Other fungi isolated from (or tested against) *Cylas* spp. include *Metarhizium anisopliae* (Metchnikoff) Sorikin from *C. formicarius* in Cuba (Castineiras *et al.* 1984b), *Fusarium* sp. from *C. formicarius* in Louisiana (Cockerham *et al.* 1954), and *Aspergillus* sp. from *C. puncticollis* in Rwanda (Allard 1990). Other pathogens have been isolated from another weevil species in Africa. Isolates of *B. brongniartii* have been found on the rough sweet potato weevil, *Blosyrus* sp., in Uganda and Kenya. Also, an unidentified *Beauveria* sp. was found on *Blosyrus* sp. in Rwanda (Allard 1990).

Four isolates of *B. bassiana* were pathogenic to adult *C. formicarius* in Cuba (Castineiras *et al.* 1984a). After 12 days at 25°C, adult mortality was 49, 48, 47, and 42% for isolates 32, 5, Villena, and 24, respectively. High levels of adult mortality (80 to 90%) were also achieved in the laboratory when spores of a *B. bassiana* isolate (JG-78) were applied to sterile soil (Diaz & Grillo 1986). An isolate of *B. bassiana* from Taiwan was very pathogenic to *C. formicarius* (Su *et al.* 1988). They showed that densities of 0, 100 and 1,000 conidia per gram of soil resulted in 0, 30, and 100% mortality, respectively. They also indicated that the population density of *B. bassiana* needed in soil to manage this weevil occurs naturally in Taiwanese soil. Despite the presence of sufficient densities of inoculum, however, epizootics from this pathogen are rare in *C. formicarius* in the field, although epizootics have been observed in the laboratory (R.K. Jansson, pers. obser.). The disease occurs naturally, albeit at low levels (enzootic). With such enzootics, effective population suppression may never be achieved unless the incidence of the disease can be increased considerably and/or the population level of the disease organism can be increased to produce an epizootic (Ignoffo 1985). In order for enzootics to develop into epizootics, several factors need to be present; the host and

Table 9.2. Fungal pathogens reported to attack *Cylas* spp.

Cylas spp.	Fungal pathogen	Location	Reference
formicarius	*Beauveria* sp. (as *Isaria* sp.)	Java Rwanda	Kemner 1924 Allard 1990
	Beauveria bassiana (Bals.) Vuill.	Cuba	Castineiras *et al.* 1984a,b, Diaz & Grillo 1986
		Taiwan	Su *et al.* 1988
		Uganda U.S.A.	Allard 1990 Sherman & Tamashiro 1954
	(as *B. globulifera* [Speg.] Pic.)	U.S.A.	Cockerham et al. 1954
	Fusarium sp.	U.S.A.	Cockerham et al. 1954
	Metarhizium anisopliae (Metchnikoff) Sorikin	Cuba	Castineiras *et al.* 1984b
brunneus	*B. bassiana*	Uganda	Allard 1990
	Beauveria sp.	Rwanda	Allard 1990
puncticollis	*Aspergillus* sp.	Rwanda	G. Allard, pers. comm.
	B. bassiana	Kenya Uganda Zimbabwe	Allard 1990 Allard 1990 Allard 1990
	Beauveria sp.	Kenya Rwanda	Allard 1990 Allard 1990

pathogen must be present in sufficient densities, and environmental conditions must be favorable. Ignoffo (1985) reviewed each of these factors and the components that are important for epizootic development.

Our knowledge of entomopathogenic fungi is limited; knowledge of fungi attacking *Cylas* spp. is minimal. More research is needed to survey for entomopathogenic fungi attacking *Cylas* spp. world-wide. Additionally, laboratory and field experiments should be conducted to more fully assess the potential of these fungi at managing these weevils world-wide. Information is needed on efficacy of fungi for control of these weevils, and on the factors that limit the success of fungi at managing weevils.

BACTERIA

With the exception of *Xenorhabdus* spp. bacteria infecting *C. formicarius* in association with entomopathogenic nematodes, there are no other reports of bacteria infecting *Cylas* spp. Numerous bacteria are known to attack insects (Bucher 1981, Krieg 1987), including members of Psuedomonadaceae, Enterobacteriaceae, Streptococcaceae, and Bacillaceae. Coleopterous insects are attacked by most types of bacteria. Of these, the Bacillaceae, which includes *Bacillus thuringiensis* Berliner, are the most important and best known bacteria for biological control of insects. *Bacillus thuringiensis* is produced on a commercial scale and has been applied for insect control in many countries. Three pathotypes of *B. thuringiensis* are known: (1) pathotype A - pathogenic or toxic to Lepidoptera (var. *thuringiensis* as the prototype and includes *B. thuringiensis* var. *kurstaki*) (Berliner 1915); (2) pathotype B - pathogenic to Nematocera (Diptera) (var. *israelensis* as the prototype) (Goldberg & Margalit 1977); and (3) pathotype C - pathogenic to Coleoptera, especially members of the family Chrysomelidae (var. *tenebrionis* as the prototype) (Krieg *et al.* 1983). These pathotypes can be further divided into subtypes (Krieg *et al.* 1983).

Because of rapid biotechnological advances, improved biological insecticides composed of bacteria have been developed. For example, genes of *B. thuringiensis* var. *morrisoni*, which are toxic to Coleoptera, have been incorporated into *B. thuringiensis* var. *kurstaki* to produce a broad spectrum biological insecticide with toxicity to both Lepidoptera and Coleoptera (e.g., Foil™, Ecogen Inc., Langhorne, Pennsylvania). Preliminary experiments were conducted to test the toxicity of Foil™ to *C. formicarius* adults. Results showed that Foil™ did not kill *C. formicarius* adults when they were fed concentrations between 2 to 50 ml Foil™ (7.5% AI/liter; 0.076 kg AI/liter) per liter of water (R.K. Jansson, unpubl. data). Similarly, weevil feeding and oviposition were not affected by the concentration applied to storage roots. The toxicity of Foil™ to immature weevils is not known. Although Foil™ had no activity against *C. formicarius*, other bacterial strains with toxicity to other weevils, such as the boll weevil, *Anthonomus grandis grandis*

Boheman, have been produced commercially and may have potential against *C. formicarius*.

It is obvious that there is a lack of information concerning pathogenicity/toxicity of bacteria to *Cylas* spp. Studies are needed to screen bacterial isolates with known toxicity to Coleoptera for their activity against *Cylas* spp. Also, surveys should be conducted world-wide to isolate *Cylas* spp. infected with bacteria. Identification of bacterial isolates that are toxic to these weevils will be useful for developing transgenic sweet potato plants that are resistant to these insects (see Chapter 10), and for determining the potential of bacteria as a component in weevil management programs.

ENTOMOPATHOGENIC NEMATODES

Perhaps the group of organisms with the greatest potential as biological control agents of *Cylas* spp. world-wide are entomopathogenic nematodes. Entomopathogenic nematodes comprise a small percentage of all entomophilic nematodes, however, their potential as biological control agents is great. This group is comprised of nematodes in the families Steinernematidae and Heterorhabditidae. Collectively, these nematodes infect over 1,000 species of insects world-wide (Nickle & Welch 1984).

The taxonomy, biology, and techniques for handling these nematodes have been described (Gaugler 1981, Nickle 1984, Poinar 1979, Woodring & Kaya 1988). Currently, there is considerable confusion regarding the taxonomy of these nematodes. Taxonomy of Heterorhabditidae is less confusing than that of Steinernematidae (Woodring & Kaya 1988). Poinar (1990) presented the most recent classification of these nematodes. In addition to binomial designations for nematodes, intraspecific strain designations are used for both types of nematodes. Strain names may refer to accession number (e.g., DD-136), geographic origin (e.g., Mexican, Breton, Italian), persons credited with isolation (e.g., All, Pye), culturing technology (e.g., G-13; refers to repeated selection for a specific trait for 13 generations on greater wax moth larvae, *Galleria mellonella* (L.)), or host from which nematodes were isolated (e.g., *Agriotes*). Differentiation of strains is important since certain strains are more pathogenic than others against certain insects (Bedding *et al.* 1983).

These nematodes have a simple life cycle that includes egg, four juvenile stages and the adult. The third-stage juvenile is the infective stage (hence, infective juvenile), which is ensheathed by the second-stage juvenile cuticle thereby enhancing the infective juvenile's resistance to environmental conditions. The third-stage juvenile is the only free-living stage of these nematodes. Third-stage juveniles do not feed, but may survive long periods of time when conditions are favorable (Woodring & Kaya 1988). Both steinernematid and heterorhabditid nematodes enter hosts through natural body openings, such as the mouth, anus, and spiracles. Heterorhabditid nematodes may also enter by using a tooth-like structure which helps to penetrate at soft intersegmental membranes of the host (Bedding & Molyneux 1982). These nematodes release mutualistic bacteria, *Xenorhabdus* spp., from their intestine into the insect host's hemolymph. Bacteria multiply rapidly, causing septicemia and death of the host. Developing nematode progeny (infective juveniles) subsequently feed on bacterial cells and host tissue. Steinernematid nematodes are associated with four subspecies of *X. nematophilus* bacteria, whereas heterorhabditid nematodes are associated with one bacterium, *X. luminescens* (Woodring & Kaya 1988). Steinernematid nematodes are amphimictic in all generations. Larger females and smaller males mate and produce progeny. Because of ample food supplies, the first generation adults are large. The second generation is usually much smaller due to depleted food reserves, and subsequently form the new infective juveniles. Sometimes a complete third generation may develop. When conditions are favorable, infective juveniles leave the host cadaver to infect new hosts (Figure 9.1). This cycle requires 10 to 14 days for *Steinernema* (= *Neoaplectana*) *carpocapsae* Weiser in *G. mellonella* larvae (Woodring & Kaya 1988). Fewer or no complete generations may occur on smaller hosts (Jackson 1985). Heterorhabditid nematodes, on the other hand, are not amphimictic in all generations. Infective juveniles develop into hermaphroditic females. These females produce a sexual generation that subsequently produces hermaphroditic infective juveniles.

Both steinernematid and heterorhabditid nematodes have been isolated from *C. formicarius*. *Heterorhabditis heliothidis* (Khan, Brooks, and Hirschmann) was isolated from larvae and pupae of *C. formicarius* in Cuba (Hernandez & Mracek 1984). Other nematodes, *Rhabditis* sp., *Aphelenchus avenae*, *Acrobeloides* sp., and *Steinernema* (= *Neoaplectana*) sp., were reportedly isolated from *C. formicarius* in Mississippi and Florida (Cockerham *et al.* 1954); however, of these, only *Steinernema* sp. is entomopathogenic. *Steinernema* sp. was shown to be pathogenic to *C. formicarius* in the laboratory (Swain 1943). These nematodes have also been shown to be very pathogenic to other root weevils, including black vine weevil, *Otiorhynchus sulcatus*

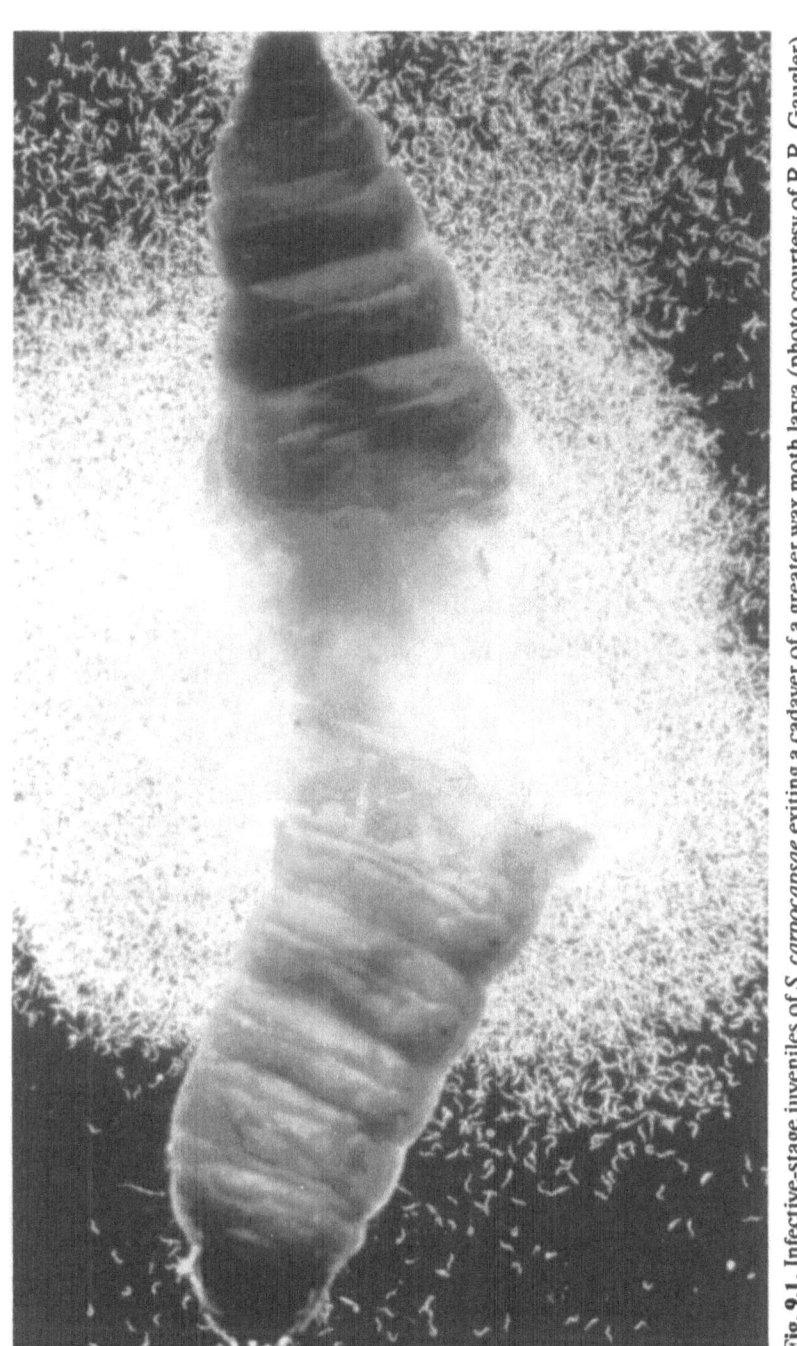

Fig. 9.1. Infective-stage juveniles of *S. carpocapsae* exiting a cadaver of a greater wax moth larva (photo courtesy of R.R. Gaugler).

(Fabricius) (Bedding & Miller 1981, Dolmans 1983, Georgis & Poinar 1984b, Rutherford *et al.* 1987), strawberry root weevil, *O. ovatus* L. (Rutherford *et al.* 1987), woods weevil, *Nemocestes incomptus* (Horn) (Georgis & Poinar 1984a), carrot weevil, *Listronotus oregonensis* (LeConte) (Belair & Boivin 1985, Boivin & Belair 1989), and citrus root weevil (sugarcane rootstalk borer), *Diaprepes abbreviatus* (L.) (Roman & Figueroa 1985, Schroeder 1987).

Steinernematid and heterorhabditid nematodes possess many of the attributes associated with an ideal biological control organism. These nematodes are safe, relatively inexpensive to mass rear *in vitro*, are capable of locating well-concealed hosts, such as root weevils, possess high virulence and reproductive rates, kill most hosts within 48 hours and subsequently reduce host feeding on the crop, and attack a wide range of host insects (Gaugler 1981, Gaugler & Kaya 1990). Additionally, these nematodes are easily applied using methods ranging from pouring suspensions of nematodes over plants to those made using high pressure boom sprayers at up to 70.3 kg/cm^2 (1,000 psi) (Kaya 1985, Woodring & Kaya 1988).

Entomopathogenic nematodes inhabit the soil. Thus, for these nematodes to be effective against *Cylas* spp., the majority of the weevil population should be below the soil surface. In a recent study, between 54-57% of the total population of *C. formicarius* in vines of sweet potato was found at or below the soil surface (Jansson *et al.* 1987). More recently, between 78-91% of the total weevil population in vines and roots of four cultivars of sweet potato was found below the soil surface (Jansson *et al.* 1990a).

Several constraints, such as soil moisture, solar radiation, and temperature, may limit the effectiveness of entomopathogenic nematodes as biological control agents of insects (Kaya 1985). However, certain characteristics of the *Cylas*/sweet potato system may minimize the negative effects of these factors. Entomopathogenic nematodes are sensitive to soil moisture levels (Gaugler 1981, Kaya 1985, 1987). Unless nematodes can enter a state of anhydrobiosis, they will dessicate quickly and die (Crowe & Madin 1975). Steinernematid nematodes, in particular, cannot tolerate rapid drying (Dutky 1959). Fortunately, *Cylas* weevils are pantropical in their distribution (Chalfant *et al.* 1990, Sutherland 1986a) and are found in parts of the world that typically receive high levels of precipitation. Thus, the climate in regions where these weevils are found may help to reduce the deleterious effects of soil moisture on nematode effectiveness. It should be noted, however, that although weevils are found in regions that receive high levels of precipitation, patterns of precipitation are typically cyclical. There are distinct wet and dry seasons throughout the tropics. In most of these regions, the weevil is more problematic during the dry season than during the wet season (see Chapter 4). Thus, nematode effectiveness against *Cylas* weevils may differ between wet and dry seasons. Recent advances in anhydrobiosis (Georgis 1987) may help improve the potential of nematodes during the dry season.

Entomopathogenic nematodes are also very susceptible to sunlight and ultraviolet radiation (Gaugler & Boush 1978). Exposure of steinernematid nematodes to direct sunlight for 60 minutes reduced their pathogenicity to *G. mellonella* larvae by 95%. Exposure of nematodes to short durations of ultraviolet radiation (254 nm for 3.5 minutes) had deleterious effects on nematode development, reproduction, and pathogenicity. This may be minimized, however, by applying nematode suspensions at dusk and irrigating after application.

These nematodes are also influenced by soil temperature (Gaugler 1981, Kaya 1985). Infection by steinernematid nematodes may occur at temperatures as low as 9°C (Dutky *et al.* 1964, Gaugler & Molloy 1981), but the host may remain alive for a considerable time (300 hours) after infection. At 30°C, death of the host may occur in 16 hours (Dutky *et al.* 1964). Infective stage juveniles are most active between 22 and 32°C (Schmiege 1963). In the tropics and subtropics, soil temperatures coincide with the temperatures at which these nematodes are most active. Thus, soil temperature is probably not an important constraint of nematode effectiveness against *Cylas* weevils.

It should be noted that other environmental factors, such as soil pH, oxygen content, and salinity, may also influence the effectiveness of nematodes as biological control agents. However, there is little information available on the effects of these factors on these nematodes.

Biological Control of *Cylas formicarius* with Entomopathogenic Nematodes: A Case Study

Several laboratory and field experiments have assessed the effectiveness of entomopathogenic nematodes as biological control agents of *C. formicarius*. The pathogenicities of three nematode strains were recently evaluated in the laboratory: *S. carpocapsae* (All and G-13 strains) and *H. bacteriophora* Poinar (HP88 strain). 'HP88' and 'All' strain nematodes were supplied by Biosys, Inc., Palo Alto, California. 'G-13' nematodes were cultured from 'All' nematodes. The host finding ability of this strain was improved by repeatedly selecting for individuals that were successful at locating *G. mellonella* larvae (Gaugler *et al.* 1989b). Various densities of each strain were tested using standard methods for bioassaying nematodes for pathogenicity to insects in a Petri dish (Woodring & Kaya 1988).

All three strains were pathogenic to *C. formicarius* larvae and pupae (Jansson *et al.* 1990b). Virulence did not differ among nematodes (Table 9.3). LD_{50} and LD_{90} values after 96 hours did not differ among nematodes within each life stage nor between life stages within each nematode. Larvae tended to be more susceptible (although not significant) to nematodes than pupae.

More recent studies tested the virulence of ten entomopathogenic nematode species or strains to *C. formicarius* (Mannion & Jansson 1992). Virulence at 48 hours differed among species and strains (Table 9.3). One unidentified heterorhabditid nematode, 'FL2122' (from Florida), and *S. feltiae* (= *S. bibionis* Steiner) (N-27 strain) were the most virulent nematodes tested. Most nematodes were more virulent to larvae than to pupae (Figure 9.2). Adults were less susceptible to nematodes than other stages, and adult males were more susceptible than females (Figure 9.2). This finding is currently being investigated more fully. The mode of entry of these nematodes into each of the various stages is not known.

Weevil immatures develop inside the vines and roots of plants. Thus, these nematodes must locate and kill weevil hosts within the plant to be effective as biological control agents. Simulated field experiments were conducted in the laboratory to determine if nematodes could locate and kill weevil hosts within storage roots (Jansson *et al.* 1990b). Roots, infested with *C. formicarius*, were buried in soil and various densities of 'All', 'G-13', and 'HP88' nematodes were then applied over the soil surface. Roots were dissected after three weeks and weevil mortality was recorded. Results showed that all three nematode strains were effective at locating and killing weevil immatures within storage roots (Figure 9.3). Percentage mortality of weevil immatures increased with an increase in nematode density applied. 'G-13' nematodes were most efficacious at locating and killing weevil hosts within roots followed by 'HP88' and 'All' nematodes (Table 9.4).

Although several studies have demonstrated the potential of entomopathogenic nematodes at killing insects in the laboratory, several reports indicate the inconsistency and poor efficacy of these nematodes in the field. Although these nematodes are adapted to the soil environment, their efficacy is not predictable and, in general, they

Table 9.3. Virulence of various entomopathogenic nematode species and strains to *C. formicarius* larvae and pupae at either 96 (Experiment 1) or 48 hours (Experiment 2).

Nematode species/strain	Life stage	LD$_{50}$ ($\pm$95%FL)[a]	LD$_{90}$ ($\pm$95%FL)[a]
Experiment 1[b]			
H. bacteriophora HP88	Larva	3.4 (2.0-4.6)	25.0 (17.2-47.3)
H. bacteriophora HP88	Pupa	4.8 (3.6-5.9)	46.2 (28.5-108.8)
S. carpocapsae All	Larva	2.6 (0.6-4.7)	21.5 (11.4-114.2)
S. carpocapsae All	Pupa	4.9 (2.8-7.6)	51.2 (24.6-292.4)
S. carpocapsae G-13	Larva	2.9 (1.6-4.0)	14.0 (9.8-27.0)
S. carpocapsae G-13	Pupa	5.9 (NE)[c]	113.3 (NE)[c]
Experiment 2[d]			
S. carpocapsae Agriotes	Larva	8.5 (6.4-11.8)	48.3 (27.9-140.2)
S. carpocapsae All	Larva	8.4 (4.9-13.7)	41.3 (21.6-334.5)
S. carpocapsae Breton	Larva	5.3 (4.4-6.2)	16.3 (13.0-23.3)
S. carpocapsae Italian	Larva	7.4 (5.1-11.0)	35.6 (20.0-134.4)
S. carpocapsae Mexican	Larva	6.4 (5.1-7.9)	20.2 (14.9-34.0)
S. feltiae N-27	Larva	2.9 (2.1-3.8)	21.3 (14.8-37.8)
S. intermedia	Larva	19.7 (NE)[c]	77.7 (NE)[c]
H. bacteriophora HP88	Larva	9.3 (5.7-15.3)	55.1 (26.3-886.7)
H. bacteriophora N. Carolina	Larva	29.7 (17.3-114.9)	1,668.7 (283.9-237,860)
Heterorhabditis sp. FL 2122	Larva	1.9 (0.9-2.8)	7.4 (5.1-13.2)

[a] LD$_{50}$ and LD$_{90}$ values did not differ among nematodes within each life stage
[b] From Jansson *et al.* (1990).
[c] NE, not estimable. Heterogeneity was high and index of significance (*g*) for potency estimation was high. Probit analysis provided a poor fit of the data.
[d] From Mannion and Jansson (1991).

are not as effective as chemical insecticides at managing insect populations (Gaugler 1988). For these nematodes to have potential against *Cylas* weevils, they must be effective at suppressing weevil populations on plants and protecting storage roots from weevil damage.

Both steinernematid, 'All', and heterorhabditid, 'HP88', nematodes have been shown to be effective at suppressing weevil populations and

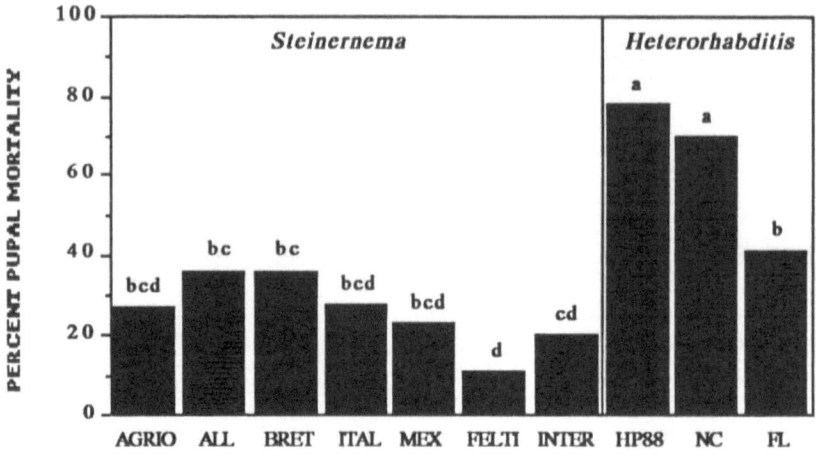

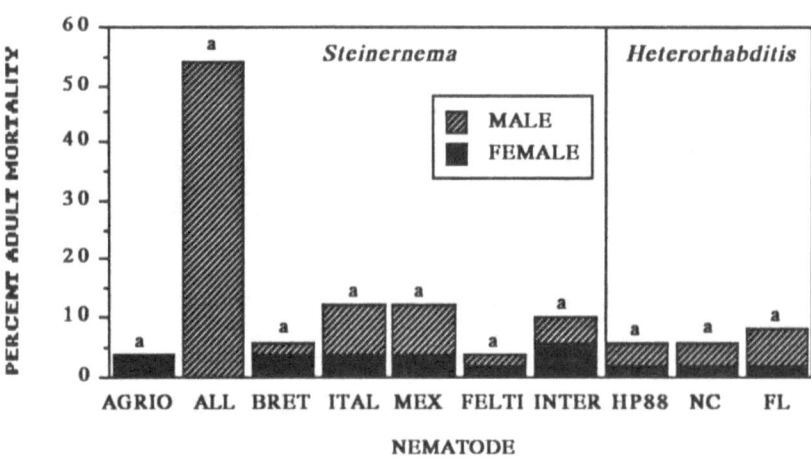

Fig. **9.2.** Percentage mortality of *C. formicarius* pupae (top) and adult males and females (bottom) when exposed to the larval LD_{50} density of various entomopathogenic nematodes in Petri dish bioassays (Mannion & Jansson 1992).

reducing weevil damage to storage roots (Jansson *et al.* 1990b). Weevil populations were reduced 68 to 83% and 45 to 81% on plants treated with 'HP88' and 'All' nematodes, respectively, than on those treated with chemical insecticides (Figure 9.4). Similarly, populations were

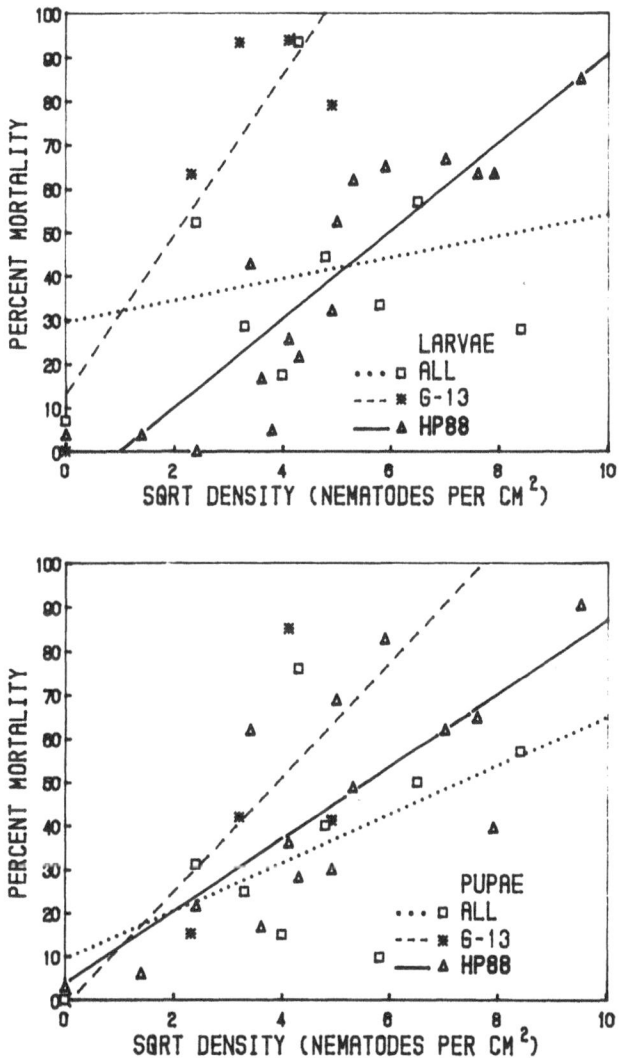

Fig. 9.3. Relationship between percentage mortality of *C. formicarius* larvae and pupae within infested storage roots and entomopathogenic nematode ('All', 'HP88', and 'G-13' strains) density applied to the soil surface under simulated field conditions.

Table 9.4. Parameter estimates ($\pm$ SEM), significance levels, and coefficients of determination (r^2) for percentage mortality of *C. formicarius* larvae and pupae within infested sweet potato roots regressed on square-root-transformed density of nematodes applied to the soil surface (from Jansson *et al.* 1990b).

Nematode species/strain	Life stage	Intercept[a]	Slope[a]	P[b]	r^2
H. bacteriophora HP88	Larva	-10.6(7.3)a,A	10.2(1.4)ab,A	***	0.80
S. carpocapsae All	Larva	29.6(19.3)a,A	2.4(3.9)b,A	ns	0.05
S. carpocapsae G-13	Larva	13.3(18.8)a,A	18.0(5.6)a,A	*	0.78
H. bacteriophora HP88	Pupa	3.2(10.2)a,A	8.4(1.9)a,A	**	0.61
S. carpocapsae All	Pupa	9.4(15.7)a,A	5.5(3.2)a,A	ns	0.30
S. carpocapsae G-13	Pupa	-1.0(21.6)a,A	12.9(6.4)a,A	ns	0.58

[a] Lowercase letters correspond to comparisons among nematodes within life stages. Uppercase letters correspond to comparisons between life stages for each nematode. Parameter estimates followed by different letters indicate significant differences by the general linear test.

[b] Significance level of regression: ns, $P > 0.05$; *, $P < 0.05$; **, $P < 0.001$; ***, $P < 0.0001$.

reduced 57 to 78% and 22 to 74% on plants treated with 'HP88' and 'All', respectively, than on nontreated plants. As found in many other studies, however, weevil densities on plants were not correlated with the nematode density applied.

'HP88' and 'All' nematodes were also effective at protecting storage roots from weevil damage (Figure 9.5). Percentages of storage root weight that were not damaged by weevils (damage rating = 1) ranged between 30 to 40% in sweet potatoes treated with 'HP88' compared with 15 to 20% in sweet potatoes treated with 'All' and 3 to 5% in nontreated sweet potatoes and those treated with chemical insecticides. Similar results were found for percentages of root weight that were slightly damaged (damage rating = 2). Overall, percentages of roots with little or no damage (damage rating = 1 or 2) ranged between 65 to 80% for sweet potatoes treated with 'HP88' compared with about 40% for those treated with 'All' nematodes and about 18% for nontreated sweet potatoes and those treated with chemical insecticides. Reverse trends were observed for percentages of root weight with severe weevil damage (damage rating = 6). Weevil damage was not correlated with the nematode density applied.

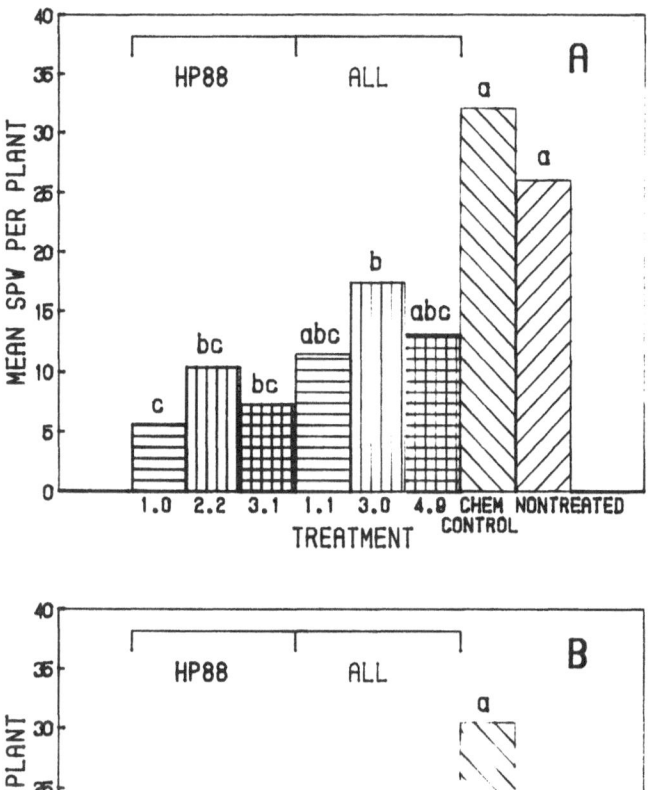

Fig. 9.4. Mean numbers of *C. formicarius* larvae, pupae, and adults on 'Jewel' sweet potato plants treated with three monthly applications of two entomopathogenic nematodes or chemical insecticides, and on nontreated plants. Three densities of nematodes were applied on each application date ('HP88', 1.0, 2.2, and 3.1 billion infective juveniles/ha; 'All', 1.1, 3.0, and 4.9 billion infective juveniles/ha) (from Jansson *et al.* 1990). Two sample dates shown: (A) one month before harvest; and (B) at harvest.

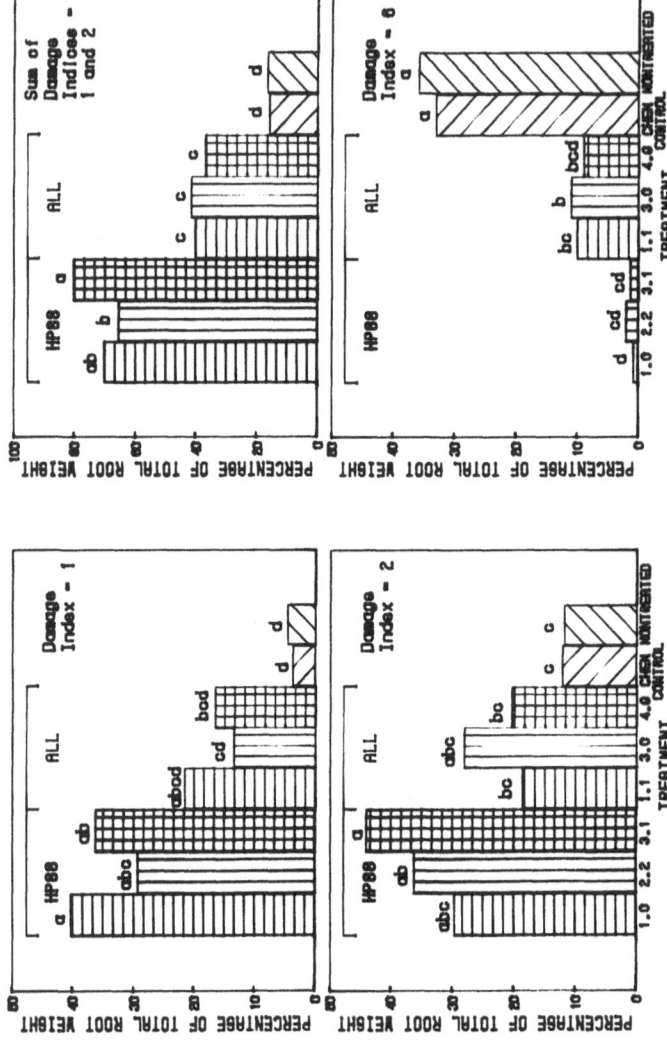

Fig. 9.5. Percentages of 'Jewel' sweet potato storage roots that were not damaged by *C. formicarius* (damage index = 1), slightly damaged (damage index = 2), not damaged or slightly damaged (damage index = 1 or 2), and severely damaged (damage index = 6) on plants treated with three monthly applications of two entomopathogenic nematode strains or monthly applications of chemical insecticides, and on nontreated plants. Three densities of nematodes were applied on each application date ('HP88', 1.0, 2.2, and 3.1 billion infective juveniles/ha; 'All', 1.1, 3.0, and 4.9 billion infective juveniles/ha) (from Jansson *et al.* 1990; reprinted with permission from J. Econ. Entomol. [Entomological Society of America, Lanham, Maryland]).

One major limitation of entomopathogenic nematodes is that they generally do not persist in the soil for long periods of time after application. Integration of these nematodes into low-input sustainable sweet potato systems in developing countries might be facilitated if these nematodes persisted in soil and remained pathogenic to weevils during the duration of the growing season. A recent study found that 'HP88' persisted (presumably by recycling) in soil for a considerable time after application (Jansson *et al.* 1992). A single application of 'HP88' (3.1 billion infective juveniles/ha) was as effective at protecting storage roots from weevil damage as two and three monthly applications (3.1 billion infective juveniles/ha at each application) (Figure 9.6). Soil sampling indicated that this nematode persisted in Rockdale soil for up to 134 days after application (Figure 9.7). A subsequent study found that this nematode persisted in soil for over 250 days after application (Jansson *et al.* 1992). Work is currently underway to examine nematode persistence in sweet potato fields in southern Florida more fully.

FUTURE OUTLOOK

Biological control of *Cylas* spp. has a very promising future. The potential of biological control agents can only be realized, however, by conducting more research. Concerning predators and parasitoids, extensive surveys should be conducted to search for other natural enemies of these weevils in India, Africa, and Madagascar, the centers of origin of *C. formicarius* and *Cylas* spp., respectively, and in South America, the center of origin of the sweet potato. Studies are also needed to characterize the impact of endemic parasitoids on weevil populations. Attempts should also be made to introduce exotic natural enemies of *Cylas* spp. and assess their success of establishment and impact on weevil populations.

Fungi have a bright future in management programs for *Cylas* spp.. The success of isolates of *B. bassiana* from Taiwanese soil against adult *C. formicarius* demonstrates their potential. Historically, however, control of insects with fungi has not been consistent. The effectiveness of fungi for insect control has been hampered by several factors, including limited dispersal of fungi, the need for optimum environmental conditions, and the need for high host and pathogen densities. Extensive surveys are needed to isolate entomopathogenic fungi attacking *Cylas* spp. Studies are also needed to determine the

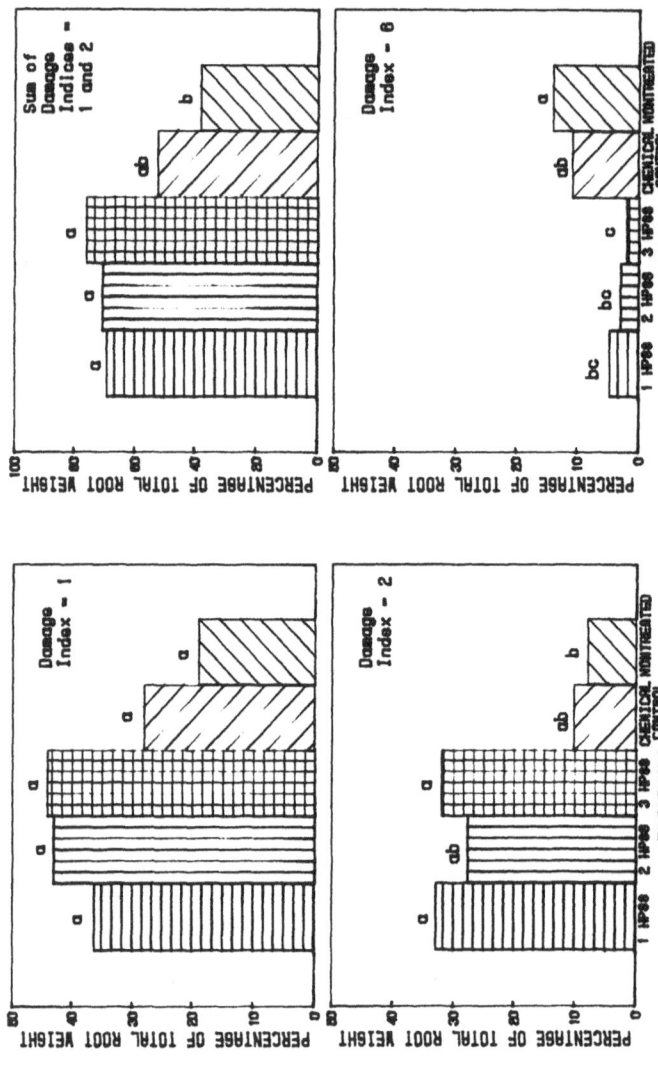

Fig. 9.6. Percentages of 'Jewel' sweet potato storage roots that were not damaged by *C. formicarius* (damage index = 1), slightly damaged (damage index = 2), not damaged or slightly damaged (damage index = 1 or 2), and severely damaged (damage index = 6) on plants with one, two, or three monthly applications of *H. bacteriophora* strain 'HP88' (1HP88, 2HP88, and 3HP88, respectively; 1.3 billion infective juveniles/ha per application), or monthly applications of chemical insecticides, and on nontreated plants (from Jansson *et al.* 1992).

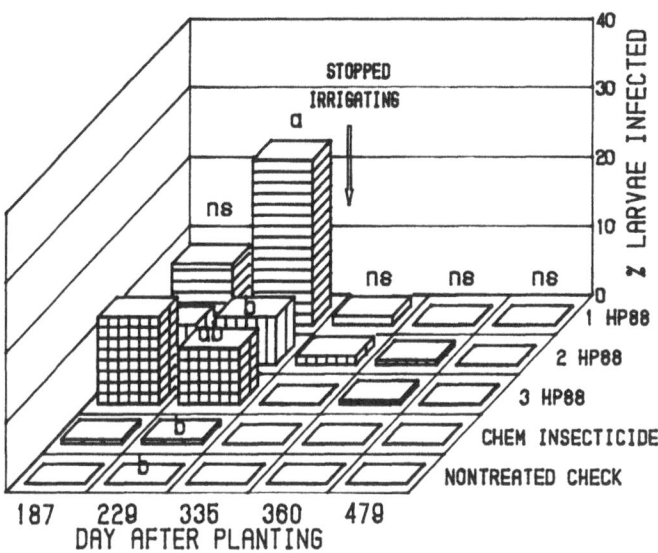

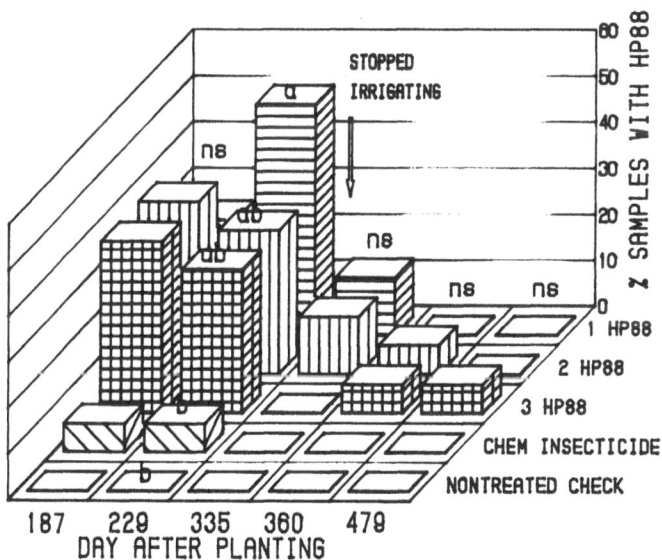

Fig. 9.7. Percentages of greater wax moth, *G. mellonella*, larvae and soil samples in which *Heterorhabditis* sp. (presumably *H. bacteriophora* strain HP88) were recovered at different times of the year. Days post-application of 'HP88' were equal to day of year minus 39, 82, and 107 for one (1HP88), two (2HP88), and three (3HP88) applications of 'HP88', respectively (from Jansson *et al.* 1992).

optimum conditions that produce fungal epizootics in *Cylas* populations.

The potential of bacteria for control of *Cylas* spp. is not known. Identification of bacteria varieties and/or strains that are toxic to these weevils is central to assessing their future potential.

Management of *Cylas* spp. with entomopathogenic nematodes has a bright future. There are several factors, however, which may hinder their integration into management programs for weevils, especially in developing countries. Current availability of commercial nematode strains is limited, especially in developing countries. Although there are several companies that mass produce these nematodes, most are located in the United States, and all are located in developed countries. Thus, nematodes must be shipped considerable distances to reach their targets in developing countries. Storage and shipping procedures for nematodes are not well developed (Bedding 1984). Considerably more research effort is needed to produce a nematode product that will remain viable, pathogenic, and virulent after being in transit for a considerable time period. Also, the costs involved to deliver nematodes from a commercial company in a developed country to a developing country might be too excessive and hence not practical for current low-input sweet potato systems. For these reason, laboratories or cottage industries should be established in developing countries for culturing entomopathogenic nematodes. Initially, surveys should be conducted to isolate nematode strains that are locally adapted to the country or region in which they will be used. Preliminary data suggest that locally-adapted nematode strains are very virulent to these weevils (C.M. Mannion & R.K. Jansson, unpubl. data). Isolated nematodes should be cultured and tested for pathogenicity, virulence, and efficacy against these weevils. The importance of using standardized protocols in all surveys and evaluations is critical (Gaugler & Kaya 1990, Kaya 1985, Woodring & Kaya 1988).

Information on the effects of environmental parameters on nematode efficacy in the field is lacking (Gaugler 1988). More research is needed on nematode ecology so that we can define the factors influencing nematode persistence, mobility, and development of epizootics. For these reasons, standardized protocols must be established when comparing nematode effectiveness against *Cylas* spp. world-wide. Of paramount importance are the effects of soil type (texture), moisture, aeration, temperature, and chemistry on nematode ecology (Kaya 1990 and references therein). Since distinct wet and dry seasons occur world-wide, information on nematode performance during both of these seasons is needed. In many developing countries, irrigation of sweet potato fields is not feasible, and this may limit nematode effectiveness, especially during the dry season. Gaugler (1988) has listed several parameters that are important to measure in all field evaluations, including the times and methods of applications, air and soil temperatures, cloud cover, thatch depth, soil type and moisture, pest density, pest-activity zone, post-treatment irrigation, and precipitation during the study. As more studies evaluate these nematodes under different soil environments, we may begin to better understand the importance that these nematodes may have in management programs for these weevils.

In summary, the microbial agents (entomopathogenic fungi, bacteria, and nematodes) appear to have the greatest potential as biological control agents of *Cylas* spp. Historically, these control strategies have been used in an augmentative or inundative manner. Such approaches can be very labor intensive because they involve the production, formulation, and application of a biological pesticide (Anonymous 1989). Production of these agents will need to done locally in order to be practical for developing countries. These agents can be produced in cottage industries as is done in China (Anonymous 1989, Greathead & Waage 1983). Integration of these agents with other methods, such as cultural controls and sex pheromone, may make these approaches more cost-effective.

ACKNOWLEDGMENTS

Research on biological control of *C. formicarius* was supported, in part, by the U.S. Department of Agriculture under CSRS Special Grant No. 88-34135-3564 (to R.K.J.) managed by the Caribbean Basin Administrative Group (CBAG). I thank F.D. Bennett, R.R. Gaugler, J. Waage, and J. Fuxa for critically reviewing an earlier version of this manuscript. This is Florida Agricultural Experiment Station No. R-01029.

REFERENCES

Akazawa, T., L. Uritani & H. Kubota. 1960. Isolation of ipomea-marone and two coumarin derivatives from sweet potato roots injured by the weevil, *Cylas formicarius elegantulus*. Arch. Biochem. Biophys. 88:150-156.

Allard, G.B. 1990. Integrated control of arthropod pests of root crops November 1988-December 1989. CAB International Institute of Biological Control Mid-term Rept., CAB International Institute of Biological Control, Nairobi, Kenya.

Allen, G.E., C.M. Ignoffo & R.P. Jacques (eds.). 1978. Microbial control of insect pests: future strategies in pest management systems. University of Florida, Gainesville.

Andreadis, T.G. 1987. Transmission, pp. 159-176. *In* J.R. Fuxa & Y. Tanada (eds.). Epizootiology of insect diseases. J. Wiley, New York.

Anonymous. 1979. Microbial processes: promising technologies for developing countries. Natl. Acad. Sci., Washington, D.C.

Anonymous. 1985. Sweet potato weevil. Visayas St. Coll. Agric. Plant Pest Clinic Advisory Bull. Visayas State College of Agriculture, Leyte, Philippines.

Anonymous. 1989. Low technology mycopesticide production for smallholder crop production: a synopsis of the scientific literature and recommendations for future development. CAB International Development Series, Wallingford, Oxon, United Kingdom.

Arnett, R.H., Jr. 1985. American insects: a handbook of the insects of America north of Mexico. Van Nostrand Reinhold, New York.

AVRDC. 1983. AVRDC progress report summaries 1982. Asian Vegetable Research and Development Center, Shanhua, Tainan, Taiwan, Republic of China.

AVRDC. 1988. Annual review 1987. Crop improvement program. Sweet potato entomology. Asian Vegetable Research and Development Center, Shanhua, Tainan, Taiwan, Republic of China.

Beardsley, J.W., T.H. Su, F.L. McEwen & D. Gerling. 1982. Field investigations on the interrelationships of the big-headed ant, the gray pineapple mealybug, and the pineapple mealybug wilt disease in Hawaii. Proc. Ha. Entomol. Soc. 24:51-67.

Bedding, R.A. 1984. Large scale production, storage and transport of insect-parasitic nematodes *Neoaplectana* spp. and *Heterorhabditis*. Ann. Appl. Biol. 101:117-120.

Bedding, R.A. & L.A. Miller. 1981. Use of a nematode, *Heterorhabditis heliothidis*, to control black vine weevil, *Otiorhynchus sulcatus*, in potted plants. Ann. Appl. Biol. 99:211-216.

Bedding, R.A. & A.S. Molyneux. 1982. Penetration of insect cuticle by infective juveniles of *Heterorhabditis* spp. (Heterorhabditidae: Nematoda). Nematologica 28:354-359.

Bedding, R.A., A.S. Molyneux & R.J. Akhurst. 1983. *Heterorhabditis* spp., *Neoaplectana* spp., and *Steinernema kraussii*: interspecific and intraspecific differences in the infectivity for insects. Exp. Parasitol. 55:249-257.

Belair, G. & G. Boivin. 1985. Susceptibility of the carrot weevil (Coleoptera: Curculionidae) to *Steinernema feltiae*, *S. bibionis*, and *Heterorhabditis heliothidis*. J. Nematol. 17:363-366.

Berliner, E. 1915. Uber die schlaffsucht der melhmottenraupe (*Euphestia kuhniella* Zell.) und ihren erreger *Bacillus thuringiensis* n. sp. Z. Angew. Entomol. 2:29-56.

Boivin, G. & G. Belair. 1989. Infectivity of two strains of *Steinernema feltiae* (Rhabditida: Steinernematidae) in relation to temperature, age, and sex of carrot weevil (Coleoptera: Curculionidae) adults. J. Econ. Entomol. 82:762-765.

Boyden, B.L. 1922. The sweet potato weevil in Florida. Fla. State Plant Bd. Quart. Bull. 6:76-87.

Boyden, B.L. 1927. Sweet potato weevil eradication in Florida and Georgia. Monthly Bull. Fla. State Plant Bd. 12:17-53.

Bucher, G.E. 1981. Identification of bacteria found in insects, *In* H.D. Burges (ed.). Microbial control of pests and plant diseases 1970-1980. Academic Press, London.

Burges, H.D. (ed.). 1981. Microbial control of pests and plant diseases 1970-80. Academic Press, London.

Carruthers, R.I. & R.S. Soper. 1987. Fungal diseases, pp. 357-416. *In* J.R. Fuxa & Y. Tanada (eds.). Epizootiology of insect diseases. J. Wiley, New York.

Castineiras, A. & A. Calderon. 1982. Susceptibilidad de *Pheidole megacephala* a tres insecticidas microbianos: Dipel, bitoxibacillin 202, y *Beauveria bassiana* en condiciones de laboratorio. Cienc. Tec. Agric., Prot. Plantas 5:61-66.

Castineiras, A., S. Caballero, G. Rego & M. Gonzalez. 1982. Efectividad tecnico economica del empleo de la hormiga leona, *Pheidole megacephala*, en el control del tetuan del boniato, *Cylas formicarius elegantulus*. Cienc. Tec. Agric., Prot. Plantas, Supl. 5:103-109.

Castineiras, A., T. Cabrera, A. Calderon & O. Obregon. 1984a. Virulencia de tres cepas de *Beauveria bassiana* sobre adultos de *Cylas formicarius elegantulus* (Coleoptera: Curculionidae). Cienc. Tec. Agric., Prot. Plantas 7:67-74.

Castineiras, A., M. Perez, O. Obregon & I. Castaneda. 1984b. Virulencia de tres cepas de *Metarhizium anisopliae* sobre adultos de *Cylas formicarius elegantulus* (Coleoptera: Curculionidae). Cienc. Tec. Agric., Prot. Plantas 7:129-136.

Chalfant, R.B., R.K. Jansson, D.R. Seal & J.M. Schalk. 1990. Ecology and management of sweet potato insects. Annu. Rev. Entomol. 35:157-180.

Cockerham, K.L. 1944. Some parasites of the sweetpotato weevil. J. Econ. Entomol. 37:546-547.

Cockerham, K.L., O.T. Deen, M.B. Christian & L.D. Newsom. 1954. The biology of the sweet potato weevil. La. Tech. Bull. No. 483.

Crowe, J.H. & K.A.C. Madin. 1975. Anhydrobiosis in nematodes: evaporative water loss and survival. J. Exp. Zool. 193:323-334.

Diaz, J. & H. Grillo. 1986. An isolate of *Beauveria bassiana* (Bals.) Vuillemin as a pathogen of *Cylas formicarius elegantulus* Summers. Centro Agric. 13:94-95.

Dolmans, N.G.M. 1983. Biological control of the black vine weevil (*Otiorhynchus sulcatus*) with a nematode (*Heterorhabditis* spp.). Med. Fac. Landbouww. Rijksumiv. Gent. 48:417-420.

Dutky, S.R. 1959. Insect microbiology. Adv. Appl. Microbiol. 1:175-200.

Dutky, S.R., J.V. Thompson & G.E. Cantwell. 1964. A technique for the mass propagation of the DD 136 nematode. J. Insect Pathol. 6:417-422.

Ferron, P. 1978. Biological control of insect pests by entomogenous fungi. Annu. Rev. Entomol. 23:409-442.

Fuxa, J.R. & Y. Tanada (eds.). 1987. Epizootiology of insect diseases. J. Wiley, New York.

Gaugler, R.R. 1981. Biological control potential of neoaplectanid nematodes. J. Nematol. 13:241-249.

Gaugler, R.R. 1988. Ecological considerations in the biological control of soil-inhabiting insects with entomopathogenic nematodes. Agric. Ecosystems Environ. 24:351-360.

Gaugler, R.R. & G.M. Boush. 1978. Effects of ultraviolet radiation and sunlight on the entomogenous nematode, *Neoaplectana carpocapsae*. J. Invertebr. Pathol. 32:291-296.

Gaugler, R.R. & H.K. Kaya (eds.). 1990. Entomopathogenic nematodes in biological control. CRC Press, Boca Raton, Florida.

Gaugler, R. & D. Malloy. 1981. Field evaluation of the entomogenous nematode, *Neoaplectana carpocapsae*, as a biological control agent of black flies (Diptera: Simuliidae). Mosq. News 41:459-464.

Gaugler, R., J.F. Campbell & T.R. McGuire. 1989a. Selection for host finding in *Steinernema feltiae*. J. Invertebr. Pathol. 54:363-372.

Gaugler, R., T. McGuire & J. Campbell. 1989b. Genetic variability among strains of the entomopathogenic nematode *Steinernema feltiae*. J. Nematol. 21:247-253.

Georgis, R. 1987. Nematodes for biological control of urban insects. Div. Environ. Chem. Am. Chem. Soc. Mtg. (New Orleans) (extended abstract).

Georgis, R. & G.O. Poinar, Jr. 1984a. Field control of strawberry root weevil, *Nemocestes incomptus*, by neoaplectanid nematodes (Steinernematidae: Nematoda). J. Invertebr. Pathol. 43:130-131.

Georgis, R. & G.O. Poinar, Jr. 1984b. Greenhouse control of the black vine weevil *Otiorhynchus sulcatus* (Coleoptera: Curculionidae) by heterorhabditid and steinernematid nematodes. Environ. Entomol. 13:1138-1140.

Goldberg, L.J. & J. Margalit. 1977. A bacterial spore demonstrating rapid larvicidal activity against *Anopheles sergentii, Uranotaenia unguiculata, Culex univittatus, Aedes aegypti*, and *Culex pipiens*. Mosq. News 37:355-358.

Gonzales, S.S. 1925. The sweetpotato weevil (*Cylas formicarius* Fbr.). Philippine Agric. 14:257-281.

Greathead, D.J. & J.K. Waage. 1983. Opportunities for biological control of agricultural pests in developing countries. World Bank, Washington, D.C.

Hernandez, E.M.A. & Z. Mracek. 1984. *Heterorhabditis heliothidis*, a parasite of insect pests of Cuba. Folia Parasitol. (Praha) 31:11-17.

Holdaway, F.G. 1941. Insects of sweet potato and their control. Ha. Agric. Exp. Stn. Prog. Notes 26:7.

Horton, D.E. 1988. World patterns and trends in sweet potato production. Trop. Agric. 65:268-270.

Ignoffo, C.M. 1985. Manipulating enzootic-epizootic diseases of arthropods, pp. 243-262. *In* M.A. Hoy & D.C. Herzog (eds.). Biological control in agricultural IPM systems. Academic Press, New York.

IITA. 1975. IITA annual report 1975. International Institute of Tropical Agriculture, Ibadan, Nigeria.

IITA. 1986. Annual report and research highlights 1985. International Institute of Tropical Agriculture, Ibadan, Nigeria.

Jackson, J.J. 1985. Parasitism of Western corn rootworm with the nematode *Steinernema feltiae*. Ph.D. Dissertation. University of Minnesota, Minneapolis, Minnesota.

Jansson, R.K. & S.H. Lecrone. 1991. *Euderus purpureas* Yoshimoto (Hymenoptera: Eulophidae): parasitoid of sweetpotato weevil (Coleoptera: Apionidae) in southern Florida. Fla. Entomol. (In review).

Jansson, R.K., H.H. Bryan & K.A. Sorensen. 1987. Within-vine distribution and damage of sweetpotato weevil, *Cylas formicarius elegantulus* (Coleoptera: Curculionidae), on four cultivars of sweet potato in southern Florida. Fla. Entomol. 70:523-526.

Jansson, R.K., R.R. Heath & J.A. Coffelt. 1989. Temporal and spatial patterns of sweetpotato weevil (Coleoptera: Curculionidae) counts in pheromone-baited traps in white-fleshed sweet potato fields in southern Florida. Environ. Entomol. 18:691-697.

Jansson, R.K., S.H. Lecrone & R.R. Gaugler. 1992. Efficacy and persistence of *Heterorhabditis bacteriophora* Poinar (Nematoda: Heterorhabditidae): a biological control agent of *Cylas formicarius elegantulus* (Summers) (Coleoptera: Apionidae). Biol. Control (In review).

Jansson, R.K., A.G.B. Hunsberger, S.H. Lecrone & S.K. O'Hair. 1990a. Seasonal abundance, population growth, and within-plant distribution of sweetpotato weevil (Coleoptera: Curculionidae) on sweet potato in southern Florida. Environ. Entomol. 19:313-321.

Jansson, R.K., S.H. Lecrone, R.R. Gaugler & G.C. Smart, Jr. 1990b. Potential of entomopathogenic nematodes as biological control agents of sweetpotato weevil (Coleoptera: Curculionidae). J. Econ. Entomol. 83:1818-1826.

Kaya, H.K. 1985. Entomogenous nematodes for insect control in IPM systems, pp. 283-302. *In* M.A. Hoy & D.C. Herzog (eds.). Biological control in agricultural IPM systems. Academic Press, New York.

Kaya, H.K. 1987. Diseases caused by nematodes, pp. 453-470. *In* J.R. Fuxa & Y. Tanada (eds.). Epizootiology of insect diseases. J. Wiley, New York.

Kaya, H.K. 1990. Soil ecology, pp. 93-115. *In* R. Gaugler & H.K. Kaya (eds.). Entomopathogenic nematodes in biological control. CRC Press, Boca Raton, Florida.

Kemner, N.A. 1924. Der batatenkafer (*Cylas formicarius* F.) auf Java und den benachbarten inseln ostindiens. Z. Angew. Entomol. 10:398-435.

Krieg, A. 1987. Diseases caused by bacteria and other Prokaryotes, pp. 323-355. *In* J.R. Fuxa & Y. Tanada (eds.). Epizootiology of insect diseases. J. Wiley, New York.

Krieg, A., A.M. Huger, G.A. Langenbruch & W. Schnetter. 1983. *Bacillus thuringiensis* var. *tenebrionis* ein neuer gegenuber larven von Coleopteren wirksamer pathotyp. Z. Angew. Entomol. 96:500-508.

Mannion, C.M. & R.K. Jansson. 1992. Comparison of ten entomopathogenic nematodes as biological control agents of *Cylas formicarius* (Fabricius). J. Econ. Entomol. (In review).

Morales, A. 1988. El cultivo del boniato (*Ipomoea batatas* (L.) Lam.) y su principal enemigo al *Cylas formicarius* var. *elegantulus*. *In* Proceedings of a sweetpotato weevil workshop, International Potato Center, Santo Domingo, Dominican Republic.

Nawale, R.N. 1981. A study on the stage of harvest of sweet potato variety, H-268 under Konkan (Maharashtra) conditions. J. Root Crops 7:29-31.

Nickle, W.R. (ed.). 1984. Plant and insect nematodes. Decker, New York.

Nickle, W.R. & H.E. Welch. 1984. History, development, and importance of insect nematology, pp. 627-653. *In* W.R. Nickle (ed.). Plant and insect nematodes. Decker, New York.

Pardales, J.R., Jr. & A.F. Cerna. 1987. An agronomic approach to the control of sweetpotato weevil (*Cylas formicarius elegantulus* F.). Trop. Pest Manage. 33:32-34.

Poinar, G.O., Jr. 1979. Nematodes for biological control of insects. CRC Press, Boca Raton, Florida.

Poinar, G.O., Jr. 1990. Taxonomy and biology of Steinernematidae and Heterorhabditidae, pp. 23-61. *In* R. Gaugler & H.K. Kaya (eds.). Entomopathogenic nematodes in biological control. CRC Press, Boca Raton, Florida.

Rajamma, P. 1980. On some natural enemies of sweet potato weevil *Cylas formicarius* Fab. (Curculionidae: Coleoptera). J. Root Crops 6:59-60.

Reimer, N.J. & J.W. Beardsley, Jr. 1990. Effectiveness of hydramethylnon and fenoxycarb for control of bigheaded ant (Hymenoptera: Formicidae), an ant associated with mealybug wilt of pineapple in Hawaii. J. Econ. Entomol. 83:74-80.

Reinhard, H.J. 1923. The sweet potato weevil. Texas Agric. Exp. Stn. Bull. No. 308.

Risbec, J. 1947. Les charancons nuisibles aux patates douces. Agron. Trop. 2:379-398.

Rolston, L.H., T. Barlow & E.G. Riley. 1983. Control of the sweetpotato weevil in planting material. La. Agric. Exp. Stn. Bull. No. 752.

Roman, J. & W. Figueroa. 1985. Control of the larva of the sugarcane rootstalk borer, *Diaprepes abbreviatus* (L.), with the entomogenous nematode *Neoaplectana carpocapsae* Weiser. J. Agric. Univ. Puerto Rico 69:153-158.

Rutherford, T.A., D. Trotter & J.M. Webster. 1987. The potential of heterorhabditid nematodes as control agents of root weevils. Can. Entomol. 119:67-73.

Sato, K., I. Uritani & T. Saito. 1981. Characterization of the terpene-inducing factor isolated from the larvae of the sweet potato weevil, *Cylas formicarius* Fabricius (Coleoptera: Brenthidae). Appl. Entomol. Zool. 16:103-112.

Schmiege, D.C. 1963. The feasibility of using a neoaplectanid nematode for control of some forest insect pests. J. Econ. Entomol. 56:427-431.

Schroeder, W.J. 1987. Laboratory bioassays and field trials of entomogenous nematodes for control of *Diaprepes abbreviatus* (Coleoptera: Curculionidae) in citrus. Environ. Entomol. 16:987-989.

Sherman, M. & M. Tamashiro. 1954. The sweetpotato weevils in Hawaii: their biology and control. Ha. Agric. Exp. Stn. Bull. 23:1-36.

Singh, B., S.S. Yazdani, R. Singh & S.F. Hameed. 1984. Effect of intercropping on the incidence of sweet potato weevil, *Cylas formicarius* Fabr., in sweet potato (*Ipomoea batatas* Lam.). J. Entomol. Res. 8:193-195.

Smee, L. 1965. Insect pests of sweet potato and taro in the Territory of Papua and New Guinea, their habits and control. Papua New Guinea Agric. J. 17:99.

Su, C.Y., S.S. Tzean & W.H. Ko. 1988. *Beauveria bassiana* as the lethal factor in Taiwanese soil pernicious to sweet potato weevil *Cylas formicarius*. J. Invertebr. Pathol. 52:195-197.

Sutherland, J.A. 1985. The chemical control of *Cylas formicarius* (Fab.) using insecticides at low volume rates, with some considerations on the economics of control. J. Plant Prot. Tropics 2:97-103.

Sutherland, J.A. 1986a. A review of the biology and control of the sweetpotato weevil *Cylas formicarius* (Fab.). Trop. Pest Manage. 32:304-315.

Sutherland, J.A. 1986b. Damage by *Cylas formicarius* (Fab.) to sweet potato vines and tubers, and the effect of infestations on total yield in Papua New Guinea. Trop. Pest Manage. 32:316-323.

Sutherland, J.A. 1986c. Evaluation of foliar sprays, soil treatment and vine dip for the control of sweetpotato weevil, *Cylas formicarius* (Fab.). J. Plant Prot. Tropics 3:95-103.

Swain, R.B. 1943. Nematode parasites of the white-fringed beetle. J. Econ. Entomol. 36:671-673.

Talekar, N.S. 1983. Infestation of a sweetpotato weevil (Coleoptera: Curculionidae) as influenced by pest management techniques. J. Econ. Entomol. 76:342-344.

Talekar, N.S. 1988. How to control sweetpotato weevil: a practical IPM approach. Asian Veg. Res. Devel. Center Bull. No. 88-292.

Uritani, I., T. Saito, H. Honda & W.K. Kim. 1975. Induction of furano-terpenoids in sweet potato roots by the larval components of the sweet potato weevils. Agric. Biol. Chem. 37:1857-1862.

Villareal, R.L. 1982. Sweet potato in the tropics - progress and problems, pp. 3-15. *In* Sweet potato. Proceedings of the 1st international symposium. Asian Vegetable Research and Development Center, Shanhua, Tainan, Taiwan, Republic of China.

Woodring, J.L. & H.K. Kaya. 1988. Steinernematid and heterorhab-ditid nematodes: a handbook of biology and techniques. Southern Coop. Ser. Bull. No. 331. Arkansas Agric. Exp. Stn., Fayetteville, Arkansas.

10. Potential Use of *Agrobacterium*-Mediated Gene Transfer to Confer Insect Resistance in Sweet Potato

John H. Dodds, Christa Merzdorf, Victor Zambrano, and
  Carmen Sigüeñas
International Potato Center
P.O. Box 5969, Lima, Peru

Jesse Jaynes
Department of Biochemistry
Louisiana State University
Baton Rouge, Louisiana 70803 U.S.A.

Agrobacterium: NATURE'S GENETIC ENGINEER

In the last five years, rapid progress has been made in the use of *Agrobacterium* spp. as gene vectors of foreign DNA into genomes of dicotyledonous plants (Fraley *et al.* 1986, Horsch *et al.* 1985, Klee *et al.* 1987, Klee *et al.* 1985, Perani *et al.* 1986, Rogers *et al.* 1986, Schardl *et al.* 1987, Schell 1988, Schell *et al.* 1983, Stiekema *et al.* 1988). The basic principles rest on the natural ability of *Agrobacterium* to act as a genetic engineer. A number of genes have become available that may have potential for conferring insect resistance to sweet potato. This chapter presents some of the basic protocols involved in transformation of the genes that are currently under study to confer insect resistance, and the potential for using this technology to improve programs for insect pests of sweet potato.

Both *Agrobacterium rhizogenes* and *A. tumefaciens* may act as gene vectors. *Agrobacterium rhizogenes* causes a proliferation of fine root hairs on the plants it infects at the point of inoculation; this phenomenon has been termed "hairy root" (Jaynes 1985). Each plant is derived from a single genetically-transformed cell. The etiology of "hairy root" is quite distinct from that of crown gall, a neoplastic cancer growth, which is caused by *A. tumefaciens*. The factor in *A. tumefaciens*

DNA, i.e., T-DNA, is transferred and integrated into the DNA of the infected plant tissue (Chilton *et al.* 1980). This T-DNA is the region responsible for tumor formation and determines two fundamental characteristics of crown gall tumor cells: 1) cell growth in the absence of phytohormones, which are normally required for cell growth in callus culture; and 2) synthesis of unusual tumor-specific compounds, called opines. The Ti plasmid also confers to the bacterium the ability to catabolize the particular opine that is produced in the tumor (Bomhoff *et al.* 1976). This represents a classic example of bacterial transformation of an eukaryotic organism. It has been shown recently that hairy root disease, caused by *A. rhizogenes*, is also due to the transfer of plasmid DNA (Ri-DNA) to the plant cell (White *et al.* 1982). The hairy root tumors produce opines as well.

One strategy for introduction of a foreign gene into a plant by means of the Ri plasmid and *A. rhizogenes* is to utilize a small recombinant plasmid suitable for cloning in *Escherichia coli*, into which a known fragment of recombinant DNA (R/DNA) has been spliced; this recombinant plasmid is cleaved at a site within the rhizogenes DNA (R-DNA). Ri-DNA is a proportion of the agrobacterial plasmid DNA that is moved to the plant R-DNA by the process of transformation. A piece of "passenger" DNA is spliced into this opening. The passenger DNA consists of the foreign gene one intends to incorporate into the plant DNA as well as a selectable marker, e.g., a bacterial gene for resistance to an antibiotic, such as kanamycin. This plasmid is then recloned into a larger plasmid and then introduced into an *Agrobacterium* strain carrying an unmodified Ri plasmid. This process allows molecular biologists to use small plasmids which are easily manipulated *in vitro*. During growth of the bacteria, a rare double-recombination sometimes occurs resulting in *Agrobacterium* whose Ri-DNA harbors an insert: the passenger of foreign R-DNA. Such bacteria can be identified and selected by their survival on media containing the selectable antibiotic, such as kanamycin. These *Agrobacterium* can then be used to insert their R/DNA (modified with passenger DNA) into a plant genome (Herrera Estrella *et al.* 1985, Matzke & Chilton 1981) (Figure 10.1).

This protocol could be used to modify plant cells *in vitro*; however, because of the presence of neoplastic cancerous genes, it has been difficult to obtain normal phenotypic plants from these tranformed plant cells (Espinoza *et al.* 1989, Hanish *et al.* 1988, Ooms *et al.* 1986, Shahin *et al.* 1986, Tepfer 1984, 1987). The same experimental strategy can be utilized by employing a disarmed strain (without neoplastic genes) of *A. tumefaciens*. This may be a better method since the

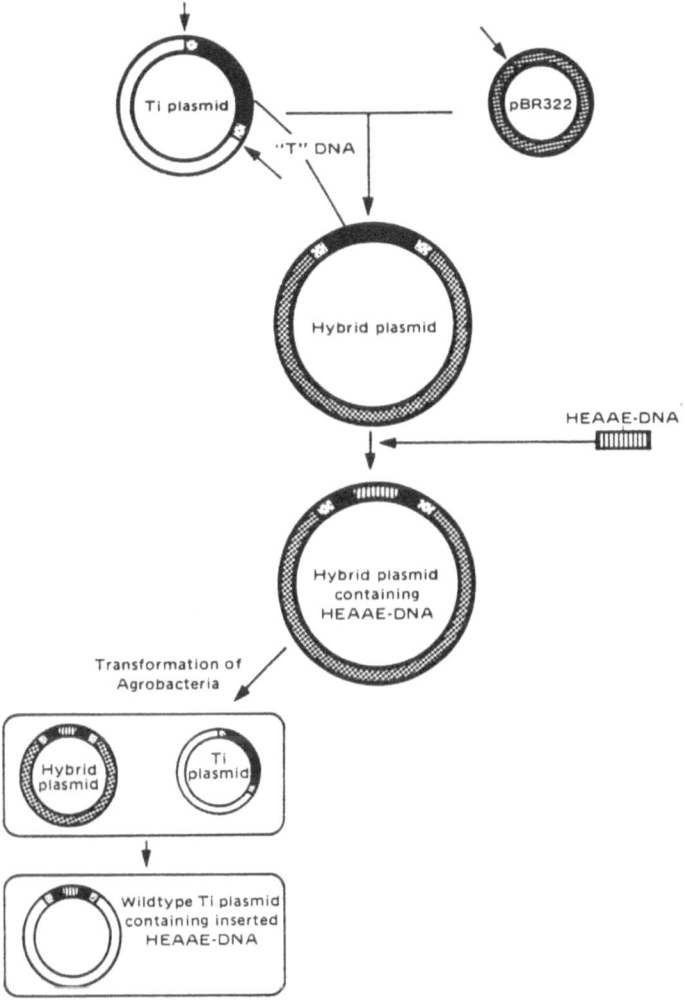

Fig. 10.1. Diagram showing how foreign DNA (HEAAE) can be included into an *Agrobacterium* plasmid vector.

disarmed Ti plasmids give rise to transformed plant cells that can regenerate into intact, healthy, fertile plants (Chilton 1983, Chilton *et al.* 1982). In this case, a selectable marker gene, such as antibiotic resistance, is included to allow selection of transformed plantlets.

In order for a gene transformation system to succeed in a crop species, a number of component parts must be functional in the system. These include: (1) appropriate *in vitro* regeneration systems, (i.e., intact plants from cells or tissues) and a host that is susceptible to infection by *Agrobacterium*; (2) genes of interest; (3) appropriate promoter sequences to drive those genes; and (4) detection methods to determine gene insertion and expression.

These component technologies are described in this review using synthetic protein production from an inserted synthetic DNA sequence (as a model). Similar types of experiments and several others using a range of potential anti-insect genes are currently underway in our laboratory.

THE SYNTHETIC PROTEIN MODEL

In Vitro Propagation and Regeneration of Sweet Potato

Sweet potatoes can be micropropagated readily by cutting *in vitro* plantlets into single node cuttings, transferring the nodes onto the surface of fresh growth medium for plantlet development, and then culturing plantlets in either glass tubes or sterile plastic boxes maintained under a 16-hour photoperiod (3,000 lux) at 28°C (Figure 10.2). A number of different culture media are available for this type of micropropagation (Frison 1981, Litz & Conover 1978).

Transformation. The transformation process involves first wounding *in vitro* plantlets of sweet potato, approximately 5 cm high, in the stem with a sterile needle and then inoculating the wound with a small amount of inoculum of either wild type or transformed *A. rhizogenes* (containing synthetic gene). The transformed *A. rhizogenes* strains contain an Ri plasmid which has the synthetic DNA under the control of the nopaline synthase promoter. Inoculated *in vitro* plantlets should be maintained under optimal lighting conditions, at 22°C. Other plantlets, wounded in a similar manner but inoculated with distilled water, should be grown as a check.

"Hairy roots," 3 to 5 cm in length (Figure 10.3), are typically induced by the various strains of *A. rhizogenes* after approximately three weeks at the inoculation site. These roots are excised, placed onto the surface of CDA and MSA agar solidified media, and incubated under 16-hour photophase (3,000 lux) at 22°C, in order to regenerate plantlets (Espinoza & Dodds 1985). A visible callus phase does not preceed plantlet regeneration. Small plantlets are apparent on some of the cultivars in about six to eight weeks (Figure 10.4). Plantlets can be excised and transferred to normal propagation

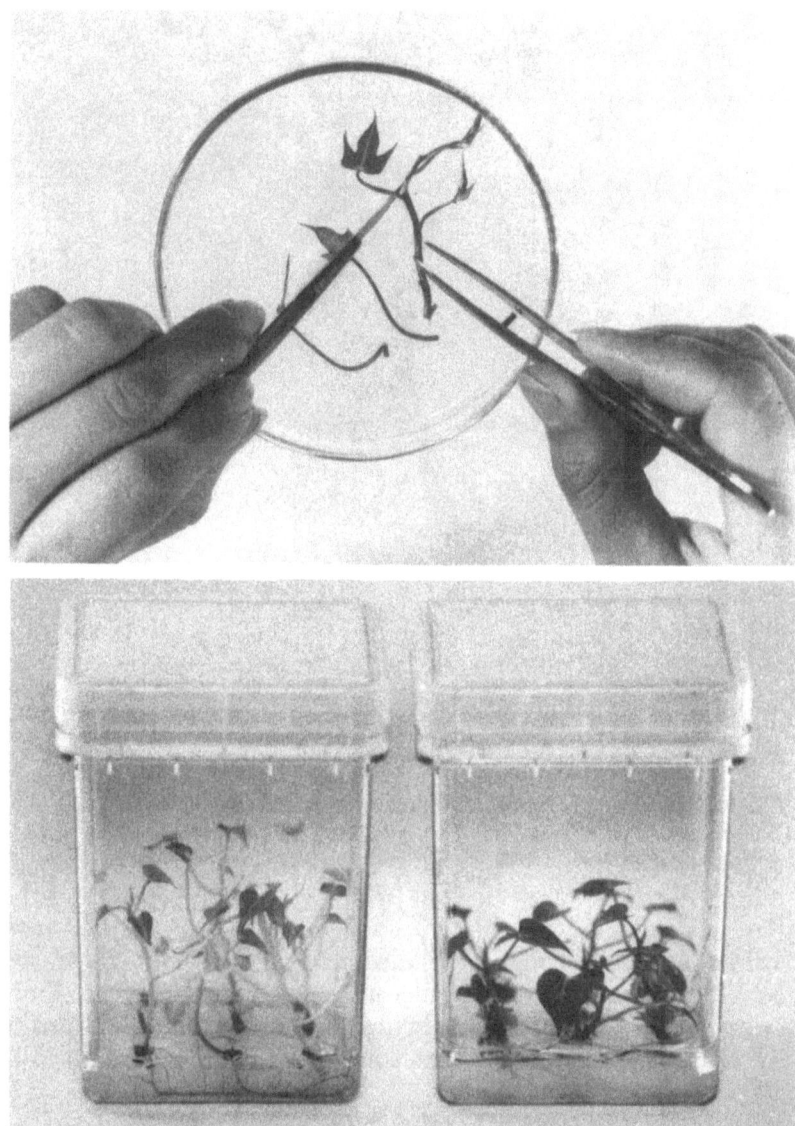

Fig. 10.2. Preparation of single node cutting for micropropagation of sweet potato plantlets *in vitro* (top); micropropagation of sweet potato plantlets on solidified agar culture medium (bottom).

medium. Fully developed and rooted plantlets can be transplanted to jiffy pots in non-sterile conditions (Figure 10.5). Control inoculations

Fig. 10.3. *In vitro* plantlet of sweet potato, wounded and inoculated with *A. rhizogenes*. After five days, hairy roots can be seen forming at the point of inoculation.

typically yield no root material indicating that the "hairy roots" were derived from transformation events induced by *A. rhizogenes*. Among cultivars which are capable of being regenerated, nearly all of the "hairy roots" yielded plantlets, with some roots yielding as many as five plants per root.

Supply of Genes of Interest

In the synthetic protein model, several different synthetic genes were constructed to enhance protein nutritional value. All constructions were coded to enrich essential amino acid content. Figure 10.6 shows a commonly used synthetic sequence (SP47)

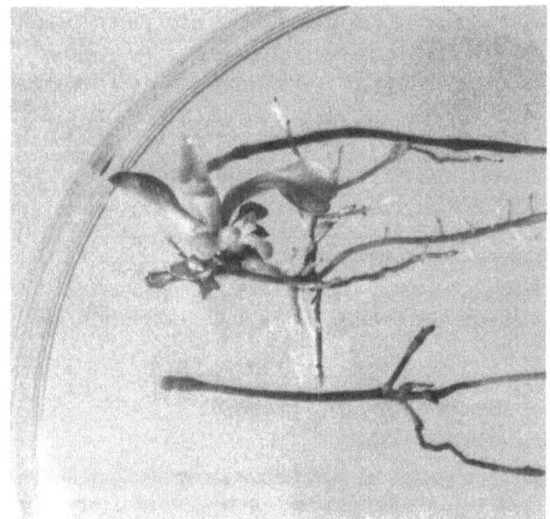

Fig. 10.4. Regeneration of sweet potato plantlet on cultured hairy roots.

Fig. 10.5. Transformed sweet potato plantlet containing genes for high essential amino acid protein, after transplanting to jiffy pot.

Gene Fragment 1

Reading direction of top strand (A)

AATTCGGGGATCGTAAGAAATGGATGGATCGTCATCCATTTCTTCATCCATTTCTTAC
GCCCCTAGCATTCTTTACCTACCTAGCAGTAGGTAAAGAAGTAGGTAAAGAATG

GATCCATCCATTTCTTAAGAAATGGATGAAGAAATGGATGACGATCCATCCATTTCTT
CTAGGTAGGTAAAGAATTCTTTACCTACTTCTTTACCTACTGCTAGGTAGGTAAAGAA

CATCCATTTCTTCATCCATTTCTTACGATCAAGAAATGGATGAAGAAATGGATGAAGA
GTAGGTAAAGAAGTAGGTAAAGAATGCTAGTTCTTTACCTACTTCTTTACCTACTTCT

AATGGATGAAGAAATGGATGCATCCATTTCTTAAGAAATGGATGAAGAAATGGATGAA
TTACCTACTTCTTTACCTACGTAGGTAAAGAATTCTTTACCTACTTCTTTACCTACTT

GAAATGGATGACGATCGATCGTAAGAAATGGATGACGATCCATCCATTTCTTACGATC
CTTTACCTACTGCTAGCTAGCATTCTTTACCTACTGCTAGGTAGGTAAAGAATGCTAG

CCCG
GGGCTTAA

Reading direction of bottom strand (B)

Sequence of protein (A)

GlyAspArgLysLysTrpMetAspArgHisProPheLeuHisProPheLeuThrIleHisProPheLeu-
- LysLysTrpMetLysLysTrpMetThrIleHisProPheLeuHisProPheLeuHisProPheLeuThr-
- IleLysLysTrpMetLysLysTrpMetLysLysTrpMetLysLysTrpMetHisProPheLueLysLys-
- TrpMetLysLysTrpMetLysLysTrpMetThrIleAspArgLysLysTrpMetThrIleHisProPhe-
- LueThrIlePro

Sequence of protein (B)

GlyAspArgLysLysTrpMetAspArgHisProPheLueThrIleAspArgHisProPheLueHisPro-
- PheLueHisProPheLueLysLysTrpMetHisProPheLueHisProPheLueHisProPheLueHis-
- ProPheLueAspArgLysLysTrpMetLysLysTrpMetLysLysTrpMetAspArgHisProPheLue-
- HisProPheMetLysLysTrpMetAspArgLysLysTrpMetLysLysTrpMetThrIleHisProPhe-
- LueThrIlePro

Fig. 10.6. Diagram of the synthetic coding sequence, SP47, (A) which codes
for synthetic proteins (B) which are enriched in essential amino acids.

together with its translational product. This synthetic sequence has
been the most commonly used gene in our laboratory and one for
which gene probes and antisera to the protein have been produced.

The above experiments and other plant transformation studies
have demonstrated that foreign genes can be inserted into both sweet
potato and potato and that inserted genes are expressed (Dodds 1987,
Espinoza *et al.* 1988, Hsiung & Hansen 1986, Jaynes & Dodds 1987,
Jaynes *et al.* 1985, Jaynes *et al.* 1985). Several other genes, such as
Bacillus thuringiensis Berliner, cowpea trysin inhibitor, and chitinase,
are currently being investigated to determine if they might confer plant
resistance to insect pests of sweet potato

Bacillus thuringiensis. *Bacillus thuringiensis* is a gram-positive
bacterium which produces endogenous crystals upon sporulation. The

crystals are proteinaceous in nature and are specifically toxic against certain insect larvae (Andrews *et al.* 1987, Aronson *et al.* 1986, Brosseau & Massow 1988, Dean 1984, Fast 1981, Kristiansen & Lewis 1986, Schnepf & Whiteley 1981, 1985, Vaeck *et al.* 1987a,b,c). The crystals dissolve when ingested, and release proteins of smaller molecular weight, which are then processed into toxic fragments in the insect gut. A wide range of *B. thuringiensis* genes are now being tested commercially against a wide range of insect pests. The *B. thuringiensis* system has been reviewed in several reports (Barton *et al.* 1987, Fischhoff *et al.* 1987, Kirschbaum 1985, Meeusen & Warren 1989, Vaeck *et al.* 1987a).

The endotoxins and exotoxins have been divided into three categories which differ in their effectiveness against Lepidoptera, Diptera, and Coleoptera. The correct gene for the toxin must be selected to be effective against a specific insect.

Bacillus thuringiensis toxins are highly specific in their activity and therefore, are a relatively safe biological insecticide. Their low toxicity to non-target organisms is often argued as one of the environmental advantages of the *B. thuringiensis* system. Because of their specificity, more screening work is needed to identify *B. thuringiensis* strains that may be active against insect pests of sweet potato, especially sweet potato weevils. A number of research groups in the United States and India are currently active in this area. Currently, no sweet potato plantlet has been transformed with genes of *B. thuringiensis*. However, several researchers have conducted similar experiments with potato. Before sweet potato plantlets are transformed, strains of *B. thuringiensis* that are toxic to these insects, especially sweet potato weevils, must be identified.

Cowpea Trypsin Inhibition. A protein (trypsin) inhibitor gene which confers high insect resistance has been isolated and characterized from a cowpea accession. This gene has been patented by a British company (AGC Ltd., Cambridge, U.K.). This gene is active against a wide range of insect pests in several crops. Similar protein inhibitors have also been reported from other sources. At the present time, little data is available on the use of this system for transforming plantlets because patents are currently pending for the researchers involved in the work. The potential for transforming sweet potato with these genes will be understood better once patent rights have been established and information is made public.

Chitinase. The exoskeleton of insects is composed mostly of chitin. Several authors suggested that weakening of the exoskeleton, foregut, and hindgut may be achieved using a chitinase enzyme. A number of

Table 10.1. Types of promoters and their activity.

Promoter	Activity
Nopaline synthesis (NOS)	Constitutive (all plant); low level of expression in all plant parts.
Cauliflower mosaic virus (CaMv)	Constitutive; medium level of expression in all plant parts.
Double cauliflower mosaic (Ca$_2$Mv)	Constitutive; probably the highest level of expression of promoter available at this time.
Rubisco	Only produced in areas of the plant that are photosynthetically active, i.e. in leaves, but not in storage roots.
Wound induced (WI)	The gene is turned on in the entire plant when a plant part is damaged, e.g., by insect attack.
Sporamin	This promoter allows production of gene product in sweet potato storage root, but not in other plant parts.

chitinase genes have been isolated and characterized (Cohen 1987, Mayer *et al.* 1980, Vogely-Lange *et al.* 1988). Currently, it is not known if sufficient quantities of chitinase could be produced to confer an effective level of resistance to insect pests. This approach may be overoptimistic, however, because few promoters are available to produce sufficient chitinase activity, and chitinase genes are fairly nonspecific in action.

Promoter Sequences

A promoter sequence is a control DNA sequence which regulates the expression of the gene under its control. During the past five years, several different promoter sequences have been isolated. A promoter sequence controls the amount, site, and time of production of a particular gene product. Some of these are listed in Table 10.1.

The localization of the gene expression and the total amount of gene product expressed can be regulated by selecting the proper promoter. Gene expression is also significantly affected by the site of insertion of the T-DNA in the plant genome and the number of copies inserted. Previously, the initial detection of transformed plants required complex and time-consuming biochemical analyses. Recently, reporter genes, or makes genes, have been developed and these accelerate this initial selection stage.

Detection Methods

Use of Reporter Genes. The use of reporter genes as monitors for detecting gene insertion is common. One of the most commonly used reporter genes, and the one we used, is the GUS (gen *E. coli* B-glucuronidase) system. In this system, a histochemical reaction is used to determine if a reporter gene in the same insect resistance gene/GUS complex is inserted and expressed in putatively transformed plants. However, the definitive test for genetic transformation has been physical detection of the inserted gene through Southern blot analysis (Southern 1975) followed by detection of the messenger RNA (mRNA) produced from the inserted DNA by Northern blot analysis (Alwine *et al.* 1977), and finally detection of the novel gene product, the protein (Western blot) (Burnett 1981, Towbin *et al.* 1979). The protocol used and results of transformations of sweet potato conducted using the synthetic protein model are presented below.

Southern Blotting for Detection of Gene Insertion in Sweet Potato. Leaf tissue was harvested from a number of regenerated sweet potato plants derived from *A. rhizogenes* infections. DNA from both control plants and plants transformed with *A. rhizogenes* containing the synthetic DNA gene fragments was treated with the restriction enzyme, Hind III. This enzyme removes the synthetic DNA contained within the Hind III-17 fragment of the wildtype Ri plasmid. This was transferred onto strips of nitrocellulose filter paper. After DNA was transferred, radioactively labeled DNA (the synthetic gene fragment) was used as a probe. The labeled probe was mixed with 4 ml of a solution containing 1.08 M NaCl, 0.108 M Na-citrate (pH 7.0), 0.02% (w/v) Ficoll Type 400, 0.02% (w/v) polyvinyl pyrollidone 10,000, 0.2% (w/v) bovine serum albumin (BSA), 0.5% (w/v) sodium dodecyl sulfate (SDS), 50 μg/ml salmon sperm DNA, 50 μg/ml yeast transfer RNA (tRNA), and 100 μg/ml calf thymus DNA. This solution was poured into a Seal-a-Bag™ containing the nitrocellulose filters and the bubbles were removed. The bag was sealed and placed in a 70°C water bath overnight. The filter was removed and washed for 15 minutes three times at 60°C in probe solution without DNAs. The filter was then washed four times in a solution (50 ml) containing 0.015 M NaCl and 0.0015 M Na-citrate (pH 7.0) for 30 minutes each at 52°C. The filter was blotted dry and placed in a vacuum oven for 10 minutes and

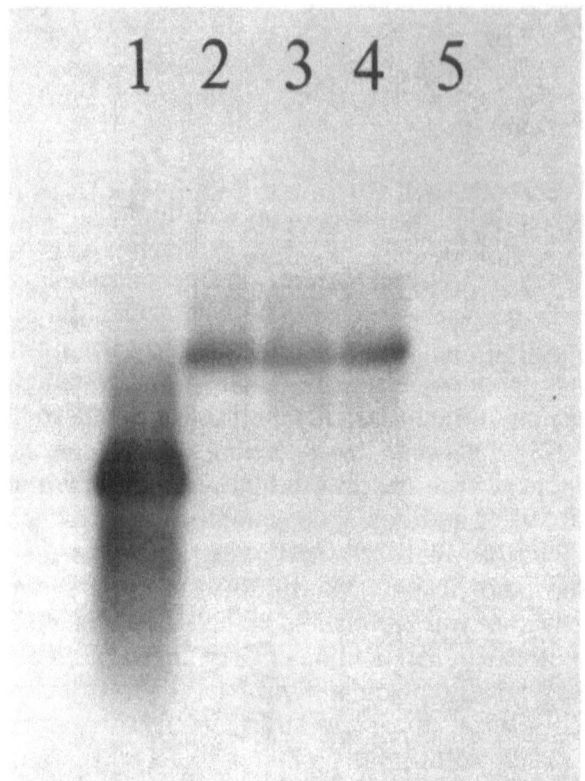

Fig. 10.7. Southern blot analysis of synthetic protein gene (lane 1, positive control; lanes 2-4, transformed plant; lane 5, negative control).

then placed in contact with a KODAK X-omat X-ray film. The autoradiogram was developed after several days. The results of this analysis showed that the synthetic protein gene was inserted into the plant genome (Figure 10.7). Similar experiments with Irish potato have also indicated production of mRNA and synthetic protein production (Dodds 1987, Jaynes *et al.* 1985).

Similar experiments can be conducted to probe for anti-insect genes because in all cases, probes and antisera are readily available. The technologies needed to produce transgenic sweet potato plants are currently available and it is likely that transgenic plants with some resistance to insect pests will be available for testing shortly.

OUTLOOK

The use of synthetic protein sequences for transforming plants demonstrates that there is potential to genetically engineer sweet potato using *Agrobacterium* plasmids as a gene vector system. In our studies, we used an artificial gene which codes for a synthetic protein and a plant-vector system employing *A. rhizogenes* as a model system (Espinoza *et al.* 1989, Jaynes & Dodds 1987). We have had good success in developing a routine system for transforming and regenerating healthy sweet potato plants from roots induced with both *A. rhizogenes* and *A. tumefaciens* vectors.

We are currently employing an *A. tumefaciens* leaf disk system to introduce the synthetic protein coding DNA into a number of sweet potato cultivars. In general, the regenerated plants produced from this method seem to grow more normally than those produced from the rhizogene system. This technique may also be expanded to insert genes for both disease and insect resistance in addition to the synthetic protein DNA. Both the synthetic protein gene and other disease and pest resistance genes may lead to the production of improved germplasm.

Most sweet potatoes are produced in developing countries where protein malnutrition is endemic (Horton 1988). These countries lack financial resources to import expensive chemical insecticides to control insects and diseases. For these reasons, any genetic improvement in germplasm might significantly improve both the quality and yield of sweet potatoes produced by these countries.

REFERENCES

Alwine, J.C. D.R. Kemp & G.R. Starket 1977. Method for detection of specific RNAs in agarose gels by transfer to diazobenzyloxymethyl-paper and hybridization with DNA probes. Proc. Natl. Acad. Sci. 74:5350-5354.

Andrews, R.E., R.M. Faust, H. Wabiko, K.C. Roymond & L.A. Bulla. 1987. The biotechnology of *Bacillus thuringiensis*. CRC Crit. Rev. Biotechnol. 6:163-232.

Aronson, A.I., W. Backman & P. Dunn. 1986. *Bacillus thuringiensis* and related insect pathogens. Microbiol. Rev. 50:1-24.

Barton, K.A., H.R. Whiteley & N. Yang. 1987. *Bacillus thuringiensis* delta-endotoxin expressed in transgenic *Nicotiana tabacum* provides resistance to lepidopteran insects. Plant Physiol. 85:1103-1109.

Bomhoff, G., P.M. Klapwijk, C.H.M. Kester, R.A. Schilperoort, J.P. Hernalsteens & J. Schell. 1976. Octopine and nopaline synthesis and break-down genetically controlled by a plasmid of *Agrobacterium tumefaciens*. Mol. Gen. Genet. 145:177-181.

Brousseau, R. & L. Masson. 1988. *Bacillus thuringiensis* insecticidal crystal toxins: Gene structure and mode of action. Biotech. Adv. 6:697-724.

Burnett, W.H. 1981. Western blotting: electrophoretic transfer of proteins from SDS-polyacrilamide gels to unmodified nitrocellulose and radiographic detection with antibody and radioiodinated protein A. Anal. Biochem. 112:195-203.

Chilton, M.D. 1983. A vector for introducing new genes into plants. Sci. Amer. 248:36-45.

Chilton, M.D., R.K. Saiki, N. Yodov, M.P. Gordon & F. Quetier. 1980. T-DNA from *Agrobacterium* Ti plasmid is in the nuclear DNA fraction of crown gall tumor cells. Proc. Natl. Acad. Sci. 77:4060-4064.

Chilton, M.D., D.A. Tepfer, A. Petit, D. Chantal, F. Casse-Delbart & J. Tempe. 1982. *Agrobacterium rhizogenes* inserts T-DNA into the genomes of the host plant root cells. Nature 295:432-434.

Cohen, E. 1987. Chitin biochemistry: synthesis and inhibition. Annu. Rev. Entomol. 32:71-93.

Dean, D.H. 1984. Biochemical genetics of the bacterial insect-control agent *Bacillus thuringiensis*: basic principles and prospects for genetic engineering. Biotechnol. Genet. Eng. Rev. 2:341-363.

Dodds, J.H. 1987. CIP collaboration research on tissue culture and genetic engineering for potato improvement. CIP Circ. 15:1-4.

Espinoza, N.O. & J.H. Dodds. 1985. Adventitious shoot formation on cultured potato roots. Plant Science 41:121-124.

Espinoza, N.O., J.H. Dodds & J. Jaynes. 1990. Regeneration of sweet potato (*Ipomoea batatas* L.) plants transformed by *Agrobacterium rhizogenes* containing a gene encoding a protein high in essential amino acids. Plant Cell Rep. (In review).

Fast, P.G. 1981. The crystal toxin of *Bacillus thuringiensis*, pp. 223-248. *In* H.D. Burges (ed.). Microbial control of pests and plant diseases, 1970-1980. Academic, London.

Fischhoff, D.A., K.S. Bowdish, F.J. Perlack, P.G. Marrone & S.M McCormick. 1987. Insect tolerant transgenic tomato plants. Biotechnol. 5:807-813.

Fraley, R.T., S.G. Rogers, R.B. Horsch & G.F. Barry. 1986. Gene transfer in plants: a tool for studying gene expression and plant development, pp. 15-26. *In* A.R. Liss (ed.). Molecular developmental biology.

Frison, E.A. 1981. Tissue culture: a tool for improvement and international exchange of tropical root and tuber crops. IITA research briefs, Vol. 2, Ibadan, Nigeria.

Hanisch Ten-Cate, Ch.H., E. Ennik, S. Roest, K. Sree Ramulu, P. Dijkhuis & B. DeGroot. 1988. Regeneration and characterization of plants from potato root lines transformed of *Agrobacterium rhizogenes*. Theor. Appl. Genet. 75:452-459.

Herrera-Estrella, L., M. De Block, P. Zambryski, M. Van Montagu & J. Schell. 1985. *Agrobacterium* as a vector system for the introduction of genes into plants, pp. 61-93. *In* J.H. Dodds (ed.). Plant genetic engineering. Cambridge Univ. Press, Cambridge.

Horsch, R.B., J.E. Fry, N.L. Hoffmann, D. Eichholtz, S.G. Rogers & R.T. Fraley. 1985. A simple and general method for transferring genes into plants. Science 227:1229-1231.

Hsiung, H.M. 1986. Use of synthetic DNA in expression of foreign proteins. Biotech. Adv. 4:1-11.

Jaynes, J.M., P. Langridge, K. Anderson, C. Bond, D. Sands, C.E. Newman & R. Newman. 1985. Construction and expression of synthetic DNA fragments coding for polypeptides with elevated levels of essential amino acids. Appl. Microbiol. Biotechnol. 21:200-205.

Jaynes, J.M. & J.H. Dodds. 1987. Crop plant genetic engineering: science fiction or science fact? Outlook Agric. 16:1-5.

Kirschbaum, J.B. 1985. Potential implication of genetic engineering and other biotechnologies to insect control. Annu. Rev. Entomol. 30:51-70.

Klee, H.J., M.F. Yanofsky & E.W. Nester. 1985. Vectors for transformation of higher plants. Bio/Technology 3:637-642.

Klee, H., R.B. Horsch & S.G. Rogers. 1987. Agrobacterium-mediated plant transformation and its further applications to plant biology. Annu. Rev. Physiol. 38:467-486.

Kristiansen, B. & C. Lewis. 1986. Bacterial insecticides: recent developments. Trends Biotech. 4:56-58.

Litz, R.E. & R.A. Conover. 1978. *In vitro* propagation of sweet potato. HortScience 13:659-660.

Matzke, A.J.M. & M.D. Chilton. 1981. Site-specific insertion of genes into T-DNA of the *Agrobacterium tumefaciens* Ti plasmid: an approach to genetic engineering of higher plant cells. J. Mol. Appl. Genet. 1:39-49.

Mayer, R.T., A.C. Chen & J.R. DeLoach. 1980. Characterization of a chitin synthase from the stable fly, *Stomoxys clacitrans* (L.). Insect Biochem. 10:549-556.

Meeusen, R.L. & G. Warren. 1989. Insect control with genetically engineered crops. Annu. Rev. Entomol. 34:373-381.

Ooms, G.D.T., M.E. Bossen, J.H. Hoge & M. Burrell. 1986. Developmental regulation of Ri TL-DNA gene expression in roots, shoots, and tubers of transformed potato (*Solanum tuberosum* cv. Desiree). Plant Mol. Biol. 67:321-330.

Perani, L., S. Radke, M. Wilke-Douglas & M. Bossert. 1986. Gene transfer methods for crop improvement: introduction of foreign DNA into plants. Physiol. Plantarum 68:566-570.

Rogers, S.G., R.B. Horsch & R.T. Fraley. 1986. Gene transfer in plants: production of transformed plants using Ti plasmid vectors. Meth. Enzymol. 118:627-641.

Schardl, C.L., A.D. Byrd, G. Benzion, M.A. Altschuler, D.F. Hildebrand & A.G. Hunt. 1987. Design and construction of a versatile system for the expression of foreign genes in plants. Gene 61:1-11.

Schell, J. 1988. Transfer of T-DNA from *Agrobacterium* into plants. Transposition, pp. 616-618. *In* A.J. Kingsman, S.M. Kingsman & K.F. Chater (eds.). Society for general microbiology symposium, Vol. 43. Cambridge Univ. Press, Cambridge.

Schell, J. & M. Van Montagu. 1983. The Ti-plasmids as natural and as practical gene vectors for plants. Bio/Technol. 1:175-180.

Schnepf, H.E. & H.R. Whiteley. 1981. Cloning and expression of the *Bacillus thuringiensis* crystal protein gene in *E. coli*. Proc. Natl. Acad. Sci. 78:2893-2897.

Schnepf, H.E. & H.R. Whiteley. 1985. Delineation of a toxin-encoding segment of a *Bacillus thuringiensis* crystal protein gene. J. Biol. Chem. 260:6273-6280.

Shahin, E.A., K. Sukhapinda, R.B. Simpson & R. Spivey. 1986. Transformation of cultivated tomato by a binary vector in *Agrobacterium rhizogenes*: sexual transmission of the transformed genotype and phenotype. Cell 37:959-967.

Southern, E.M. 1975. Detection of specific sequences among DNA fragments separated by gel electrophoresis. J. Mol. Biol. 98:503-571.

Stiekema, W.J., F. Heidekamp, J.D. Louwerse, H.A. Verhoeven & P. Dijkhuis. 1988. Introduction of foreign genes into potato cultivars 'Bintje' and 'Desiree' using *Agrobacterium tumefaciens* binary vector. Plant Cell Rep. 7:45-50.

Tepfer, D. 1987. Ri T-DNA from *Agrobacterium rhizogenes*, a semiochemical that alters morpholicial plasticity, pp. 565-571. *In* D. von Wettstein & N.-H. Chua (eds.). Plant molecular biology. Plenum, New York.

Towbin, H., T. Staehelin & J. Gordon. 1979. Electrophoretic transfer of proteins from polyacrylamide gels to nitrocellulose sheets. Procedure and some applications. Proc. Natl. Acad. Sci. 76:4350-4354.

Vaeck, M., H. Hofte, A. Reynaerts, J. Leemans, M. Van Montagu & M. Zabeau. 1987a. Engineering of insect resistance plants using a *B. thuringiensis* gene, pp. 355-366. *In* Molecular strategies for crop protection.

Vaeck, M., A Reynaerts, H. Hofte, H. Vanderbruggen, S. Janses & J. Leemans 1987b. Insect resistance in transgenic plants expressing *Bacillus thuringiensis* toxin genes. Ann. Entomol. Soc. Brasil 16:427-435.

Vaeck, M., A. Reyaerts, H. Hofte, S. Jansens, M. De Beuckeleer, C. Dean, M. Zabean, M. Van Montagu & J. Leemans. 1987c. Transgenic plants protected from insect attack. Nature 327:33-37.

Vogeli-Lange, R., A. Hansen-Gehri, T. Boller & F. Meins, Jr. 1988. Induction of the defense-related glucanohydrolases, B-1 3-glucanase, and chitinase, by tobacco mosaic virus infection of tobacco leaves. Plant Science 54:171-176.

White, F., G. Ghidossi, M. Gordon & E. Nester. 1982. Tumor induction by *Agrobacterium rhizogenes* involves the transfer of plasmid DNA to the plant genome. Proc. Natl. Acad. Sci. 79:3193-3197.

11. Oviposition Stimulant for *Cylas formicarius* in Sweet Potato: Isolation, Identification, and Development of Analytical Screening Method

David D. Wilson, Ray F. Severson,[1] *and Stanley J. Kays*
Department of Horticulture
University of Georgia
Athens, Georgia 30602 U.S.A.

Sweetpotato weevil, *Cylas formicarius* (Fabricius) (= *C. f. elegantulus* [Summers] and *C. f. formicarius* [Fabricius]), is a major pest of sweet potato worldwide, both in the field and in storage. The development of cultivars resistant to this weevil has been a major objective of plant breeders for some time (e.g., see Chapter 20). The criteria used to assess and select individual traits in sweet potato breeding programs often represent a formidable obstacle, which has hampered the development of resistant cultivars. In addition, inconsistent and clumped infestation levels of weevils in the field have often made it difficult to differentiate resistance and escape.

Currently, no cultivars have been developed that are completely immune to feeding and oviposition by these weevils. However, cultivars show varying degrees of susceptibility. Thus, an understanding of the basis of susceptibility could greatly enhance the incorporation of desired characters into new cultivars. An analytical approach based on the levels of kairomones involved in the interaction of weevils with sweet potato would be an important tool to assist plant breeders. Such an approach would enable researchers to screen large numbers of clones for specific phytochemical(s), and then test promising clones for weevil resistance under controlled conditions in the laboratory, greenhouse, and in the field. We report here the development of a rapid bioassay technique for screening sweet potato clones for the

---

[1] U.S. Department of Agriculture, Agricultural Research Service, Tobacco Quality and Safety Research Unit, Athens, Georgia 30613 U.S.A.

presence of an oviposition stimulant, its isolation and identification, and its use in the development of an analytical screening technique.

DEVELOPMENT OF RAPID BIOASSAY TECHNIQUE

Accuracy of selection represents a weak link in many breeding programs. The main factor limiting improvement of pest resistances in sweet potato seems to be the development of reliable evaluation techniques (Martin & Jones 1986). This problem may be compounded by variations in natural field populations of the target insects, and often requires supplemental infestations in field trials to maintain adequate selection pressures. These large variations in infestation levels of weevils in the field also make it difficult to determine the type(s) and levels of resistance present (Barlow & Rolston 1981). Martin (1984) noted that the most reliable test for resistance was to expose a uniform amount of sweet potato roots to a controlled number of weevils at a given level of hunger under controlled temperature and humidity. These conditions are difficult to obtain in the field and costly in time and effort. Thus, controlled, reproducible laboratory and greenhouse assays are essential for rapid prescreening of lines to be tested in the field. A laboratory feeding assay was developed by Mullen et al. (1980); however, Wilson et al. (1988) suggested that an oviposition assay would be a better indicator of plant resistance to the weevil since oviposition is a behavioral event further in the host selection process by the weevil and is more accurately quantified. Wilson et al. (1988) modified the previous assay (Mullen et al. 1980), and used storage root cores in tissue culture plates (24-well), along with one to three female weevils per core per cage. This technique has been used to demonstrate the presence and characteristics of a chemical oviposition stimulant in the periderm of sweet potato roots (Wilson et al. 1988, 1989). It has also been used to demonstrate the feeding and oviposition preferences of the sweetpotato weevil for the inner core and periderms of different sweet potato cultivars (Nottingham et al. 1987, 1989).

The objective of the study by Wilson et al. (1988) was to develop a simple, fast, and reliable bioassay for an oviposition stimulant, and to demonstrate its presence in the periderm of sweet potato storage roots. In initial tests with small cages and one tissue culture plate with an array of potato cores per cage, the number of eggs laid per core was examined at different combinations of root extract volume and weevil densities tested to select an optimum combination for the bioassays.

There were significant ($P < 0.05$) main effects for weevil densities and treatment (i.e., control, surface extracted, and surface extracted and then reapplied), but not for extract volume. A combination of 20 μl of extract, which sufficiently covered a core surface, with 40 weevils gave significantly higher oviposition on both control and extracted roots with extract reapplied than on extracted (surface washed) cores, and this combination was adopted for further tests. Additional tests with larger cages, two arrays of cores per cage (about 5 cm apart), 20 μl of extract per core and 80 weevils per cage confirmed that the weevils laid significantly more eggs on the control and extracted roots on which the extract had been reapplied than on extracted cores. Various concentrations of the extract were then assayed. Cores with extracts always contained more eggs than cores with solvent only, however, numbers of eggs laid did not differ among extract dosages applied to root cores. In order to obtain a more neutral substrate for the bioassays, Wilson *et al.* (1989) tested sweet potato root cores with and without periderm and found that the inner core (without periderm) was more inert. The test chemical applied to filter paper disks placed on the inner core was a better test substrate than placing the chemical on the core itself because of solvent-induced rotting. The filter paper/core assembly was chosen for subsequent bioassays and has proved to be the best substrate thus far in our tests with the surface chemicals.

The weevils used in the bioassays were reared on sweet potato storage roots in plastic cages at a temperature of 28°C and 55 to 75% RH. Weevils were allowed to feed and oviposit on sweet potato roots for about 7 days before they were transferred to fresh storage roots. The old roots were then incubated until a new generation of weevils emerged in about 4 to 5 weeks. Weevils were collected as they emerged and held for at least 3 weeks before females were used in the bioassays to ensure adequate egg-laying capability (Wilson *et al.* 1988).

The experimental set-up after the initial tests used 48 female weevils per tissue culture plate (24-well; Falcon) to give a weevil/core ratio of 2:1. Females were held for three to six hours in plastic cages (32 x 24 x 10 cm) under test conditions before being presented with an array of 24 cores. Each core, which was cut with a cork borer (no. 11), had either its periderm intact or removed depending on the experiment. Cores were pushed into each well such that only the top surface (1.6 cm diam., 2 cm² area) was exposed to the weevils. The surface area was the same as that used by Mullen *et al.* (1980) and therefore permitted a direct comparison of results in cultivar preference tests, in which the root periderms were left intact. For the tests with sweet potato surface extracts, the test chemicals, silicic acid fractions and boehmeryl acetate, were applied in 5 μl aliquots to small (4 mm diam.) filter paper discs and the solvent was allowed to evaporate. The filter paper discs were then stapled in the middle of each interior core, leaving a peripheral core surface on which the weevils oviposited (Wilson *et al.* 1989). The number of eggs laid in each core was counted after a 48 hour period in total darkness at 28°C and 90 to 100% RH.

EXTRACTION AND GC PROFILES OF SURFACE COMPONENTS

The surface extracts from six different sweet potato cultivars with differing levels of susceptibility were obtained after a 9 minute dip of the storage roots in methylene chloride in an ultrasonic bath (Son *et al.* 1987). After solvent reduction on a roto-evaporator, a portion equivalent to 15 cm² surface area was dried under nitrogen and treated with a silylation reagent, trimethylsilylimidazol-pyridine (1:3), for one hour at 76°C. The samples were then analyzed by capillary gas chromatography on a 0.3 mm inside diameter x 25 m thin-coated (0.1 μm) SE-54 fused silica capillary column (temperature program 140 to 290°C at 4°C/min). Gas chromatograph profiles were obtained, and extract component levels were calculated using an internal standard method (hexacosanol). Root surface areas were determined by peeling periderm from storage roots and passing it through a LI-COR 3000 leaf area meter. The gas chromatograph profiles of the root surface extracts of six lines showed qualitative and quantitative differences (Figure 11.1, Table 11.1). The three more susceptible lines ('Centennial', 'W-115', and '11-49-19') had a major component, a triterpenol acetate (peak no. 5), that was absent in the other three lines. A second peak (no. 2) which was present only in the more susceptible lines, was the free alcohol of this compound.

ISOLATION AND IDENTIFICATION OF OVIPOSITION STIMULANT

To isolate and identify the components stimulating oviposition, the extract from the cultivar 'Centennial' was fractionated on silicic acid (10 g, 5 psi under N₂) and eluted with hexane, methylene chloride-hexane (1:3), methylene chloride-hexane (1:1), and methylene

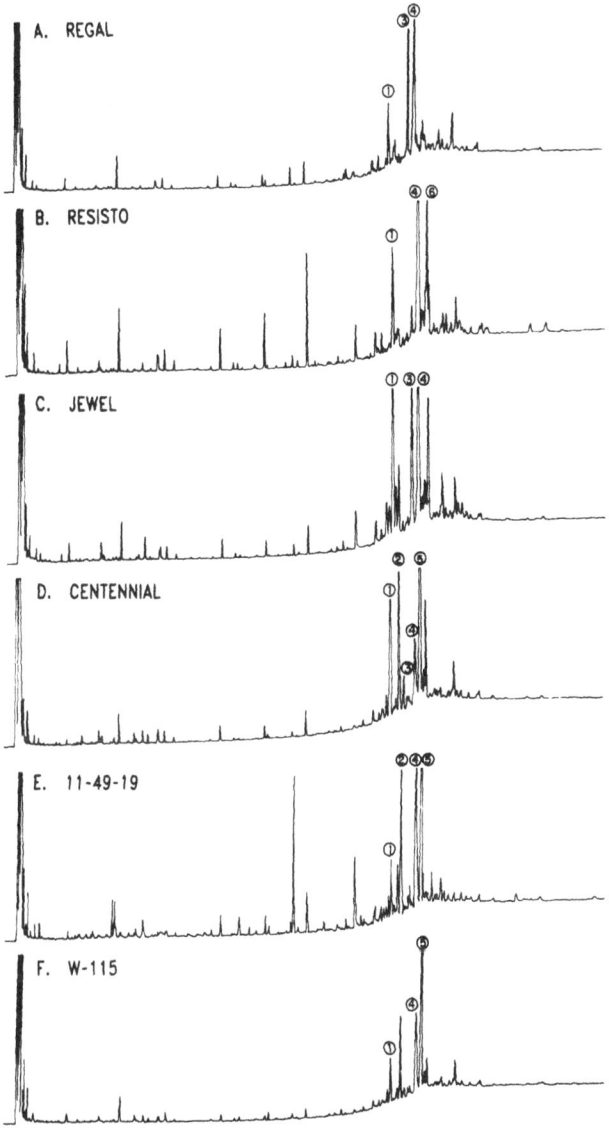

Fig. 11.1. Gas chromatograph profiles of storage root surface components of six sweet potato cultivars (peak 1, unknown triterpenol silyl ether; 2, alcohol of boehmerol silyl ether; 3, unknown [molecular weight=526]); 4 unknown triterpenol silyl ether; 5 boehmeryl acetate; and 6 unknown [molecular weight=426]).

Table 11.1. Levels of storage root surface components from six sweet potato cultivars.

	Peak number [a]					
	1	2	3	4	5	6
	Relative retention time [b]/Molecular weight [c]					
	1.24/498	1.27/498	1.30/526	1.32/498	1.33/468	1.34/426
Cultivar	Levels (μg/cm^2) [d]					
Centennial	0.51	0.66	-[e]	0.27	3.67	-
W-115	0.33	0.54	-	0.65	3.25	-
11-49-19	0.14	0.54	-	0.85	1.94	-
Jewel	0.36	-	0.59	1.54	-	-
Resisto	0.22	-	-	1.53	-	0.30
Regal	0.20	-	0.64	1.38	-	-

[a] Numbers correspond to peaks on chromatograms in Figure 11.1
[b] Relative to internal standard, hexacosanol silyl ester.
[c] Molecular weight based upon GC/MS analysis of silylated extracts.
[d] Calculated assuming chromatographic response of boehmeryl acetate.
[e] Absent or below detection limits.

chloride-methanol (19:1) (Son *et al.* 1987). Elution with methylene chloride-hexane (1:3) produced a highly pure isolate (>96%) which represented the most dominant compound present. Characterization of this compound by GC-MS, IR, ^{13}C, and ^{1}H NMR and other physical data (Son *et al.* 1990) established its identity as boehmeryl acetate ($C_{32}H_{52}O_2$), a pentacyclic triterpene, the alcohol of which was recently isolated from the bark of *Boehmeria excelsa* (Oyarzun *et al.* 1987) (Figure 11.2).

BIOLOGICAL ACTIVITY

Different Cultivars

Nottingham *et al.* (1987, 1989) examined the feeding and oviposition preferences of *C. formicarius* for four different cultivars and found that a highly susceptible cultivar, 'Centennial', in field-plot

227

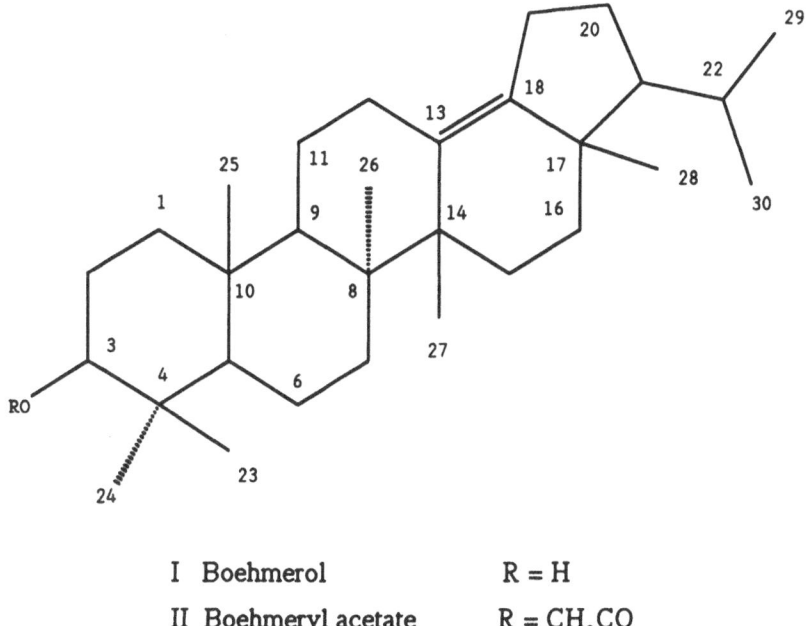

I Boehmerol   R = H

II Boehmeryl acetate $R = CH_3CO$

Fig. 11.2. The chemical structures of boehmerol and boehmeryl acetate.

experiments, was preferred for feeding and oviposition in choice bioassays and for oviposition in no-choice bioassays. No differences in feeding or oviposition were observed for any of the cultivars when inner cores (without periderm) were exposed to the weevils. Oviposition was drastically reduced to almost zero on the inner cores, and was observed only on cores with periderm intact, when compared with those without the periderm (Nottingham *et al.* 1987). When 'Centennial' roots were extracted (washed) with methylene chloride, the level of oviposition was significantly reduced (Wilson *et al.* 1988). Oviposition was restored when the extract was reapplied to the washed surface. This might indicate the presence of a chemical oviposition stimulant in the periderm of 'Centennial'.

Whole Extract and Fractions

Both whole extracts and fractions significantly increased oviposition compared to solvent controls (Wilson *et al.* 1989). Dose-response studies with the whole extract showed a sharp rise in oviposition after a low threshold concentration followed by a relatively dose-independent oviposition response (Wilson *et al.* 1988, 1989). The 1:3 methylene chloride-hexane fraction, which contained > 96% of boehmeryl acetate, had consistently higher oviposition than other fractions (Wilson *et al.* 1989). High levels of oviposition in fractions other than that which contained the dominant peak in the GC profiles suggested that more than one compound might be involved in ovipositional stimulation. Research on preference mechanisms of resistance in plants indicates that a blend of chemicals is usually involved, rather than a single specific compound (Maxwell 1980). In these cases, however, breeding becomes more difficult because each chemical is usually controlled by different genes. The use of key chemical markers may enhance the breeding of plants for lower or higher concentrations of biologically active chemicals.

Boehmeryl Acetate

To assess the effect of boehmeryl acetate on oviposition, recrystallized boehmeryl acetate, solubilized in methylene chloride, was applied to the filter paper discs and presented to the weevils in choice tests along with filter paper discs alone or with solvent. Various dilutions of the compound, relative to the normal concentration found per unit surface area of the roots, were used in dose-response studies to determine the optimum level of activity (Wilson *et al.* 1990). Boehmeryl acetate applied to filter paper discs elicited significantly higher ($P < 0.05$) oviposition by female weevils on the cores compared with filter paper discs alone and those treated with solvent.

There were no significant differences ($P > 0.05$) in feeding punctures at 24 hours, in response to different doses of boehmeryl acetate, indicating that boehmeryl acetate may not be important in feeding stimulation. Oviposition was significantly lower ($P < 0.05$) at 24 hours compared with that at 48 hours (about 20% of that at 48 hours). Oviposition at 48 hours increased ($P < 0.05$) with an increase in dose of boehmeryl acetate, peaked at a dose of about 0.04 μg per core (0.01 sweet potato equivalents), and then decreased with further increases in dose (Figure 11.3). Ovipositional stimulation peaked at the same dose (0.04 μg/core) at both 24 and 48 hours after treatment. The compound appears to be relatively stable, activating the oviposition response through direct contact. All data to date indicate that no special methods are required to maintain it in a pure state (Son *et al.* 1990). This is consistent with the fact that the root is an acceptable host for many months after harvest. These results agree

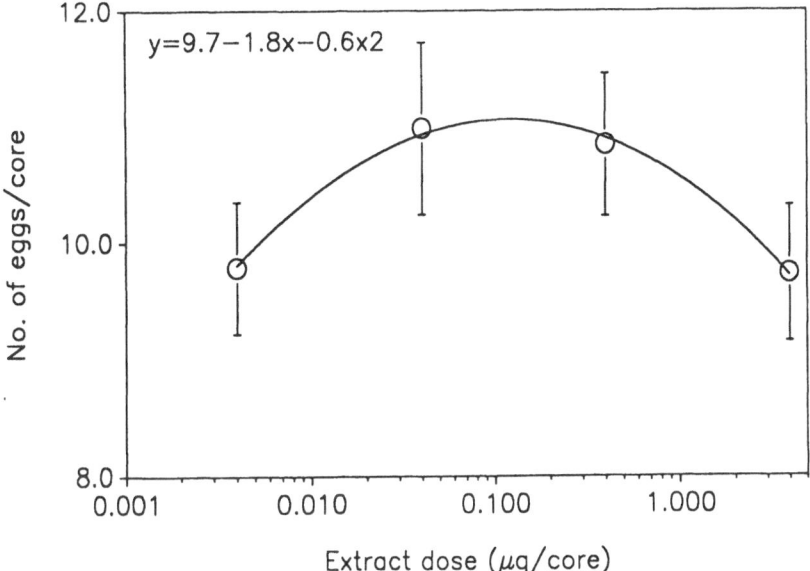

Fig. 11.3. Oviposition by female sweetpotato weevils in response to boehmeryl acetate. Each data point represents a mean ± s.e. ($n = 112$). Mean oviposition for the solvent control (dose = $0\,\mu g$) was 8.5 ± 0.5.

with data for whole extracts (Wilson *et al.* 1989) indicating a high level of activity with a low response threshold.

DEVELOPMENT OF ANALYTICAL SCREENING TECHNIQUE

An understanding of the chemical factors which impart the various levels of resistance/susceptibility of present sweet potato cultivars to weevils would facilitate the rate at which meaningful levels of resistance could be incorporated into new cultivars. Specialist herbivores, such as the sweetpotato weevil, usually respond to specific secondary plant compounds and feed and/or oviposit if the right stimulant is present; generalist herbivores respond in the absence of deterrents (Renwick 1983). The potential for breeding plants that discourage oviposition is gaining widespread recognition. In the past, the basis for improving plant resistance to insects has been to increase levels of allomones (deterrents) rather than to reduce levels of kairomones (stimulants) (Jackson & Lewis 1981). Reduction in the levels of kairomones in crops attacked by specialists might enhance the development of resistance to these pests.

Breeding of less preferred plants may be accomplished by reducing levels of attractants, arrestants, and/or stimulants within plant parts. Chemical assay methods developed for the identified attractants or stimulants, could then be used as a primary screening device for detecting less susceptible plants (Stadler 1983). The potential of using the content of attractant chemicals in breeding programs has been presented for cucurbitacins (Chambliss & Jones 1966, Metcalf *et al.* 1982) and for mustard oils (Ellis *et al.* 1976). Metcalf *et al.* (1982) suggested that elimination of cucurbitacins might produce cucurbits that are less susceptible to attack by *Diabrotica* spp. and this was a useful objective for promoting host plant resistance by antixenosis (nonpreference).

An example of a chemical approach to incorporating insect resistance can be found in tobacco. Leaf surface chemistry differed between selections displaying some resistance to tobacco budworm, *Heliothis virescens* (Fabricius), and tobacco hornworm, *Manduca sexta* (L.), and the green peach aphid, *Myzus persicae* (Sulzer), and selections that were susceptible to these pests (Severson *et al.* 1985). Tobacco leaves with low concentrations of duvane and total diterpene showed some resistance to these insects. Using controlled cage studies, they found that the major mode of budworm resistance in tobacco was oviposition nonpreference. When leaf washes from susceptible lines were applied onto resistant lines, the resistant lines became susceptible. Duvatriene diols were found to be the active chemical components that stimulated oviposition. These results underscore both the validity and functional utility of this type of approach for sweet potato breeding programs.

The development of an analytical selection program for weevil resistance based on the chemistry of sweet potato cultivars was suggested by Kays and Horvat (1983) who listed the following steps: (1) isolate and identify the sweet potato surface and/or internal chemical constituents responsible for resistance or susceptibility; (2) ascertain the nature and level of resistance imparted by these compounds; and (3) develop appropriate analytical procedures for rapid screening of potential parent lines and crosses. The methodology for the rapid, qualitative and quantitative screening of sweet potato germplasm for oviposition stimulants has been developed by Son (1989). He developed an efficient method for measuring the concentration of boehmeryl acetate and other surface components in field-sampled sweet potato roots. He also determined the minimum surface area that could be sampled to give accurate results, differences in the distribution of surface components between proximal and distal ends of storage roots, differences in the cross-sectional distribution, and the stability of boehmeryl acetate in samples stored under various conditions.

Three periderm discs (surface area = 11.4 cm²) were removed using a cork borer (no. 14). Discs (3 mm thick) were put in a 20 ml scintillation vial containing 10 ml of methylene chloride and 30 μl of an internal standard (hexacosanol), sealed, and then extracted using an 8 minute ultrasonication. A portion of extract equivalent to that removed from a section of root surface (4 cm² area) was silylated and analyzed on a SE-54 fused silica capillary column using a splitless injection mode. The GC profile of extracts derived from the discs was similar to the surface extract from intact roots, indicating that the extract equivalent to 4 cm² of surface area (i.e., one disc) was adequate for quantitative analysis of surface components. Boehmeryl acetate existed only in the surface 1 to 1.2 mm of the periderm and not in the tissue beneath the periderm. The position on the root from which the disk was removed did not affect the concentration of boehmeryl acetate. Boehmeryl acetate present in root extracts did not degrade when held under ambient laboratory conditions for more than one month.

This technique for chemical analysis of root surface components is accurate, fast, relatively inexpensive, and amenable to assessment of a large number of genetic crosses. Thus, an analytical approach has the potential to be a powerful tool for sweet potato breeders, and may help to develop resistant cultivars more rapidly since technique development, especially in the use of chemical methodologies, has been one of the major restraints to progress in the field (Maxwell 1980). A possible plan for the use of chemical analyses would be to first prescreen a large number of clones and then select promising clones based on the analyses. Selections should then be tested against weevils under controlled laboratory conditions using the core/tissue culture plate assembly as described previously. Selected clones should then be evaluated in the field and results compared to laboratory tests to correlate laboratory and field resistance.

REFERENCES

Barlow, T. & L.H. Rolston. 1981. Types of host plant resistance to the sweet potato weevil found in sweet potato roots. J. Kansas Entomol. Soc. 54: 649-657.

Chambliss, O.L. & C.M. Jones. 1966. Cucurbitacins: specific insect attractants in Cucurbitaceae. Science 153:1392-1393.

Ellis, P.R., J.A. Hardman, P. Crisp, A.G. Johnson & R.A. Cole. 1976. Resistance of brassicas and radish to cabbage root fly. Ann. Rep. Natl. Veg. Res. Stn. (U.K.) for 1975 26:88-90.

Jackson, R.D. & W.J. Lewis. 1981. Summary of significance and employment strategies for semiochemicals, pp. 283-295. In D.A. Nordlund, R.L. Jones & W.J. Lewis (eds.). Semiochemicals: their role in pest control. J. Wiley, New York.

Kays, S.J. & R.J. Horvat. 1983. Insect resistance and flavor chemistry: integration into future breeding programs. Am. Soc. Hort. Sci. Trop. Reg. 27(B):97-106.

Martin, F.W. 1984. Variation in the sweet potato reaction to the sweetpotato weevil, pp. 36-44. In M.A. Mullen & K.A. Sorensen (eds.). Proceedings of a sweetpotato weevil workshop. Department of Entomology, North Carolina State University, Raleigh.

Martin, F.W. & A. Jones. 1986. Breeding sweet potatoes, pp. 313-345. In J. Janick (ed.). Plant breeding reviews IV, Avi, Westport, Connecticut.

Maxwell, F. 1980. Future opportunities and directions, pp. 535-538. In F.G. Maxwell & P.R. Jennings (eds.). Breeding plants resistant to insects. J. Wiley, New York.

Metcalf, R.L., A.M. Rhodes, R.A. Metcalf, J. Ferguson, E.R. Metcalf & P.-Y. Lu. 1982. Cucurbitacin contents and diabroticite (Coleoptera: Chrysomelidae) feeding upon *Cucurbita* spp.. Environ. Entomol. 11:931-937.

Mullen, M.A., A. Jones, R. Davies & G.C. Pearman. 1980. Rapid selection of sweet potato lines resistant to the sweetpotato weevil. HortScience 15:70-71.

Nottingham, S.F., D.D. Wilson, R.F. Severson & S.J. Kays. 1987. Feeding and oviposition preferences of the sweetpotato weevil, *Cylas formicarius elegantulus* (Summers), on the outer periderm and inner core of selected sweet potato cultivars. Entomol. Exp. Appl. 45:271-275.

Nottingham, S.F., K.-C. Son, D.D. Wilson, R.F. Severson & S.J. Kays. 1989. Feeding & oviposition preference of sweetpotato weevil, *Cylas formicarius elegantulus* (Summers) on storage roots of sweet potato cultivars with differing surface chemistries. J. Chem. Ecol. 15:895-903.

Oyarzun, M.L., J.A. Garbarino, V. Gambaro, J. Guilhem & C. Pascard. 1987. Two triterpenoids from *Boehmeria excelsa*. Phytochem. 26:221-223.

Renwick, J.A.A. 1983. Nonpreference mechanisms: plant characteristics influencing insect behavior. Am. Chem. Soc. Symp. Ser. 183:199-213.

Severson, R.F., A.W. Johnson & D.M. Jackson. 1985. Cuticular constituents of tobacco: factors affecting their production and role in insect and disease resistance and smoke quality. Recent Adv. Tob. Sci. 11:105-174.

Son, K.-C. 1989. Methodology for the rapid screening of sweet potato genotypes for oviposition stimulants to the sweetpotato weevil, *Cylas formicarius elegantulus* (Summers). Ph.D. Dissertation, University of Georgia, Athens.

Son, K.-C., R.F. Severson & S.J. Kays. 1987. Extraction and isolation of surface components on sweet potato storage roots and leaves. J. Ga. Acad. Sci. 45:78.

Son, K.-C., R.F. Severson, R.F. Arrendale & S.J. Kays. 1990. Isolation and characterization of pentacyclic triterpene ovipositional stimulant for the sweetpotato weevil from *Ipomoea batatas* (L.) Lam.. J. Agric. Food Chem. 38:134-137.

Stadler, E. 1983. Attractants, arrestants, feeding and oviposition stimulants in insect-plant relationships: application for pest control, pp. 243-258. *In* D.L. Whitehead & W.S. Bowers (eds.). Natural products for innovative pest management. Pergamon, London.

Wilson, D.D., R.F. Severson, K.-C. Son & S.J. Kays. 1988. Oviposition stimulant in sweet potato periderm for the sweetpotato weevil, *Cylas formicarius elegantulus*. Environ. Entomol. 17:691-693.

Wilson, D.D., K.-C. Son, R.F. Severson & S.J. Kays. 1990. Role of a pentacyclic triterpene found on the surface of sweet potato storage roots in oviposition by the sweetpotato weevil, *Cylas formicarius elegantulus* (Summers). Environ. Entomol. (In press).

Wilson, D.D., K.-C. Son, S.F. Nottingham, R.F. Severson & S.J. Kays. 1989. Characterization of an oviposition stimulant from the surface of sweet potato storage roots for the sweetpotato weevil, *Cylas formicarius elegantulus*. Entomol. Exp. Appl. 51:71-75.

12. Volatile Chemicals from Sweet Potato and Other *Ipomoea*: Effects on the Behavior of *Cylas formicarius*

Christopher K. Starr, Ray F. Severson[1], and Stanley J. Kays
Department of Horticulture
University of Georgia
Athens, Georgia 30602 U.S.A.

The most prominent component of chemical ecology has to do with the effects of plant chemicals on the behavior and development of phytophagous insects (Bell & Cardé 1984, Harborne 1988). Most of the chemicals involved are secondary metabolites -- presumed to have originated as by-products of plant metabolism -- of which there exist an estimated 50,000 to 400,000 (Metcalf 1987, Schoonhoven 1982). These include both volatile and non-volatile compounds which may mediate plant/insect interactions as attractants, repellents, stimulants or deterrents to feeding and/or oviposition. The subject of this chapter is the volatile compounds given off by one group of plants, the genus *Ipomoea*, and their influence on the behavior of one insect, the sweetpotato weevil, *Cylas formicarius* (Fabricius).

Volatility is an important property in the action of communicative compounds, or semiochemicals, both within and between species. Molecules that can be transported in air usually have fewer than 20 carbons and molecular weights of less than 300 (Wilson & Bossert 1963). Insects are known to respond to 64 plant volatiles that vary in molecular weight from 99 to 222 and in boiling point from 250° to 340°C (Finch 1980, Metcalf 1987, Visser 1986).

The genus *Ipomoea* (Convolvulaceae) comprises about 500 species worldwide (see Chapter 3), including the sweet potato, *I. batatas* (L.) Lam., water spinach, *I. aquatica* Forsk., and an array of species known

---

[1] U.S. Department of Agriculture, Agricultural Research Service, Tobacco Quality and Safety Research Unit, Athens, Georgia 30613 U.S.A.

as "morning glories". A majority of species are climbing or creeping vines, but shrub-like plants and trees are also included in the genus.

Sweet potato is an important staple crop in much of the world, while water spinach is cultivated as a vegetable in some areas. Although the nutritional composition of these two edible species has been the subject of considerable study (e.g., Collazos 1967, Watt & Merrill 1975), the volatile chemistry of *Ipomoea* spp. has until recently received little attention. The practical value of such research is in understanding how some varieties and species attract insects which may become pests. Of particular interest are beetles of the genus *Cylas* (Brentidae: Apionidae; Kuschel, in press, also see Chapter 2), the most serious insect pests of sweet potato.

The best known member of the genus, *C. formicarius*, feeds and reproduces on a variety of *Ipomoea* species worldwide (Chalfant *et al.* 1990, also see Chapters 2 and 3) and is the most important pest of cultivated sweet potato. Adults feed on the leaves, stems, and roots of the plant and lay eggs in the stems and roots, where larvae develop to maturity. Sutherland (1986) and Chalfant *et al.* (1990) have reviewed the biology and management of *C. formicarius*.

We report here on responses of *C. formicarius* to plant volatiles and on recent advances in the volatile surface chemistry of *I. batatas*. The *Ipomoea/Cylas* relationship appears to be a promising model for studying interactions between oligophagous insects and their host-plants. Also, a better understanding of *Ipomoea*-derived volatile compounds that modulate the behavior of *C. formicarius* may help to improve management programs for this pest.

RESPONSES OF *Cylas formicarius* TO *Ipomoea* ODOR

No sweet potato cultivar is known to be immune to *C. formicarius*, but cultivars are not equally susceptible to damage (Barlow & Rolston 1981, Mullen *et al.* 1980, 1981, 1982, 1985, Rolston *et al.* 1979). In what follows, we use "susceptible" and "resistant" to refer to cultivars with relatively high and low susceptibility to sweetpotato weevil, respectively.

Based upon other plant/insect interactions, differences in susceptibility to the weevil may be largely due to chemical differences among cultivars, which could result in variation in initial attraction, host choice, the weevils' success in utilizing different plants, or a combination of these factors. This possibility was first explored by Barlow & Rolston (1981); however, they found no evident inhibition of feeding or oviposition among various cultivars. They implied that resistance was probably due to lower levels of attractant and/or feeding-stimulant chemicals.

Nottingham *et al.* (1989a) used a dual-choice olfactometer (Figure 12.1) to compare responses of *C. formicarius* to leaf and storage-root volatiles from the highly susceptible sweet potato cultivar, Centennial, the moderately susceptable cultivar, Jewel, the moderately resistant cultivars, Regal and Resisto, and a non-host plant, *Plectranthus tuberosus* Poir Chev. et. Perrot. They tested female and male *C. formicarius* separately in both choice (*I. batatas* vs. *P. tuberosus*) and no-choice (*P. tuberosus* vs. a blank) tests, with the following results:

1. In a leaf-odor choice test of 'Jewel' vs. *P. tuberosus*, both adult female and male *C. formicarius* showed significantly greater attraction to 'Jewel'. Using an attraction index (AI) consisting of the percentage of weevils responding to the preferred treatment, 'Jewel', minus the percentage responding to the non-preferred treatment, *P. tuberosus*, they obtained AI values of 41.6 and 45.2 for females and males, respectively.
2. In a no-choice test of *P. tuberosus* leaf odor vs. a blank control, neither female nor male weevils showed a significant preference.
3. In leaf-odor choice tests of each of the four sweet potato cultivars vs. *P. tuberosus*, female weevils showed significantly greater attraction to sweet potato cultivars in each case (AI = 19.3 for 'Regal', 34.0 for 'Centennial', 40.7 for 'Resisto', and 46.7 for 'Jewel').
4. In no-choice tests of a methylene chloride extract from leaf surfaces of 'Jewel' vs. a solvent (methylene chloride) control, both female and male weevils showed a significantly greater attraction to a 1-leaf and a 0.2-leaf equivalent than to the control, but response to a 0.02-leaf equivalent extract did not differ from that to the control.
5. In no-choice tests of root odor of each of the four sweet potato cultivars vs. a blank control, female weevils showed significantly greater attraction to sweet potato root volatiles in each case (AI = 26.7 for 'Resisto', 30.0 for 'Centennial', 33.3 for 'Regal', and 52.0 for 'Jewel'). In similar tests of 'Centennial' and 'Jewel' root odor, male weevils showed no preference for either (AI = 7.2 and 6.8, respectively).

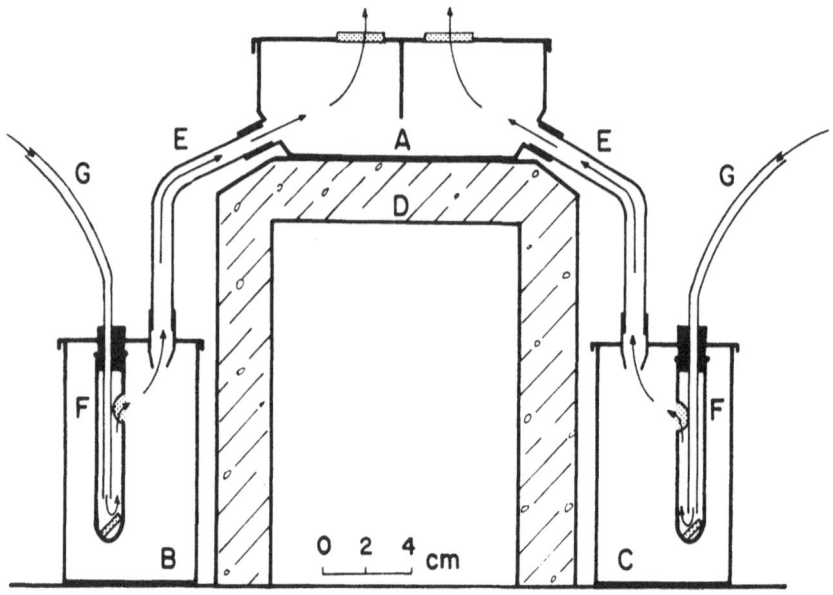

Fig. 12.1. Diagrammatic cross-section of dual-choice olfactometer used to compare attraction of *C. formicarius* to odor from various substrates. In each trial 25 female weevils were starved for 2 hours and then released in the main chamber. After an additional 2 hours, the numbers in the side chambers (B and C) were counted. Weevils found in the tubes that connected to the side chambers were regarded as having made a choice and were counted along with those in their respective side chambers. Arrows indicate the direction of air flow. The arrangement shown here was used for tests of extracts with treated filter paper placed in the test tubes in the side chambers. For tests of leaves, the test tubes were detached and placed on the floor of the side chambers, where they served as vases. For tests with roots, the test tubes were removed and the roots were placed on the floor of the side chamber. A. Main chamber. B. Treatment chamber. C. Control chamber (from Nottingham *et al.* 1989a; reprinted with permission from J. Chem. Ecol. [Plenum Publishing Corp., New York]).

6. In no-choice tests of a methylene chloride extract from 'Jewel' root surfaces vs. a solvent control, female weevils showed a significantly greater response to a 10-root and a 1-root equivalent than to the control, but their response to a 0.1-root equivalent extract did not differ from their response to the control. The attraction of males did not differ between 10-root or 1-root equivalents and the control.

These results suggest the following: 1) females respond to leaf and root volatiles from all tested sweet potato cultivars; 2) female response differs among cultivars; and 3) males respond to volatiles from leaves but apparently not those from roots.

In assays of feeding and oviposition preferences on storage roots and leaves, Nottingham *et al.* (1987, 1988, 1989b) found similar differences among cultivars. Thus, initial attraction and subsequent utilization may each contribute to variation in susceptibility. It remains to be seen whether *C. formicarius* shows comparable variation in its ability to reproduce in different sweet potato cultivars in the absence of a choice.

The range of host plants acceptable to *C. formicarius* has yet to be established, but present indications are that the weevil is restricted predominately to the family Convolvulaceae (see Chapter 3). It has been recorded feeding and/or breeding on at least 27 species of *Ipomoea* and 8 species from related genera (Austin *et al.* 1990, see Chapter 3).

In a modification of the olfactometer assay devised by Nottingham *et al.* (1989a), no-choice assays were conducted to determine the response of female weevils to volatiles from leaves ten *Ipomoea* spp. and a blank control. Of these species, nine have been recorded as host to the weevil: *I. alba, I. aquatica, I. batatas, I. hederifolia, I. nil, I. pandurata, I. pes-caprae, I. purpurea,* and *I. setosa* (Austin *et al.* 1990, Chalfant *et al.* 1990, Jansson *et al.* 1989, also see Chapter 3). The host status of *I. carnea* is unknown. Our experimental method differred from that of Nottingham *et al.* (1989a) in two ways: 1) we conducted assays under dim red light, rather than in total darkness; and 2) in tests with small-leaved plant species, we used two or three leaves together in each trial. These modifications were intended to simplify the experiment and it is unlikely that they affected the outcome. For *I. batatas*, we used the moderately susceptible cultivar, Jewel.

Figure 12.2 shows response data from two of the ten species, in which dosage was computed as [leaf area per side (cm^2)/rate of air flow (ml. min.$^{-1}$)]. The data illustrate two general results. First, contrary

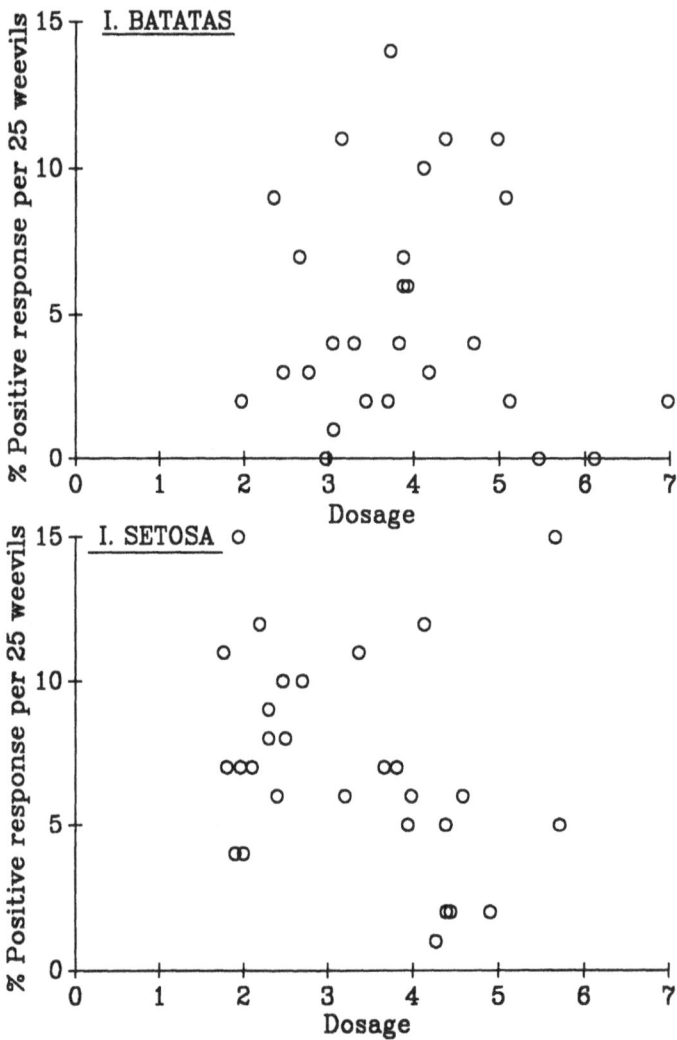

Fig. 12.2. Number of female *C. formicarius* (out of 25) responding to leaf odor from two *Ipomoea* species (*I. batatas* [top] and *I. setosa* [bottom]) in relationship to leaf volatile dosage. For simplicity, response to the blank control is omitted. Each data point corresponds to one trial.

to our expectation, we found no significant increase in weevil response with an increase in dosage of leaf volatiles (analysis of covariance, $P > 0.05$) in any of the species. Variation in dosage could therefore be discounted as a factor within the range of dosages that were tested. Second, the response of weevils to a given species was not uniform among trials. The variance in response, which is illustrated in Figure 12.2, was typical of all ten species (Table 12.1).

As expected, the weevils did not respond equally to leaf volatiles from all plant species (Table 12.1; analysis of variance of ranked AI data; $F = 6.91$; $P < 0.05$). Four species, *I. batatas*, *I. carnea*, *I. nil*, and *I. pes-caprae*, elicited responses significantly greater than the control (Bonferroni *t*-test of mean AI ranks; $P < 0.05$). Furthermore, eight of 45 pairwise comparisons between species were significant (Bonferroni *t*-test of mean AI ranks, $P < 0.05$). These were: *I. batatas* > *I. aquatica*; *I. carnea* > *I. alba*; *I. carnea* > *I. aquatica*; *I. carnea* > *I. hederifolia*; *I. carnea* > *I. purpurea*; *I. carnea* > *I. setosa*; *I. nil* > *I. aquatica*; and *I. pes-caprae* > *I. aquatica*. It is our expectation that similar data from a broader range of species within the Convolvulaceae, coupled with analysis of surface volatiles, will help to understand what makes some plants more attractive to this particular insect.

Adult male and female *C. formicarius* were not equally attracted to root and leaf volatiles. Females were strongly attracted to both leaf and root volatiles, whereas males were attracted to leaf volatiles, but were not attracted to root volatiles. There are considerable morphological differences between the sexes of *C. formicarius* in their putative olfactory apparati (C.K. Starr *et al.*, unpubl. data) which may correlate with host finding behavior.

VOLATILES OF *Ipomoea batatas*

At present no class of chemicals is definitely known to be involved in the differential attraction of *C. formicarius* to *Ipomoea* spp. There are several lines of evidence, however, that point towards terpenoids as possible candidates. In contrast to the leaf surfaces of many plant species, *I. batatas* has copious amounts of triterpenes (S.J. Kays & R.F. Severson, unpubl. data). In addition, the ovipositional stimulant for *C. formicarius* found on the surface of the storage roots is a pentacyclic triterpene and weevil larvae induce the synthesis of a series of furanoterpenoids within the roots. It would appear, therefore, that terpene chemistry might be involved in the coevolution of the two species.

Table 12.1. Attractiveness of leaf volatiles from ten *Ipomoea* species and a blank control to *C. formicarius.*

Ipomoea species	Number of trials	Mean dosage ($\pm$ SE)	Response to plant, % ($\pm$ SE)	Response to control, % ($\pm$ SE)	AI ($\pm$ SE) [a]
I. alba	20	3.69 $\pm$ 0.26	21.6 $\pm$ 3.3	4.2 $\pm$ 1.2	17.4 $\pm$ 3.8
I. aquatica	32	3.92 $\pm$ 0.15	14.6 $\pm$ 2.8	7.0 $\pm$ 1.8	7.6 $\pm$ 3.5
I. batatas	30	3.23 $\pm$ 0.22	29.1 $\pm$ 2.6	2.9 $\pm$ 0.8	26.2 $\pm$ 2.4
I. carnea	29	3.11 $\pm$ 0.13	44.6 $\pm$ 3.4	3.0 $\pm$ 1.2	41.5 $\pm$ 3.9
I. hederifolia	30	3.30 $\pm$ 0.13	30.4 $\pm$ 3.2	11.6 $\pm$ 1.8	18.8 $\pm$ 4.2
I. pandurata	15	3.53 $\pm$ 0.27	27.7 $\pm$ 5.5	4.0 $\pm$ 1.6	23.7 $\pm$ 5.8
I. nil	30	4.53 $\pm$ 0.23	36.8 $\pm$ 3.3	14.4 $\pm$ 2.7	22.4 $\pm$ 5.4
I. pes-caprae	27	2.36 $\pm$ 0.12	34.4 $\pm$ 3.4	8.4 $\pm$ 1.8	25.9 $\pm$ 4.2
I. purpurea	30	3.52 $\pm$ 0.12	16.7 $\pm$ 3.0	5.3 $\pm$ 1.6	11.4 $\pm$ 3.6
I. setosa	32	3.78 $\pm$ 0.20	18.9 $\pm$ 2.8	3.1 $\pm$ 1.2	15.8 $\pm$ 3.0
Blank [b]	15	0.00 $\pm$ 0.00	5.6 $\pm$ 1.6	3.2 $\pm$ 1.4	2.4 $\pm$ 2.0

[a] Attraction index (see text for explanation).

[b] Both choices chambers were empty; one side was randomly designated as control and the other plant in each trial.

Terpenoids are a large, heterogeneous group of organic compounds based on a branched C_5 building block. Those of both lower [hemiterpenes (C_5), monoterpenes (C_{10}) and sesquiterpenes (C_{15})] and higher molecular weights [biterpenes (C_{20}) and triterpenoids (C_{30})] are widely distributed among plants and are frequently involved in modifying insect behavior (Harborne 1988, Metcalf 1987, Pant & Rastogi 1979). Several of these compounds are known to deter leaf-feeding insects (e.g., Howard *et al.* 1989). More importantly, several sesquiterpenes are known to act as kairomones, and help insect herbivores locate their host plants (Metcalf 1987).

The volatile fractions of storage roots and leaf surfaces of *I. batatas* contain several sesquiterpenes (Figure 12.3). Five of these have been identified as copaene, trans-caryophyllene, gamma-humulene, gamma-cadinene, and gamma-elemene (Nottingham *et al.* 1989a). There are distinct differences between roots and leaves in the concentrations of these compounds (Nottingham *et al.* 1989a). Nottingham *et al.* (1988) and Son (1989) also found large differences in the leaf-surface chemicals between cultivars with differing susceptibilities to *C.*

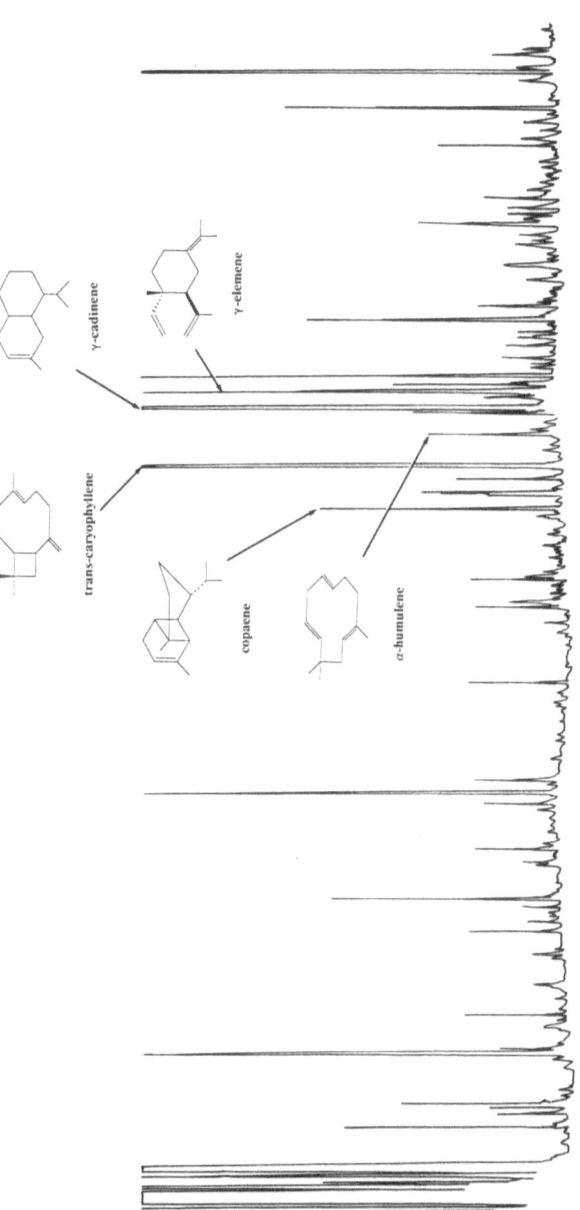

Fig. 12.3. Gas chromatogram of methylene–chloride extract from the surface of *I. batatas* (cv. Jewel) leaves. Sesquiterpenes identified in the volatile component are identified adjacent to their respective peaks.

formicarius. More work is needed, however, to determine the chemical(s) that attract the weevil to its host plants.

CONCLUSIONS

Investigations of the volatile chemistry of *Ipomoea* and its impact on insect behavior can be largely characterized as basic research. How an insect chooses among several potential host plants with differing apparency and unequal value as breeding sites is a compelling question. Comparative assays and chemistry in the *Ipomoea/C. formicarius* system can contribute to the elucidation of this more general problem. From a practical standpoint, volatile attractants from *Ipomoea* may be useful for trapping *C. formicarius* as part of an integrated pest management program. Current work with volatile attractants utilizes the female-produced sex pheromone of *C. formicarius* (see Chapters 5 and 6); however, this chemical only traps adult males. Since certain leaf volatiles from *I. batatas* attract both males and females, their identification and use may enhance the sex pheromone-based trapping systems currently in use world-wide.

ACKNOWLEDGMENTS

This research was funded by U.S.A.I.D. (Project no. 88-CSRS-2-3236 [to S.J.K. and R.F.S.). We thank D.F. Austin for identification of plants and T. Richardson for statistical advice.

REFERENCES

Austin, D.F., R.K. Jansson & G.W. Wolfe. 1990. Convolvulaceae and *Cylas*: a proposed hypothesis on the origins of this plant: insect relationship. Trop. Agric. (In press).

Barlow, T. & L.H. Rolston. 1981. Types of host plant resistance to the sweetpotato weevil found in sweet potato roots. J. Kansas Entomol. Soc. 54:649-657.

Bell, W.J. & R.T. Cardé (eds.). 1984. Chemical ecology of insects. Sinauer Assoc., Sunderland, Massachusetts.

Chalfant, R.B., R.K. Jansson, D.R. Seal & J.M. Schalk. 1990. Ecology and management of sweet potato insects. Annu. Rev. Entomol. 35:157-180.

Collazos Ch., C. 1967. La composición de los alimentos peruanos. 2nd ed. Ministerio de la Salud Pública y la Asistencia Social, Lima, Peru.

Finch, S. 1980. Chemical attraction of plant-feeding insects to plants, pp. 23-63. *In* J.R. Miller & T.A. Miller (eds.). Insect-plant interactions. Springer, New York.

Harborne, J.B. 1988. Introduction to ecological biochemistry. 3rd ed. Academic, New York.

Howard, J.J., T.P. Green & D.F. Wiemer. 1989. Comparative deterrency of two terpenoids to two genera of attine ants. J. Chem. Ecol. 15:2279-2288.

Jansson, R.K., A.G.B. Hunsberger, S.H. Lecrone, D.F. Austin & G.W. Wolfe. 1989. *Ipomoea hederifolia*, a new host record for the sweetpotato weevil, *Cylas formicarius elegantulus* (Coleoptera: Curculionidae). Fla. Entomol. 72:551-553.

Kuschel, G. 1990. A phylogenetic classification of Curculionoidea to families and subfamilies. Mem. Entomol. Soc. Wash. (In press).

Metcalf, R.L. 1987. Plant volatiles as insect attractants. Crit. Rev. Plant Sci. 5:251-301.

Mullen, M.A., A. Jones, D.R. Paterson & T.E. Boswell. 1982. Resistance of sweet potato lines to the sweetpotato weevil. HortScience 17:931-932.

Mullen, M.A., A. Jones, D.R. Paterson & T.E. Boswell. 1985. Resistance in sweet potatoes to the sweet potato weevil, *Cylas formicarius elegantulus* (Summers). J. Entomol. Sci. 20:345-350.

Mullen, M.A., A. Jones, R.T. Arbogast, D.R. Paterson & T.E. Boswell. 1981. Resistance of sweet potato lines to infestations of sweetpotato weevil, *Cylas formicarius elegantulus*. HortScience 16:539-540.

Mullen, M.A., A. Jones, R.T. Arbogast, J.M. Schalk, D.R. Paterson, T.E. Boswell & D.R. Earhart. 1980. Field selection of sweet potato lines and cultivars for resistance to the sweetpotato weevil. J. Econ. Entomol. 73:288-290.

Nottingham, S.F., D.D. Wilson, R.F. Severson & S.J. Kays. 1987. Feeding and oviposition preferences of the sweet potato weevil, *Cylas formicarius elegantulus*, on the outer periderm and exposed inner core of storage roots of selected sweet potato cultivars. Entomol. Exp. Appl. 45:271-275.

Nottingham, S.F., K.-C. Son, D.D. Wilson, R.F. Severson & S.J. Kays. 1988. Feeding by adult sweet potato weevils, *Cylas formicarius elegantulus*, on sweet potato leaves. Entomol. Exp. Appl. 48:157-163.

Nottingham, S.F., K.-C. Son, R.F. Severson, R.F. Arrendale & S.J. Kays. 1989a. Attraction of adult sweet potato weevils, *Cylas formicarius elegantulus* (Summers) (Coleoptera: Curculionidae), to sweet potato leaf and root volatiles. J. Chem. Ecol. 15:1095-1106.

Nottingham, S.F., K.-C. Son, D.D. Wilson, R.F. Severson & S.J. Kays. 1989b. Feeding and oviposition preferences of sweet potato weevil, *Cylas formicarius elegantulus* (Summers), on storage roots of sweet potato cultivars with differing surface chemistries. J. Chem. Ecol. 15:895-903.

Pant, P. & R.P. Rastogi. 1979. The triterpenoids. Phytochem. 18:1095-1108.

Rolston, L.H., T. Barlow, T. Hernandez & S.S. Nilahke. 1979. Field evaluation of breeding lines and cultivars of sweet potato for resistance to the sweet potato weevil. HortScience 14:634-635.

Schoonhoven, L.M. 1982. Biological aspects of antifeedants. Entomol. Exp. Appl. 31:57-69.

Son, K.-C. 1989. Phytochemistry of the sweet potato, *Ipomoea batatas* (L.) Lam., storage root in relation to susceptibility to the sweet potato weevil, *Cylas formicarius elegantulus* (Summers). Ph.D. Dissertation, University of Georgia, Athens.

Sutherland, J.A. 1986. A review of the biology and control of the sweetpotato weevil *Cylas formicarius* (Fab.). Trop. Pest Manage. 32:304-315.

Visser, J.H. 1986. Host odor perception in phytophagous insects. Annu. Rev. Entomol. 31:121-144.

Watt, B.K. & A.L. Merrill. 1975. Composition of foods: raw, processed, prepared. U.S.D.A. Agric. Handb. No. 8.

Wilson, E.O. & W.H. Bossert. 1963. Chemical communication among animals. Rec. Prog. Hormone Res. 19:673-716.

13. Quarantine Programs for *Cylas formicarius* in the United States

Shashank S. Nilakhe
Texas Department of Agriculture
P.O. Box 12847
Austin, Texas 78711 U.S.A.

The objectives of insect pest management are generally achieved by preventing the establishment or spread of insect pests, by controlling established pest infestations, or by keeping infestations at levels at which little or no damage occurs (NAS 1969). Quarantine regulations and actions may utilize one or all of these three approaches including eradication efforts in some special situations.

The sweetpotato weevil, *Cylas formicarius* (Fabricius), was first reported in the United States from New Orleans, Louisiana in 1875 (Newell 1917), and by 1923, it had spread to most of the southern United States (Edmond & Ammerman 1971, Reinhard 1923). As information on the insect's biology was collected in the early 1900's (e.g., Conradi 1907), it was noted that weevils dispersed primarily by transporting infested sweet potato storage roots and plant parts for propagation into weevil-free areas. It was also recognized that areawide efforts were needed to halt or reduce the spread of the weevil, and participation of entire communities was essential. As a result, quarantine regulations for the weevil were enacted.

FEDERAL REGULATIONS

The Plant Protection and Quarantine (PPQ) section of the Animal and Plant Health Inspection Service branch of the United States Department of Agriculture is responsible for regulating the sweet potato shipments from other countries into the United States. PPQ classifies plant products in two groups: propagative and non-propagative. Propagative sweet potatoes include plants, vines, cuttings, draws, slips, storage roots, or true seed used for propagation, and non-propagative sweet potatoes include primarily storage roots, and perhaps leaves, used for consumption.

Sweet Potatoes for Propagation

The entrance of sweet potato from all countries and localities, except Canada, into the United States is prohibited under the authority of the Code of Federal Regulations (CFR), Title 7, Chapter 3, part 319.37. This rule was formulated to prevent a possible introduction of a diversity of diseases. One exception to this rule is that sweet potato true seed from other countries is permitted entrance into the United States. PPQ officers are responsible for examining imported true seed; seed that enters through the Canadian border ports is inspected by state or county officials. A phytosanitary certificate from the country of origin should accompany the seed shipment. The certificate should attest that the seed is free of any pests entering the United States as stipulated under the authority of CFR, Title 7, Chapter 3, part 319.37-2. Information on where the sweet potatoes were grown and the treatments that the seed were subjected to should also be included on the certificate.

PPQ's Plant Import Manual (USDA 1987a) lists seeds of plant species that are prohibited or have restrictions in addition to inspection and a written permit. There are restrictions on importation of water-spinach morning glory, *Ipomoea aquatica* Forskal, and little bell morning glory, *I. triloba* L., but not on sweet potato seed.

Non-Propagative Sweet Potatoes

Fruits and vegetables produced in other countries that are permitted entrance into the continental United States are listed in PPQ's "Non-propagative Manual" (USDA 1987b). Because sweet potato is not included on the list, it is not permitted to enter into the continental United States.

Separate regulations apply for importing sweet potato storage roots into Puerto Rico, U.S. Virgin Islands and the other United States territories. No permit is required for roots imported into Guam and the Commonwealth of Northern Mariana Islands from Japan and South Korea. Likewise, roots can be imported into the U.S. Virgin Islands from Antigua, Bahamas, Barbados, Cayman Islands (U.K.),

Cuba, Dominica, Dominican Republic, Grenada, Guadeloupe, Haiti, Jamaica, Martinique (France), Montserrat (U.K.), Nevis, St. Barthelemy (France), St. Eustatius (Neth.), St. Kitts, St. Lucia, St. Vincent, St. Martin (Neth. and France) and Virgin Islands (U.K.). Storage roots imported into Puerto Rico from most of these island countries require a specified fumigation treatment. However, no fumigation treatment is required when roots are shipped from the Dominican Republic to Puerto Rico.

Storage roots should be cured and free from surface moisture, prior to fumigation. Roots should be fumigated in a chamber at normal atmospheric pressure using methyl bromide at the rate of 40 g/m³ at 32-35.5°C, 48 g/m³ at 26.5-31.5°C, 56 g/m³ at 21-26°C, or 64 g/m³ at 15.5-20°C for four hours (USDA 1985). Storage roots should be held at the fumigation temperature for 24 hours post-treatment. Currently, few storage roots are fumigated because of phytotoxicity and high costs. Storage roots produced in Hawaii, Puerto Rico, and the U.S. Virgin Islands can be shipped to the continental United States provided that roots are fumigated with the approved treatment and certified accordingly; in some situations, a PPQ officer may issue a permit to fumigate storage roots upon arrival at a designated northern port of the United States.

Hawaii, Puerto Rico, and the U.S. Virgin Islands are quarantined to prevent spread of the West Indian sweetpotato weevil or scarabee, *Euscepes postfasciatus* (Fairmaire), and the sweet potato vineborer, *Omphisa anastomosalis* Guenee. There are no restrictions for movement of roots between Puerto Rico and the U.S. Virgin Islands. Storage roots grown in Puerto Rico according to conditions outlined in CFR, Title 7, Chapter 3, Part 318.30a can be shipped to certain parts of the northern United States without a fumigation treatment. Shipment of plant parts other than storage roots from Hawaii, Puerto Rico, and the U.S. Virgin Islands into the continental United States is prohibited.

Current information regarding import and export of sweet potato plant parts can be obtained at the PPQ office in Hyattsville, Maryland. The customs personnel at the ports of entry in the United States check tourists' baggage and/or inquire about transport of any plant materials. Sweet potato plants are refused entry; this process helps prevent entrance of sweet potato pests into the United States, especially sweetpotato weevil.

STATE REGULATIONS

Movement of sweet potatoes within the continental United States generally falls under the jurisdiction of state governments. Each state that has concern about sweetpotato weevils, has identified counties and localities within the state that are weevil-infested or weevil-free (noninfested) and enacted regulations to prevent the spread of the insect to localities where it is not present. Regulations have been established by either a State Plant Board, State Department of Agriculture, or other appropriate state authority. The United States Department of Agriculture cooperates fully with the state authorities in controlling and preventing the spread of the weevil. State quarantines are rather lengthy documents and only salient features will be presented. Appropriate state authorities should be contacted when storage roots are shipped either within or between states.

History of the Sweetpotato Weevil Quarantines

Florida was the first state in the United States to form quarantine regulations against the sweetpotato weevil. Careful studies conducted by the Florida State Plant Board showed that the weevils were constantly introduced into Florida through shipments of sweet potato plants and storage roots from other states and countries. Quarantine regulations were initiated on July 23, 1917, which prohibited the importation of sweet potato and morning glory storage roots, plants, vines, and cuttings into Florida (Newell 1917). Eight weevil-infested counties of Florida were quarantined on September 13, 1917. Storage roots produced in these counties had to be fumigated in order to ship or sell these roots in noninfested counties of the state. Sweet potato plants produced in the noninfested areas were certified. Alabama, Arizona and Georgia formed quarantine regulations against the weevil in 1918; Virginia formed quarantine programs in 1919; Arkansas and Mississippi initiated programs in 1920; and South Carolina formed quarantines in 1922 (Reinhard 1923). Currently, 13 states have specific quarantines or regulations against the sweetpotato weevil. Because state quarantines are legal documents, each term used is defined meticulously. Definitions of key terms used in quarantines are given in Table 13.1.

Those states not having specific weevil quarantines examine sweet potato storage roots and plants shipped under a general pest

Table 13.1. Definitions of key terms used in quarantines for *C. formicarius*.

Pest: the sweetpotato weevil in any stage of development.

Infestation: the presence of the weevil or existence of circumstances that make it reasonable to believe that the weevil is present in any life stage.

Inspector: any authorized employee of a state who enforces the quarantine regulations.

Regulated articles: sweet potato plant parts including whole plants, roots, vines, and cuttings and containers used in handling these articles; vines or roots of other *Ipomoea* plants and any other host plants of the weevil.

Certificate: a document issued by an inspector to allow movement of regulated articles to any destinations.

Quarantined area/regulated area/infested area: counties/civil divisions within each state in which the weevil is known to occur.

Limited permit: a permit issued by an inspector to indicate that sweet potato storage roots were grown in a regulated area, but have been inspected and found free of the weevils.

Compliance agreement: a written agreement between those engaged in growing, dealing in, processing or moving regulated articles and the state authority to comply with certain conditions to prevent spread of the weevil.

Control area: that portion of a regulated area where control measures against the weevil may be used but eradication is not the objective.

Eradication area: that portion of a regulated area where active work and control measures are applied and eradication is the immediate objective.

Scientific permit: a document issued by the state authority to allow the movement to a specific destination of the regulated articles for scientific purposes.

---

quarantine or plant import regulation (e.g., Maryland, New Jersey and Virginia).

Weevil Distribution in the United States

The distribution of the weevil in the continental United States is shown in Figure 13.1. Edmond and Ammerman (1971) presented a similar map showing the weevil-infested areas. Significant additions to the infested areas over the last 18 years include ten counties in Mississippi (Mississippi Dept. Agric. Comm. 1976), a portion of Union County in Arkansas (Ark. State Pl. Bd. 1983), Carter County in Kentucky (California Dept. Food Agric. 1987), and portions of

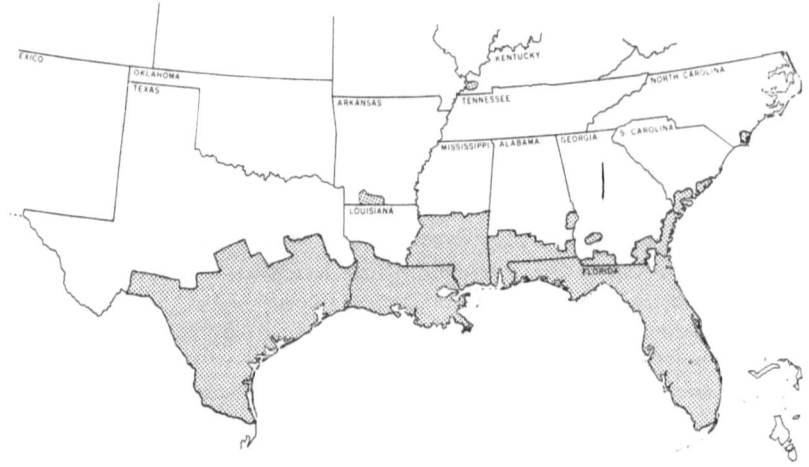

Fig. 13.1. Distribution of *C. formicarius* (shaded areas) in the continental United States.

Brunswick and New Hanover Counties of North Carolina (North Carolina Administrative Code 1989, W. Dickerson, pers. comm.).

Quarantines of Weevil-Free States

California, New Mexico, Oklahoma, and Tennessee are weevil-free states (Calfornia Dept. Food Agric. 1987, New Mexico Dept. Agric. 1969, Oklahoma Dept. Agric. 1978, Tennessee Dept. Agric 1957). Moreover, California lists Tennessee as a quarantined state, whereas other state quarantines consider it weevil free. Nevertheless, these four states do not allow entry of any sweet potato plant parts into their states except storage roots. Roots originating from infested areas of other states are permitted entrance provided they are fumigated with an approved treatment and protected from reinfestation thereafter. Storage roots from any noninfested areas may enter without fumigation provided that a certificate stating that the shipment originated in a noninfested area and precautions were taken during shipping to prevent the weevil infestation accompanies the shipment.

Quarantines of Weevil-Infested States

Infested and noninfested areas occur in Alabama, Arkansas, Georgia, Louisiana, Mississippi, North Carolina, South Carolina and Texas. The entire state of Florida is considered infested (Figure 13.1). Propagative sweet potato storage roots and other plant parts grown in infested areas of other states are not permitted entrance into these states (Georgia Dept. Agric. 1986, North Carolina Administrative Code 1989). Storage roots grown in infested areas for consumption must be fumigated using an approved treatment and shipped immediately after fumigation. The roots and the truck should be thoroughly cleaned before loading. Each sack, package, or crate of roots should be properly sealed. The truck body should also be sealed with tarpaulin or other appropriate materials. Care should be taken during transportation to avoid reinfestation. A certificate must accompany the shipment.

Storage roots grown in weevil-free areas of other states are permitted entrance without fumigation, but a certificate must accompany the shipment. Some states require colored tags to denote the origin of the storage roots. For example, roots produced in a weevil-free area are identified with a green tag on a container, and a manilla tag indicates that the roots originated from a regulated area (Alabama Dept. Agric. Indus. 1983, Mississippi Dept. Agric. Comm. 1976).

Movement of sweet potato storage roots for consumption within regulated areas of a state is allowed using a limited permit. Movement of roots from a regulated to a nonregulated area within a state requires a fumigation treatment; however, this treatment requirement may be waived in some states (e.g., North Carolina). Within a state, roots originating from a nonregulated area can be transported through a regulated area provided they are properly sealed and protected against weevil infestation during transit. Sweet potato roots produced in a weevil-free area can be moved anywhere within a state without a certificate, but the seller and buyer must know the origin of the shipment (Alabama Dept. Agric. Indus. 1983, Mississippi Dept. Agric. Comm. 1976). However, storage roots produced in a weevil-free area of Louisiana require a certificate (Louisiana Dept. Agric. 1985). Generally, persons or firms involved with the sweet potato commerce in a quarantined area need to be registered with the state. Louisiana and Texas require that such persons and firms file a bond of $1,000 with the state (Louisiana Dept. Agric. 1985, Texas Dept. Agric. 1976).

In Georgia, roots grown in areas infested with weevils must be harvested before December 15 of the production year, and the sweet potato fields must be cleaned thoroughly by February 15 (Georgia Dept. Agric. 1982). Similar restrictions are enforced in Arkansas, Louisiana, Mississippi, and Texas. To prevent spread and reduce weevil numbers, North Carolina prohibits the growing of sweet potatoes in regulated areas of the state (North Carolina Administrative Code 1989). Louisiana also prohibits sweet potato production in areas highly conducive for weevil abundance. Such areas are designated as non-sweet potato areas of the state (Louisiana Dept. Agric. 1985).

Alabama and Georgia prohibit growing sweet potato seed and plants used for propagation on properties infested with weevils and within one mile of such properties (Alabama Dept. Agric. Indus. 1983, Georgia Dept. Agric. 1986). Owners of other properties in the regulated areas may produce seed and plants for use only on the properties where produced. Sweet potato seed that is free of weevils should be selected and treated with approved insecticides before storage, and before bedding. All host material in and around the bed where sweet potatoes are grown must be destroyed immediately after use of the plants and not later than August 1 (Alabama Dept. Agric. Indus. 1983).

Sweet potato seed and plants can be "state certified" provided they meet certain conditions (e.g., they should be grown in weevil-free areas). These sweet potatoes are inspected four times: two inspections (at least 30 d apart) are conducted when plants are growing in fields, and the remaining two are conducted when roots are in storage. Plants and roots are inspected for both weevils and several diseases. An inspector may deny a crop certification if any pests are found during the inspection. The state certified sweet potato seed and plants guarantee growers of a quality product. Producers of the certified seed and plants have to follow the recommended sanitary, cultural, and chemical control practices. Details of these practices and the inspection procedures are given in sweet potato plant regulations of various states (e.g., Georgia Dept. Agric. 1982, Mississippi Dept. Agric. Comm. 1976).

Weevil Eradication

The first efforts to eradicate *C. formicarius* were undertaken in Florida under the leadership of the state and the federal authorities in

1919 (Boyden 1927). Thereafter, various states have conducted eradication and control work with varying degrees of success. Between 1937 and 1951, control and eradication work was done in 106 infested counties in seven states. By 1951, 33 counties, and 12,327 of the 16,169 infested properties were reportedly free of weevils (Roberts 1952). However, according to the current state quarantines, the weevil occurs in some 262 counties of the continental United States, which is a substantial increase in the weevil-infested area.

Of 29 counties quarantined in Mississippi, 26 are in the eradication area and three in the control area. Likewise, Texas has nine counties in the eradication area and 90 in the control area. Regulations for growing and handling sweet potatoes in eradication areas are stricter than in control areas (Mississippi Dept. Agric. Comm. 1976, Texas Dept. Agric. 1976).

Some state officials have indicated an interest to develop a nationwide weevil eradication program. An Advisory Committee made up of representatives from industry (sweet potato growers, and packers), state and federal research organizations (land-grant universities, U.S. Department of Agriculture, State Departments of Agriculture), and extension specialists could be formed to address the feasibility of such a program.

Removal of Areas from Quarantines

Typically, most growers that produce sweet potatoes in regulated counties or areas are interested in changing the status of their areas to "weevil-free." The benefits of producing sweet potatoes in a weevil-free area include no restrictions on movement of the roots and easy access to more markets, ability to produce certified storage roots and plants for propagation, and no economic loss due to the weevil and associated control costs.

An area may be removed from the quarantine provided that no weevils are found for one full crop year in a regulated area (Arkansas State Plant Board 1983, South Carolina Crop Pest Comm. 1976). In Mississippi, no weevils can be found within a five mile radius of the property or area during two consecutive years (Mississippi Dept. Agric. Comm. 1976). In Alabama, properties located within a regulated area may be removed from the quarantine if no weevils are found during three consecutive years, and the minimum area for removal is four square miles (Alabama Dept. Agric. Indus. 1983). The regulated areas may change from year to year. Information on such changes should be published annually in state registers.

The distance the weevils can fly has an important bearing on quarantine regulations. Cockerham *et al.* (1954) marked adults of *C. formicarius* and found that some of the marked weevils traversed as much as 2 km. The authors did not mention the time frame weevils required to traverse this distance. Nevertheless, more studies are needed to determine dispersal potential of the weevil. Based on available information, it appears that absence of weevils for two consecutive years within an area with a radius of 3 km from an infested property is an adequate criterion for declaring a property free of weevils.

Penalties for Violation of Weevil Quarantines

A person or a firm that violates the sweetpotato weevil quarantine is penalized by law. Penalties may include cancelation of license to participate in sweet potato commerce, a maximum fine of $100 and/or imprisonment up to 30 days. Regulated articles found in violation of quarantine regulations are destroyed and/or disposed of at the expense of the person or the firm responsible for those articles (Louisiana Dept. Agric. 1985, Texas Dept. Agric. 1976).

SWEET POTATO TRADE AND WEEVIL QUARANTINES

Trade Within the United States

In 1988, the United States produced 536,599 metric tons of sweet potatoes on 36,073 ha (USDA 1989). The majority of sweet potato production is located in weevil-free areas, except in Louisiana where about 80 percent of its 6,883 ha is produced in infested areas (C. Roussel, pers. comm.). Sweet potato trade in the United States is quite extensive. For example, North Carolina and Louisiana, the two largest producers of sweet potato, ship their sweet potatoes to most of the major cities and metropolitan areas of the country, except Los Angeles and San Francisco. During July 1986 through June 1987, 62,129 metric tons of sweet potatoes were shipped to the major cities in the United States (North Carolina Dept. Agric. 1987).

Certificates accompanying the shipments facilitate sweet potato entry into a state. However, attempts have been made to gain entry into a state without documentation. For example, two truckloads of white-fleshed sweet potatoes grown in southern Florida were denied entry into California in 1988 (S. Brown, pers. comm.). During the past 15 years, regulatory measures (destruction of the shipment or refusal of entrance) were undertaken on several occasions: two each in Alabama, North Carolina and Texas; numerous occasions in Louisiana; 10 in Mississippi (on one occasion 72.6 metric tons of seed sweet potatoes certified as weevil-free were destroyed because they were infested with *C. formicarius* (B. Graves, pers. comm.). During the past seven years, sweet potato storage roots or plants have been intercepted in California on at least 30 occasions (S. Brown, pers. comm.).

In Texas, two sweet potato farmers from Wood and Van Zandt Counties (counties that are located in a weevil-free area) imported and bedded 3.6 metric tons of Louisiana certified sweet potato seed in April 1989. These seed sweet potatoes were produced in the Bienville Parish of Louisiana, a weevil-free parish. The Louisiana grower later found some *C. formicarius* larvae and one pupa in sweet potato storage roots in his storage facility. He immediately informed the growers in Texas of his finding. To avoid a possible introduction of the weevil, the Texas growers immediately fumigated the bedded sweet potato seed, and then buried the seed in soil to a depth of four feet. One pupa and one adult weevil were found when 100 bedded storage roots were examined for weevils post-fumigation. This rapid action helped prevent the spread of the weevil into these two major sweet potato producing counties in Texas.

On numerous occasions, roadside vendors and small grocery stores pesonnel have bought sweet potatoes from an infested area and sold them in a weevil-free area. Thus, the spread of the weevil could be curtailed by inspecting such operations in weevil-free areas, and by educating vendors about quarantine regulations.

International Trade

Annual production of the sweet potato is 115 million tons worldwide (FAO 1984); however, only 132,000 metric tons are imported and 48,000 metric tons are exported (Horton 1988). Europe and Canada import 125,000 metric tons. It is unlikely that weevils would survive in these countries (if introduced accidently through imported sweet potatoes) because of the cold winters. Hong Kong, Saudi Arabia, and Trinidad and Tobago also import sweet potatoes

(Horton 1988). It is possible that weevils have been introduced in some of these countries, although never reported. Most countries, however, have regulations for importation of plant materials. For this reason, the chance of the weevil increasing its distribution through international trade is small. Some sweet potato seed and plants are traded between countries in the the Caribbean (G. Pollard, pers. comm.). Enforcement of existing plant import regulations (Pollard 1986) should help reduce possible introduction of weevils.

On several occasions, illegal shipments of sweet potatoes from various Caribbean islands, such as the Dominican Republic, have been intercepted at ports in Florida, especially in Miami (R.K. Jansson, pers. comm.). Also, many exotic pests, such as *Euscepes postfasciatus* (Fairmaire), have been intercepted at ports of entry into Florida. These shipments originated from various Caribbean island countries (R.K. Jansson, pers. comm.).

FUTURE OUTLOOK

Pheromone and its Impact on Quarantines

Coffelt *et al.* (1978) recognized that the sweetpotato weevil females produce a sex pheromone. This chemical was later isolated, identified, and synthesized by Heath *et al.* (1986). This sex pheromone has been very effective in attracting the weevils under field conditions (Jansson *et al.* 1989, Proshold *et al.* 1986,). The pheromone will be of a major help in enforcing the weevil quarantines.

The North Carolina Department of Agriculture is currently using about 1,000 pheromone traps to determine the distribution and abundance of the weevil in seaside morning glory, *I. pes-caprae* (Linnaeus) in Brunswick and New Hanover Counties; thus, the control measures could be concentrated at locations where they are needed most. Additionally, approximately 100 traps, one each at a packing and storage facility, have been used for two years to detect weevils (W. Dickerson, pers. comm.).

In Texas, one pheromone trap was placed near the storage site where the infested sweet potatoes were held briefly, and nine traps were placed in the two fields where the potatoes were bedded. These traps helped determine the extent of the infestation and the effectiveness of the eradication program in Wood and Van Zandt Counties (J. Robinson, J. Cates & J. Noe, pers. comm.). Farmers that produce certified sweet potato seed could use pheromone traps on their properties to reduce chances of shipping weevil-infested seed and plants.

Guidelines for conducting inspections for weevils in fields and at storage sites vary from state to state. Despite the best efforts, infestations can often remain undetected due to human error, as well as time and cost constraints. Pheromone traps might aid in routine inspection procedures. However, an absence of weevils in traps placed in fields, and in storage and packing facilities may not mean that fields and other sites are weevil-free because the pheromone attracts only males and plants and/or roots might be infested with immature weevils. Thus, certifying sweet potatoes as weevil-free based on negative trap-catch data alone may not be justified. In some situations, inspections may not detect weevils, whereas pheromone traps will. Thus, use of pheromone traps may complement the findings of routine inspections. Nevertheless, trap-catch data gathered using the pheromone might be the best method for detecting weevils in a quarantine program. Currently, sweet potato fields, storage facilities, seed beds, home gardens, and host plants of the weevil are inspected intensively. Areas are removed from quarantine if weevils are not found. Use of pheromone traps may complement inspections in such a program.

A pheromone-trap, field-monitoring system for the weevil in the southern United States has been developed (see Chapter 6). However, this system might need to be modified for quarantine programs. Also, the level of the weevil infestation that traps can detect needs to be determined. Seldom is the source of infestation known with any certainty after an infestation is detected in a weevil-free area. Research is also needed to determine the maximum distance that different pheromone dosages will attract weevils over various durations. The sampling range of traps baited with four different dosages of the pheromone was determined 16 and 40 hours after marked males were released (Mason *et al.* 1990); however, more studies are needed to assess recapture over longer periods and further release distances. Ultimately, pheromone traps may help to determine the present status of weevil-free and weevil-infested areas nationwide.

ACKNOWLEDGMENTS

I thank W. Dickerson, R.K. Jansson, R. Mulder, C. Roussel, and M. Woodfin for reviewing the manuscript.

REFERENCES

Alabama Department of Agriculture and Industries. 1983. Sweetpotato weevil quarantine. Agricultural Chemistry and Plant Industry Division. Plant Protection Section. Chapter 80-10-5, Ala. Dept. Agric. Indus., Montgomery.

Arkansas State Plant Board. 1983. Sweetpotato weevil quarantine. Ark. State Plant Bd., Little Rock.

Boyden, B.L. 1927. Sweetpotato weevil eradication in Florida and Georgia. Monthly Bull. Fla. State Plant Bd. 7:17-55.

California Department of Food and Agriculture. 1987. Sweetpotato weevil exterior quarantine. California Plant Quarantine Manual. Ca. Dept. Food Agric., Sacramento.

Cockerham, K.L., O.T. Dean, M.B. Christian & L.D. Newsom. 1954. The biology of the sweetpotato weevil. La. Agric. Exp. Stn. Bull. No. 483.

Coffelt, J.A., K.W. Vick, L.L. Sower & W.T. McClellan. 1978. Sex pheromone of the sweetpotato weevil, *Cylas formicarius elegantulus*: laboratory bioassay and evidence for a multicomponent system. Environ. Entomol. 7:756-758.

Conradi, A.F. 1907. The sweetpotato borer. Texas Agric. Exp. Stn. Bull. No. 93.

Edmond, J.B. & G.R. Ammerman. 1971. Sweetpotatoes: production, processing, and marketing. Avi, Westport, Connecticut.

FAO. 1984. FAO production yearbook. 1983. Food and Agricultural Organization, Rome.

Georgia Department of Agriculture. 1982. Sweetpotato plant regulations. Entomology and plant industry. Chapter 40-4-8, Ga. Dept. Agric., Atlanta.

Georgia Department of Agriculture. 1986. Sweetpotato weevil quarantine. Entomology and plant industry. Chapter 40-4-15, Ga. Dept. Agric., Atlanta.

Heath, R.R., J.A. Coffelt, P.E. Sonnett, F.I. Proshold, B. Dueben & J.H. Tumlinson. 1986. Identification of sex pheromone produced by female sweetpotato weevil *Cylas formicarius elegantulus* (Summers). J. Chem. Ecol. 12:1489-1503.

Horton, D. 1988. Underground crops: long-term trends in production of roots and tubers. Winrock, Arlington, Virginia.

Jansson, R.K., R.R. Heath & J.A. Coffelt. 1989. Temporal and spatial patterns of sweetpotato weevil (Coleoptera: Curculionidae) counts in pheromone-baited traps in white-fleshed sweet potato fields in southern Florida. Environ. Entomol. 18:691-697.

Louisiana Department of Agriculture. 1985. Sweetpotato weevil quarantine. Office of Agricultural and Environmental Sciences. La. Reg. 11:320-322.

Mason, L.J., R.K. Jansson & R.R. Heath. 1990. Sampling range of male sweetpotato weevils (*Cylas formicarius elegantulus*)(Summers) (Coleoptera: Curculionidae) to pheromone traps: inflence of pheromone dosage and lure age. J. Chem. Ecol. 16:2493-2502.

Mississippi Department of Agriculture and Commerce. 1976. Sweetpotato weevil quarantine. Division of Plant Industry. Miss. Dept. Agric. Commerce, Mississippi State.

NAS. 1969. Principles of plant and animal pest control. Vol. 3. Insect-pest management and control. Natl. Acad. Sci. Publ. No. 1695. Washington, D.C.

New Mexico Department of Agriculture. 1969. Exterior quarantine no. 2: sweetpotato weevil. Division of Agricultural and Environmental Services, Bureau of Entomology and Nursery Industries, N.M. Dept. Agric., Las Cruces.

Newell, W. 1917. Sweetpotato root weevil. Fla. State Plant Bd. Quart. Bull. 2:81-100.

North Carolina Department of Agriculture. 1987. Marketing North Carolina sweet potatoes, 1986 crop. Federal-State Marketing News Service. Division of Marketing, N.C. Dept. Agric., Raleigh.

North Carolina Administrative Code. 1989. Sweetpotato weevil. Agriculture-Plant Industry. Document T02:48A.0900, N.C. Dept. Agric., Raleigh.

Oklahoma Department of Agriculture. 1978. Sweetpotato weevil quarantine. Standard State Quarantine Order No. 12. Ok. Dept. Agric., Oklahoma City.

Pollard, G.V. 1986. Plant quarantine in the Caribbean: a retrospective view and some recent pest introductions. FAO Plant Prot. Bull. 34:145-152.

Proshold, F.I., J.L. Gonzales, C. Asencio & R.R. Heath. 1986. A trap for monitoring the sweetpotato weevil (Coleoptera: Curculionidae) using pheromone or live females as bait. J. Econ. Entomol. 79:641-647.

Reinhard, H.J. 1923. The sweetpotato weevil. Texas Agric. Exp. Stn. Bull. No. 308.

Roberts, R.A. 1952. Sweetpotato weevil. USDA yearbook of agriculture, Insects. U.S. Dept. Agric., Washington, D.C.

South Carolina Crop Pest Commission. 1976. Sweetpotato weevil quarantine. Chapter 27, Article 6. Clemson Univ., Clemson.

Tennessee Department of Agriculture. 1957. Sweetpotato weevil quarantine. Division of Plant Industries. Elington Agricultural Center. Quarantine no. 1, Tenn. Dept. Agric., Nashville.

Texas Department of Agriculture. 1976. Sweetpotato weevil quarantine. Texas Dept. Agric., Austin.

USDA. 1985. Plant protection and quarantine treatment manual. U.S. Dept. Agric., Plant Prot. Quar., Animal and Plant Health Inspection Service, Hyattsville, Maryland.

USDA. 1987a. Plant import: propagative. U.S. Dept. Agric., Plant Prot. Quar., Animal and Plant Health Inspection Service, Hyattsville, Maryland.

USDA. 1987b. Plant import: non-propagative. U.S. Dept. Agric., Plant Prot. Quar., Animal and Plant Health Inspection Service, Hyattsville, Maryland.

USDA. 1989. Crop production 1988 summary. Natl. Agric. Stat. Serv. and Agric. Stat. Board., U.S. Dept. Agric., Washington, D.C.

14. Biology and Management of the West Indian Sweet Potato Weevil, *Euscepes postfasciatus*

Kandukuri V. Raman
International Potato Center
P.O. Box 5969
Lima, Peru

Eslie H. Alleyne
Ministry of Agriculture, Food and Fisheries
P.O. Box 505
Graeme Hall, Christ Church
Barbados

The sweet potato ecosystem is inhabited by many pest species of insects and mites. Talekar (1988) has reported at least 270 species of insects and 17 species of mites feeding on sweet potato in the field and in storage world-wide. All plant parts, namely roots, stems, foliage, and even seeds, are damaged by these pests. The greatest diversity of arthropods associated with sweet potato occurs in South and Central America, where the crop and its many related wild species originated (Austin 1987, Raman 1988a).

There are only certain key pests that must be routinely or specifically targeted for control in each region and production system. On a global scale, two weevils, *Cylas formicarius* (Fabricius) and *C. puncticollis* (Boheman) (Coleoptera: Curculionidae) are most destructive. However, the West Indian sweet potato weevil, *Euscepes postfasciatus* (Fairmaire), is also a major pest of sweet potato in the South Pacific, Caribbean basin, and some countries of Central and South America. These weevils cause extensive root damage. *Euscepes postfasciatus* populations and damage increased considerably in certain areas where farmers have changed from subsistence to more intensive commercial farming. The presence of these weevils on islands in the South Pacific has resulted in strict quarantine regulations banning the export of sweet potato from these islands and concomitantly limited the potential of sweet potato as an important export commodity. Unlike *C. formicarius*, little research has been conducted on management of *E. postfasciatus*. The successful cultivation of sweet potato on certain islands of the Caribbean basin and South Pacific depends almost entirely on minimizing damage by *E. postfasciatus*. This small weevil, known locally as "scarabee", is the most serious pest of sweet potato on certain islands in these regions and in the Caribbean it has caused extensive damage to storage roots in the field and in storage for as long as sweet potato has been grown (Ballou 1912, Tucker 1937).

Within the last decade, sweet potato has become more economically important in local agriculture of the Caribbean. This new thrust in sweet potato production has not only altered the economic status of the crop, but also the economic importance of *E. postfasciatus*. This chapter discusses the distribution, biology, and prospects for managing this pest.

DISTRIBUTION

Euscepes postfasciatus is more widely distributed than *C. formicarius* in the Caribbean basin and has been reported as a pest of sweet potato in Barbados, Cuba, Dominican Republic, Grenada, Guadaloupe, Bermuda, Guyana, Haiti, Jamaica, Martinique, Puerto Rico, St. Lucia, Virgin Islands, Caroline Islands, St. Vincent, and St. Kitts (Suah 1981, Raman 1988b). In the Pacific, it occurs in Vanuatu, New Caledonia, Fiji, Kiribati, Guam, Northern Mariana Islands, Federated States of Micronesia, Tonga, Western Samoa, French Polynesia, Cook Islands, Niue, Hawaii, Wallis and Futuna, Japan, Papua New Guinea, and New Zealand (Dumbleton 1954, Messenger 1954, Pole 1988, Sherman & Tamashiro 1954). In South America, *E. postfasciatus* has been as a pest in Surinam, Venezuela, Peru, and Brazil (Reyes 1986, Raman 1988a)

In the Pacific and the Caribbean, *E. postfasciatus* usually co-occurs with *C. formicarius*. On St. Croix, about 95% of the weevils present were *C. formicarius* (Proshold 1986). However, in other areas excluding the U.S. Virgin Islands no specific research has been done to determine the relative importance of both these species. Recently, the Caribbean Agricultural and Development Research Institute (CARDI) and the International Potato Center (CIP) conducted field surveys in the Caribbean to determine the importance of these two species and assess current control strategies for these species. In the Pacific, *E. postfasciatus* has been present for a longer time than *C. formicarius* and on seven of the islands, New Caledonia, Fiji, Guam, Northern Mariana Islands, Federated States of Micronesia, Tonga, and French Polynesia, this species was reported earlier than *C. formicarius* (Sherman & Tamashiro 1954). It is likely that in some countries identification of *E. postfasciatus* damage has been confused with that of *C. formicarius*. The adults of both these species are very distinctive and easy to distinguish. In the field however, adults are less conspicuous, and immature stages and damage of these two species are more difficult to distinguish. Identification of these pests in many of these countries is most often done by non-specialists, many of whom usually associate weevil damage with *C. formicarius*, since information and education on *E. postfasciatus* is lacking. For this reason, distribution records of these pests may not be accurate. For example, *E. postfasciatus* and *C. formicarius* both reportedly occur in Western Samoa, but recent surveys by Pole (1988) showed that only *E. postfasciatus* was present. Further field surveys are needed to clarify the distribution of this species.

LIFE CYCLE

Few comprehensive studies have been conducted on the biology of this weevil (Alleyne 1982a, Sherman & Tamashiro 1954, Tucker 1937, Yoshida 1985). We report here recent research conducted in Peru and in Barbados on the biology of this pest. These results are compared with other published work.

Egg

Females lay grayish-yellow to yellow eggs singly in a cavity excavated in either roots or stems and seal the cavity with a fecal plug (Sherman & Tamashiro 1954). Eggs are oval (about 0.12mm long x 0.10mm wide) and transparent when freshly laid. When leaves and stems were provided as oviposition substrates, females oviposited exclusively in the scarred area of the nodes located at the junction of the petiole and the stem. Eggs were not found in the internodal region even when these were the only oviposition sites available to adult females. A similar observation was made in the field. Storage roots were the preferred oviposition site. More eggs were laid on storage roots with cracks or crevices than on non-cracked roots. Females prefer to oviposit on roots within the first 2 cm of soil surface, and this concurs with observations of Aramaki *et al.* (1987).

All eggs, except for some laid on the surface of leaves, were inserted into punctures in the plant tissue made by the ovipositor of the adult female. Eggs are laid singly in each cavity located just below the surface of storage root, and then covered with a transparent cement. Eggs laid on leaves are not covered. The cement protects eggs from predators, such as ants, mites, and insect parasitoids, and also prevents dessication. The cement may become darker and appear stellar as eggs mature. Tucker (1937) used this feature as a quick method to recognize eggs. This star-like appearance of the cement, however, is not always apparent. Eggs are easily recognized using a hand lens (10X) or a binocular microscope. Sherman and Tamashiro (1954) reported a mean period of egg hatch of 8.3 ± 0.4 days at $26.8°C$. In present studies, fertilized eggs hatched at room temperature ($24\pm2°C$) in 7 to 9 days. At $18\pm2°C$, the incubation period lasted 11 to 15 days. Eggs laid on leaves did not hatch. Infertile eggs are noticeably softer and smaller than fertilized ones. The pre-oviposition period for virgin females ranged between 30 to 45 days after emergence, as compared with 12.8 ± 2 days for fertilized females. At $24°C$, newly emerged mated females laid an average of 178.6 (range, 95 to 362) eggs over a six month period. Females laid an average of 1 egg per day during the egg laying phase. Tucker (1937), after studying the reproductive habits, concluded: 1) fertilization was required for egg hatching; 2) the average time between adult emergence from sweet potato and oviposition was 15 days; 3) the average rate of oviposition was 106 eggs per female per month; and 4) females oviposited for 4 to 6 months.

Larva

The placement of the egg by adult females within the sweet potato tissue ensures that the newly emerged larva is in contact with its food source. Legless larvae move deeper and deeper into the plant tissue as they feed. Thus, the surface of the stem or storage root may appear unblemished while internally the tissue may be severely damaged. Larval feeding induces production of terpenoid compounds (Sato *et al.* 1978). Internal damage of roots by larvae results in discoloration and a foul smell of terpene. Such roots are unfit for human or animal consumption. The inward movement of larvae into the plant tissue makes it difficult to detect damaged storage rots, and also reduces the effectiveness of chemical control. Soil-applied systemic insecticides which are capable of penetrating the plant tissue as well as the soil are needed for effective control.

Five larval instars have been reported. The dimensions of head capsule were in close agreement with the data of Sherman and Tamashiro (1954). Observed mean head widths (mm) were: first instar, 0.182 ± 0.008; second, 0.251 ± 0.014; third, 0.376 ± 0.021; fourth, 0.558 ± 0.030; and fifth, 0.778 ± 0.032. Larvae develop into pupae after 20 to 30 days (Tucker 1937). Larval development at $24\pm2°C$ lasted 25 ± 5 days (range, 18 to 31); at $18\pm2°C$, it was 45 ± 3.0 days (range, 40 to 53).

Pupa

A prepupal stage has been observed (Sherman & Tamashiro 1954). The pupal stage lasts 7 to 9 days (Tucker 1937). Pupation occurs in the root or stem within a small pupal chamber prepared by larvae. The pupa is exarate and whitish in color. It rarely moves, but it twitches vigrously when touched. At $24\pm2°C$, the pupal stage lasted 8 ± 1.6 days (range, 7 to 9); at $18\pm2°C$, the pupal stage lasted 10 ± 0.9 days (range, 10 to 12).

Adult

Adult weevils are inconspicuous and easily overlooked because they resemble soil particles. The body varies from reddish brown to grayish black, and is covered with short stiff erect bristles and scales, (Figures 14.1 and 14.2). Adult weevils are approximately 3.5 mm long x 1.6 mm wide, and have light yellow spots near the apex of each forwing (Fennah 1947). Adults are formed within the pupal chamber; teneral adults are pale yellow, but they gradually harden and darken to grayish black. Adults feed on sweet potato tissue and emerge by chewing exit holes. Large exit holes of adults are visible on the surface of weevil-infested roots or stems. Adults may live for up to 6 months in the laboratory (Tucker 1937) and often feed gregariously on leaves and exposed roots. This species is thigmotactic, and like *C. formicarius*, this weevil also feigns death when disturbed (Sherman & Tamashiro 1954).

Fig. 14.1. Adult female (top) and male (bottom) of *E. postfasciatus* as observed under scanning electron microscope (31.4X). The last ventral segment of abdomen in females is flat, whereas in males it curves upward.

Fig. 14.2. Side view of female (top) and male (bottom) *E. postfasciatus* (31.4X). The elytra completely covers the last abdominal segment in females. In males, the last abdominal segment is curved and not covered by the elytra.

Tucker (1937) used the shape of the posterior ventral segment to differentiate males from females. In females, the segment is nearly flat, whereas in males, it curves upward (Figure 14.1). Females are typically larger than males. Food plant nutritional status may mask these differences. Sexes could be differentiated using this parameter with approximately 80% accuracy for individuals reared on plants of good nutritional quality. However, separation of sexes can be made more accurately by allowing adults to mate, or by preparing slide mounts of genitalia. Sex ratios of populations were approximately 1:1 (male:female).

Dispersal

One of the major factors influencing the method of control of any pest is the means by which it disperses. Because adults of *E. postfasciatus* possess well developed hind wings, we hypothesized that this insect dispersed by flying. However, none of several thousand adults observed in our laboratory flew or attempted to fly. Others (Sherman & Tamashiro 1954, Tucker 1937) also stated that this weevil did not fly. Several experiments were conducted using a wide range of stimuli to promote flight. These included exposing adults to high temperatures and various wind currents, and by dropping adults (free fall) from various heights; all stimili were not successful at enhancing flight (Alleyne 1987).

Several trap designs were also tested under field conditions in an attempt to collect adults in flight. Moericke water pan traps were tested as described by (Heathcote 1957), placed in weevil-infested fields for the entire growing season, and compared with standard black light traps for four years. No adult weevils were collected (Alleyne 1987). In another study, plastic funnel traps, which were developed for monitoring *C. formicarius* (Proshold *et al.* 1986), were compared with plastic pit fall traps buried at soil surface. Adult males and females were used as live bait in these traps. Adults (1,000) were marked with non-toxic luminous paint and released 5 and 10 m downwind from the traps. Weevil captures were compared between traps weekly for one month. No adult weevils were recaptured (K.V. Raman, unpubl. data).

In other field studies, clean stem cuttings were grown in a screenhouse and planted in sterilized soil in cut oil drums spaced 3 m apart and fully exposed to wind movements on all sides to enhance weevil infestation. Storage roots were sampled for 48 weeks at two week intervals beginning sixteen weeks after planting, and examined for weevil damage. No damage was observed. Stems were also removed periodically and examined for weevil damage; no stem damage was observed. These results suggest that *E. postfasciatus* might not fly and this has altered the strategy presently being used to manage this pest in Barbados (Alleyne 1987).

Survival

Adult survival not only influences the number of eggs laid, but also their spread, because the longer adults live, the greater the opportunity for dispersal. Adult survival averaged 243 ± 25 days ($n = 20$) (range, 216 to 252 days) in the laboratory at $24\pm2°C$ with sufficient food. Under adverse conditions without food, 75% of adults died ($n = 50$) in moistened soil in less than 30 days while the remaining 25% survived for 45 days. Similar results were obtained in the absence of soil, food, and water.

Under field conditions, this species may often be subjected to occasional flooding. The effects of flooding were tested in the laboratory. Adults were submerged in test tubes for different time periods with distilled water. Most adults (79%) survived one week in water.

SOURCES OF INFESTATION

Planting Material

Sweet potatoes are usually grown from propagative cuttings or "slips" which are 20 to 30 cm in length and usually taken from above-ground vine tissues. The distribution of weevils was characterized within different stem sections of 21 cultivars in Barbados. All life stages, except eggs, were recorded. Stems were divided into three 20 to 30 cm sections, apical (tip region), middle (the region between tip and bottom), and basal (bottom region). Stems were sampled ($n = 100$) 27 weeks after planting. The results indicate that all portions of the stem were infested, but the apical section was less infested (mean infestation varied from 0 to 10%). The basal sections were more infested (mean infestation varied from 4 to 42%). Stem sections immediately above the crown are usually more susceptible to infestation. Similar results were found for *C. formicarius* in southern Florida (Jansson *et al.* 1987,

1990). These "slips" are undoubtedly a major source of infestation, and the data indicate that relatively clean planting material is central to minimizing weevil damage in storage roots.

Alternate Hosts

This weevil utilizes wild hosts in the genus *Ipomoea* (Alleyne 1982a, Bonfils & Bart 1967, Muruvanda *et al.* 1984, Pemberton 1943, Sherman & Tamashiro 1954, Shiroma & Kunishi 1977, Swezey 1925, 1946). During the last five years, surveys have been conducted to identify wild hosts of this pest. Several *Ipomoea* spp. are found in fields throughout the Caribbean and Pacific. Fifteen wild *Ipomoea* spp. were tested in the screenhouse and in the field to determine if this weevil survives and reproduces on some wild *Ipomoea* spp.. *Ipomoea nil, I. spiralis*, and *I. tiliacea* were colonized both in the field and in the screenhouse. *Ipomoea obscura* was not infested in the field, but it was colonized in the screenhouse. Hence, it is likely that this species is an alternate host of *E. postfasciatus*.

Tucker (1937) and Fennah (1947) indicated that *I. pes-caprae* was an alternate host of *E. postfasciatus*. They also listed *I. biloba* as an alternate host; however, *I. biloba* is a junior synonym of *I. pes-caprae* (Austin *et al.* 1990). In the Caribbean and Central America, this species is found in areas where sweet potato is not grown, such as on beaches, the banks of canals and ponds, and in very sandy soils. Other alternate hosts reportedly include *I. pentaphylla* (Swezey 1925), *I. triloba* (Pemberton 1943), *I. horsfalliae* (Swezey 1946), and *I. reptans* in Hawaii (Shiroma & Kunishi 1977). Recently, Muruvanda *et al.* (1984) reported that roots of radish, *Raphanus sativus* L., could serve as an alternate food for adults in the laboratory. This crop is not important and is rarely grown in close proximity to sweet potato. Austin *et al.* (1990) listed all known hosts of *Cylas* spp. It is likely that these plants are also suitable hosts for *E. postfasciatus* (see Chapter 3).

Field Workers and Plant Debris

An interesting behavior of adult *E. postfasciatus* is that they tend to cling to objects. Adults have been observed clinging to fine mesh cloth in the laboratory (K.V. Raman, pers. obser.). It is likely, therefore, that they also cling to clothing of field laborers, who may provide a vehicle for dispersing this pest between fields. The contribution of this type of dispersal to overall infestation levels is not known; however, it may be an important means of dispersal. For this reason, laborers should be discouraged from moving between infested and noninfested fields.

This species, like *C. formicarius*, can survive in plant debris left in fields after harvest. Crop rotation, and the concomitant avoidance of planting continuous sweet potato in the same field, will help reduce infestations of *E. postfasciatus*.

PROSPECTS FOR MANAGEMENT OF *E. postfasciatus*

Host Plant Resistance

Host plant resistance to this pest has only recently been studied. Field screening for resistant clones has been conducted in Tonga, Peru, and Barbados. Pole (1988) screened 30 clones from 14 breeding families during two dry seasons in Tonga. Weevil damage was measured in the stems and storage roots at harvest. A significant relationship between stem thickness and stem damage by *E. postfasciatus* and *C. formicarius* was observed. The number of adults, larvae, and pupae of both species, and the percentage of damaged stems were significantly lower in clones with thin stems than in those with medium and thick stems. Stem damage may reduce yields considerably. Thus, in Tonga, identification of clones with little stem damage is desired. Although thin stems are less damaged, they are also poor planting material. For this reason, plant breeders in Tonga selected planting material with medium stems (3.6 to 5.5 mm diam.) since these stems establish well and are intermediate in weevil damage. Pole (1988) also found a negative correlation between percent stem thickness and root volume. Larger roots were produced in thin-stemmed clones than in thick-stemmed clones. Unfortunately, such roots are usually closer to the soil surface and more accessible to weevils. Thus, storage roots on thin vines tend to have greater damage than those on thick vines. It is possible that thin vines do not provide adequate breeding sites for adults, and enhance adults to move to storage roots. If cultivars with thin or medium vine thicknesses are planted, storage roots should be hilled to minimize access to roots.

Euscepes postfasciatus also causes major losses in storage. In Peru, losses of up to 80% have occurred in sweet potato germplasm stored under ambient conditions for four months (K.V. Raman, unpubl. data). Recently, 54 clones, which were selected for resistance to this weevil in storage roots in the laboratory, were evaluated in the field using procedures described by Talekar (1982). All clones were susceptible to stem damage in the field. Storage root resistance found in the laboratory did not correlate with stem resistance in the field. Further research is needed to determine the stem thickness that optimizes good establishment and high yields, and minimizes weevil damage. Recent studies for resistance to *C. formicarius* have concentrated on utililzing *I. trifida*. This species has been crossed to *I. batatas* (Iwanaga *et al.* 1989, Talekar 1987). The progeny of these crosses have also been tested in the laboratory in Peru to select for resistance to *E. postfasciatus*. Several resistant lines have been selected. However, these lines need to be backcrossed with *I. batatas* to improve yield and quality since most of the resistant lines have fibrous or woody roots. Studies are underway to assess if such manipulations increase susceptibility to weevils.

In Barbados, 21 varieties were tested for resistance to *E. postfasciatus* at two locations for several years and a highly acceptable variety, 'B-63-726' was selected for resistance. Studies relied on natural infestations, and in some years, infestations did not develop to desired levels. Thus, inconsistent levels of resistance were found between seasons and locations (Alleyne 1982b). Storage root bioassays have been developed in the laboratory to identify resistant germplasm (Raman 1988a). A total of 1,306 clones from the sweet potato gene bank maintained at the International Potato Center (CIP) (Huaman & De la Puente 1988) have been screened for damage and population development of *E. postfasciatus*. Damage was assessed 60 days after infestation by estimating the percentage of weevil damage on the external root periderm. Roots were then dissected in half and internal damage and weevil populations were estimated using an index from 1 to 5, where 1 = no development of larvae, pupae, and adult weevils (highly resistant [HR]); 2 = population of immature stages and adults ranged from 1 to 10 weevils (resistant [R]); 3 = population ranged from 11 to 20 weevils (moderately resistant [MR]); 4 = population ranged from 21 to 50 weevils (susceptible [S]); and 5 = population > 50 weevils (highly susceptible [HS]). A total of 118 clones were selected from these bioassays for further testing.

Currently, no plant breeding studies are being conducted to incorporate resistance to this pest. An international effort should be made to collect all germplasm reportedly resistant to this weevil. Germplasm should then be evaluated systematically. Although it is unlikely that completely resistant germplasm may occur, germplasm with high, moderate, or even low levels of resistance may be useful in integrated management programs for this weevil. A data base on germplasm resistant to this weevil and *Cylas* spp. needs to be established. Evaluations of germplasm should use standardized screening methods and employ the expertise of entomologists and plant breeders in developing countries. (Raman 1988b).

Biological Control

Several strategies are available for biogical control of *C. formicarius* (see Chapter 9). Similar approaches could be used for *E. postfasciatus*. Little information is available on biological control for this pest. Recently, two larval ectoparasitoids (Hymenoptera: Pteromalidae) were found in Peru (K.V. Raman, unpubl. data). These were identified as *Eurydinoteloides* sp. and *Cerocephala* sp. The potential of these and other unidentified natural enemies at killing this weevil is currently underway.

The fungus *Beauveria* sp., isolated from infected adult weevils (K.V. Raman, unpubl. data), has been mass reared using techniques described by Torres *et al.* (1989). An experiment was conducted in the greenhouse to assess its potential as a biological control agent of this weevil in Peru. Barley husk (4 g), in which the fungus was grown, was added to the soil at three different time periods (planting and 30 and 60 days after planting). Five pairs (1:1, male:female) of *E. postfasciatus* were caged on plants immediately after application. Adult mortality and weevil abundance was assessed after 90 days by dissecting stems and storage roots. Weevil abundance was lower and adult mortality was higher on plants treated with *Beauveria* sp. than on nontreated plants (Table 14.1). In the field, weevil damage is usually greater during the dry season (Bonfils & Bart 1967). Thus, current emphasis is on identifying strains of *Beauveria* sp. that persist and remain pathogenic to this pest in dry soils. In a preliminary field test conducted during the dry season of 1989 on the coast of Peru, this fungus reduced weevil damage in storage roots by 20% (K.V. Raman, unpubl. data).

Chemical Control

Chemical insecticides have not been used widely for the control of *E. postfasciatus*. This might be due, in part, to the low economic value of the crop. Chemical control of this pest is difficult because immature

Table 14.1. Percentage adult mortality of *E. postfasciatus* and numbers of immature weevils within plants 90 days after application of *Beauveria* sp.

Time of application	Treatment	% adult mortality (± SEM)	Mean total no. larvae + pupae (± SEM)
At planting	*Beauveria* sp.	85.5(10.2)a	5.2(2.7)b
	Nontreated	22.2(7.2)b	13.4(14.7)a
30 days	*Beauveria* sp.	100.0(0.0)a	11.4(4.8)b
	Nontreated	3.3(1.7)b	17.6(3.7)a
60 days	*Beauveria* sp.	96.7(6.2)a	0.5(0.5)b
	Nontreated	9.0(3.7)b	1.4(1.0)a

Means within the same column for the same time period followed by different letters are significantly different ($P < 0.05$).

stages develop inside the vines and storage roots, and adults do not fly. Dipping infested planting material in chemical insecticides, such as carbofuran (0.01% AI), has been effective. This chemical was also effective against *C. formicarius* when used as a dip for planting material (Talekar 1989). Planting material should be dipped for at least 30 minutes. Others (Sherman & Tamashiro 1954, Tucker 1937) also reported on chemical control of *E. postfasciatus*. They found that foliar sprays of chlordane and lead arsenate were effective at controlling this pest; however, these compounds persist in the environment and their use is now prohibited in most countries.

Integrated Pest Management (IPM)

The results of recent biological and behavioral studies as well as varietal resistance trials showed that considerable control of this pest can be achieved if certain practices are followed. These include removal of infested sweet potato vines and storage roots from the field

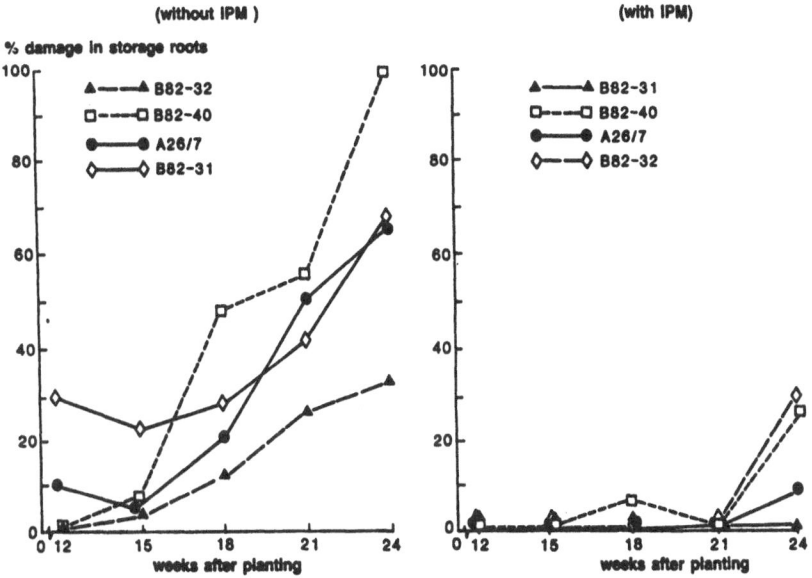

Fig. 14.3. *Euscepes postfasciatus* damage to storage roots of four cultivars of sweet potato in plots managed with or without IPM.

after harvest, removal of alternate hosts, use of non-infested planting material, frequent ridging (hilling), prompt harvesting, crop rotation, and use of less susceptible varieties. These practices were integrated and tested in three trials at three different locations in Barbados in 1985, 1986, and 1987. Each trial consisted of two field plots in which four local varieties ('A26/7', 'B82-31', 'B82-32', and 'B82-40') were planted. In one plot, normal farmer practices were used, whereas in the second plot, integrated pest management (IPM) methods were used. Results are shown in Figure 14.3. All varieties were attacked early in the growing season in plots where IPM was not used (Figure 14.3). Percentage of storage roots damaged by weevils ranged between 10 to 50% by the third sample period (18 weeks after planting). On the fifth sample period, damage ranged between 0 to 30% in plots where IPM was used, compared with 30 to 100% in plots where IPM was not used. Thus, IPM may help to reduce weevil damage to storage roots. The IPM tactics used in this study were similar to those proposed by Talekar (1989) for control of *C. formicarius* (see Chapter 7). In the Caribbean and the South Pacific, *E. postfasciatus* and *C. formicarius* often co-occur. The biologies and damage for these two species are similar. Thus, control tactics for both of these species are similar (Chalfant *et al.* 1990). Weevil management will improve when the IPM tactics mentioned above are adopted by most farmers and modified to meet their local conditions. Education of farmers in IPM technology is central to facilitating grower acceptance. Sponsored pest surveys, training programs, and on-farm demonstration trials will be important to facilitate this process. National and international programs should encourage such activities.

ACKNOWLEDGMENTS

We thank J.E. Wilson, for providing the information from the South Pacific islands, M. Palacios, O. Fabian, J. Alcazar, and V. Cañedo for help in data collection in Peru, and the Taxonomic Services Unit, Systematic Entomology Laboratory, U.S. Department of Agriculture, Agricultural Research Service, Beltsville, Maryland for identification of insect specimens.

REFERENCES

Alleyne, E.H. 1982a. Studies on the biology and behavior of the West Indian sweet potato weevil, *Euscepes postfasciatus* (Fairmaire) (Coleoptera: Curculionidae). Proc. Caribb. Food Crops Soc. 18:236-243.

Alleyne, E.H. 1982b. Varietal resistance as a control strategy against West Indian sweet potato weevil, *Euscepes postfasciatus* (Fairmaire), on sweet potatoes in Barbados. Proc. Caribb. Food Crops Soc. 18:54-62.

Alleyne, E.H. 1987. Dispersal of the West Indian sweet potato weevil, *Euscepes postfasciatus* (Fairmaire) (Coleoptera: Curculionidae), in Barbados. Proceedings of the 2nd Caribbean Regional Workshop on Tropical Root Crops, St. Vincent.

Aramaki, Y., T. Yoshida & K. Tashiro. 1987. An ecological study of the two weevils, the sweet potato weevil *Cylas formicarius* (Fabricius) (Coleoptera: Brentidae) and the small sweet potato weevil *Euscepes postfasciatus* (Fairmaire) (Coleoptera: Curculionidae), in the Amani Islands. On the ovipositional behavior of the two species. Res. Bull. Plant Prot. Serv. Japan 23:67-69 (In Japanese).

Austin, D.F. 1987. The taxonomy, evolution and genetic diversity of sweet potatoes and related wild species, pp. 27-60. *In* Exploration, maintenance, and utilization of sweet potato genetic resources. International Potato Center, Lima, Peru.

Austin, D.F., R.K. Jansson & G.W. Wolfe. 1990. Convolvulaceae and *Cylas*: a proposed hypothesis on the origin of this plant:insect relationship. Trop. Agric. (In press).

Ballou, H.A. 1912. Insect pest of the Lesser Antilles. Imperial Dept. Agric. West Indies, Barbados. Pamphlet Series No. 71.

Bonfils, J. & A.A. Bart. 1967. Distribution of sweet potato weevils in the French West Indies, pp. 27-29. *In* E.A. Tai, W.R. Charles, E.F. Iton, P.H. Haynes & K.A. Leslie (eds.). Proceedings of a symposium for the International Society for Tropical Root Crops., Vol. 2., Guadaloupe.

Chalfant, R.B., R.K. Jansson, D.R. Seal & J.M. Schalk. 1990. Ecology and management of sweet potato insects. Annu. Rev. Entomol. 35: 157-180.

Dumbleton, L.J. 1954. A list of insect pests recorded in South Pacific territories. Tech. Paper No. 79. South Pacific Commission, Novomea Cedex, New Caledonia.

Fennah, R.G. 1947. Insect pests of food crops in the Lesser Antilles. St. George's Grenada: Windward and Leeward Islands Dept. Agric.

Heathcote, G.D. 1957. The comparison of yellow cylindrical, flat, and water traps and of Johnson suction traps for sampling aphids. Ann. Appl. Biol. 45:133-139.

Huaman, Z., F. del Puente. 1988. Development of a sweet potato gene bank at CIP. CIP Circ. 16(2): 1-6.

Iwanaga, M., J.Y. Yoon, N.S. Talekar & Y. Umemura. 1989. Evaluation of the breeding value of 5X interspecific hybrids between sweet potato cultivars and 4X. *I. trifida*. Proceedings of an international sweet potato workshop. Visca, Baybay, Philippines. (In Press).

Jansson, R.K., H.H. Bryan & K.A. Sorensen. 1987. Within-vine distribution and damage of the sweet potato weevil, *Cylas formicarius elegantulus* (Coleoptera: Curculionidae), of four cultivars of sweet potato in southern Florida. Fla. Entomol. 70:523-526.

Jansson, R.K., A.G.B. Hunsberger, S.H. Lecrone & S.K. O'Hair. 1990. Seasonal abundance, population growth, and within plant distribution of the sweetpotato weevil (Coleoptera: Curculionidae) on sweet potato in southern Florida. Environ. Entomol. 19:313-321.

Messenger, A.P. 1954. Organization of a plant quarantine system in Okinawa. J. Econ. Entomol. 47:703-704.

Muruvanda, D.A., J.W. Beardsley & W.C. Mitchell. 1984. Additional alternate hosts of sweet potato weevils *Cylas formicarius elegantulus* and *Euscepes postfasciatus* (Coleoptera: Curculionidae) in Hawaii. Proc. Hawaii Entomol. Soc. 26:93-96

Pemberton, C.E. 1943. A note. Proc. Hawaii Entomol. Soc. 12:9.

Pole, F.S. 1988. Vine thickness in sweet potato (*Ipomoea batatas*): its inheritance and relationship to weevil damage. M.S. Thesis, University of the South Pacific, Western Samoa.

Proshold, F.I. 1986. Development of weevil populations on sweet potatoes in St. Croix, U.S. Virgin Islands. Trop. Pest Manage. 32:5-10.

Proshold, F.I., J.L. Gonzales, C. Ascencio & R.R. Heath. 1986. A trap for monitoring the sweetpotato weevil (Coleoptera: Curculionidae) using pheromone or live females as bait. J. Econ. Entomol. 79:641-647.

Raman, K.V. 1988a. Major pests of sweet potato and selection for resistance to sweet potato weevil *Euscepes postfasciatus* (Fairmaire), pp. 233-242. *In* Proceedings of a workshop on development of sweet potato (*Ipomoea batatas*) in Latin America. International Potato Center, Lima, Peru (In Spanish).

Raman, K.V. 1988b. Strategies to develop sweet potatoes with weevil resistance in developing countries, pp. 203-211. *In* Improvement of sweet potato (*Ipomoea batatas*) in Asia. International Potato Center, Lima, Peru.

Reyes, M.V. 1986. Bioecological aspects of *Cylas formicarius elegantulus* (Summers) (Coleoptera: Curculionidae) and factors determining resistance in different varieties of *Ipomoea batatas* (L.) Lam. (Convolvulaceae). Ph.D. Dissertation, Central University of Venezuela. Maracay, Venezuela (In Spanish).

Sato, K., I. Uritani, T. Saito & H. Honda. 1978. Factor causing terpene induction in sweet potato roots extracted from the West Indian sweet potato weevil, *Euscepes postfasciatus* (Fairmaire) (Coleoptera: Curculionidae). Appl. Entomol. Zool. 13:227-228.

Sherman, M. & M. Tamashiro. 1954. The sweetpotato weevils in Hawaii: their biology and control. Ha. Agric. Exp. Sta. Tech. Bull. No. 23.

Shiroma, E. & J. Kunishi. 1977. Occurence of *Euscepes postfasciatus* on swamp cabbage. *Ipomoea reptans*. Ha. Coop. Econ. Pest. Rpt. June 17.

Suah, J.R. 1981. Major pests of root crops in the Caribbean. Paper delivered at a short course on integrated pest management of tropical crops. Fac. Agric., Univ. West Indies, St. Augustine, Trinidad.

Swezey, O.H. 1925. A note. Proc. Ha. Entomol. Soc. 6:19.

Swezey, O.H. 1946. A note. Proc. Ha. Entomol. Soc. 13:4

Talekar, N.S. 1982. A search for sources of resistance to sweet potato weevils, pp. 147-156. *In* R.L. Villarreal & T.D. Griggs (eds.). sweet potato. Proceedings of the 1st international symposium. Asian Vegetable Research and Development Center, Shanhua, Taiwan, Republic of China.

Talekar, N.S. 1987. Feasibility of the use of resistant cultivars in sweet potato weevil control. Insect Sci. Applic. 8:815-817.

Talekar, N.S. 1988. Insect pests of sweet potato in the tropics. *In* Proceedings of the 11th International Congress of Plant Protection, Manila, Philippines (In press).

Talekar, N.S. 1989. Development and testing of an integrated pest management technique to control sweet potato weevil, pp. 117-126. *In* Improvement of sweet potato (*Ipomoea batatas*) in Asia. International Potato Center, Lima, Peru.

Torres, M., J. Alcazar & C. Vitorelli. 1989. Mass rearing of *Beauveria* sp.: regulator of the Andean potato weevil. *In* The 14th meeting of the Potato Association of Latin America. Mar del Plata, Republica Argentina. (Abstract) (In Spanish).

Tucker, R.W. 1937. The control of scarabee (*Euscepes batatae*) (Waterhouse) in Barbados. Barbados Dept. Sci. Agric. J. 6:133-156.

Yoshida, T. 1985. An ecological study of the two weevils, *Cylas formicarius* (Fabricius) and the small sweet potato weevil, *Euscepes postfasciatus* (Fairmaire) (Coleoptera: Curculionidae) in the Amani Islands. On life spans of the two species. Res. Bull. Plant Prot. Japan 21:55-59. (In Japanese).

15. Approaches to the Control of Multiple Insect Problems in Sweet Potato in the Southern United States

James M. Schalk, Alfred Jones, Philip D. Dukes, and Joseph K. Peterson
U.S. Department of Agriculture, Agricultural Research Service
U.S. Vegetable Laboratory
Charleston, South Carolina, 29414 U.S.A.

A complex of soil insect pests is a major limiting factor in sweet potato production in the southern United States. The most destructive stage of these pests is the larva which feeds on roots and reduces the quality and yield of the product. Historically, insecticides have been the first line of defense in reducing root damage by these pests; however, since the suspension of the persistent chlorinated hydrocarbon insecticides, alternative methods for insect control have been sought. Alternatives to insecticides include the development of insect resistant plants, and the use of biological control agents and a sex attractant.

INSECT PEST COMPLEX

There are over nineteen species of insect pests of sweet potatoes in the United States, however, this chapter reviews only ten. Cuthbert (1967) identified insects responsible for certain kinds of damage to sweet potato. Injury incurred by several different species during root growth appears similar at harvest and it is difficult to differentiate damage associated with each pest species. For this reason, damage resulting from several different species of wireworms (*Conoderus* spp.), rootworms, *Diabrotica* spp., and flea beetles, *Systena* spp., has been grouped into a complex called the wireworm-*Diabrotica-Systena* complex (WDS) (Cuthbert 1967, Schalk & Jones 1985). This complex is composed of seven root-damaging species: southern potato wireworm, *C. falli* Lane; the tobacco wireworm, *C. vespertinus*

(Fabricius); the spotted and banded cucumber beetle, *D. undecimpunctata howardi* Barber and *D. balteata* LeConte, respectively; the pale-striped flea beetle, *S. blanda* Melsheimer; the elongate flea beetle, *S. elongata* (Fabricius); and another flea beetle species *S. frontalis* Fabricius.

When holes are first made by larvae of the WDS complex, they are usually shallow, but when larvae penetrate into the vascular cambium, the damage can increase considerably by subsequent root growth (Figure 15.1).

BIOLOGY OF INSECT PESTS

Wireworm biology and management will not be discussed in this chapter because it is reviewed in another chapter (see Chapter 16). The biology and management of the cucumber beetles and flea beetles, however, are presented below. In addition, the biology and management of the sweetpotato flea beetle, *Chaetocenema confinis* Crotch, white grubs, *Plectris aliena* Chapin and *Phyllophaga ephilida* Say, and the white-fringed beetles, *Graphognathus* spp., are presented below.

Cucumber Beetles

The distribution of *D. balteata* is from Columbia, Venezuela, and Cuba to the southern United States, whereas *D. undecimpunctata* is found from Central Mexico to east of the Rocky Mountains in the United States. Both species feed on several different host plants. The larvae eat small holes through the periderm of storage roots and form irregular cavities under the skin. Feeding holes are in groups and may enlarge as roots develop. The roots are often attacked during early development, which results in many unsightly healed holes at harvest. Adult feeding on sweet potato foliage produces irregular holes. Adults, which feed on plants in the family Convolvulaceae, lay their eggs in the soil where they hatch in one to two weeks, depending on temperature. Larvae of these species are nearly indistinguishable. The larval stage may last 8 to 30 days, depending on food supply. Pupae are found in cells just below the soil surface and emerge as adults in about one week. Adults can overwinter in warm climates in the United States (Krysan & Miller 1986, Schalk & Jones 1985).

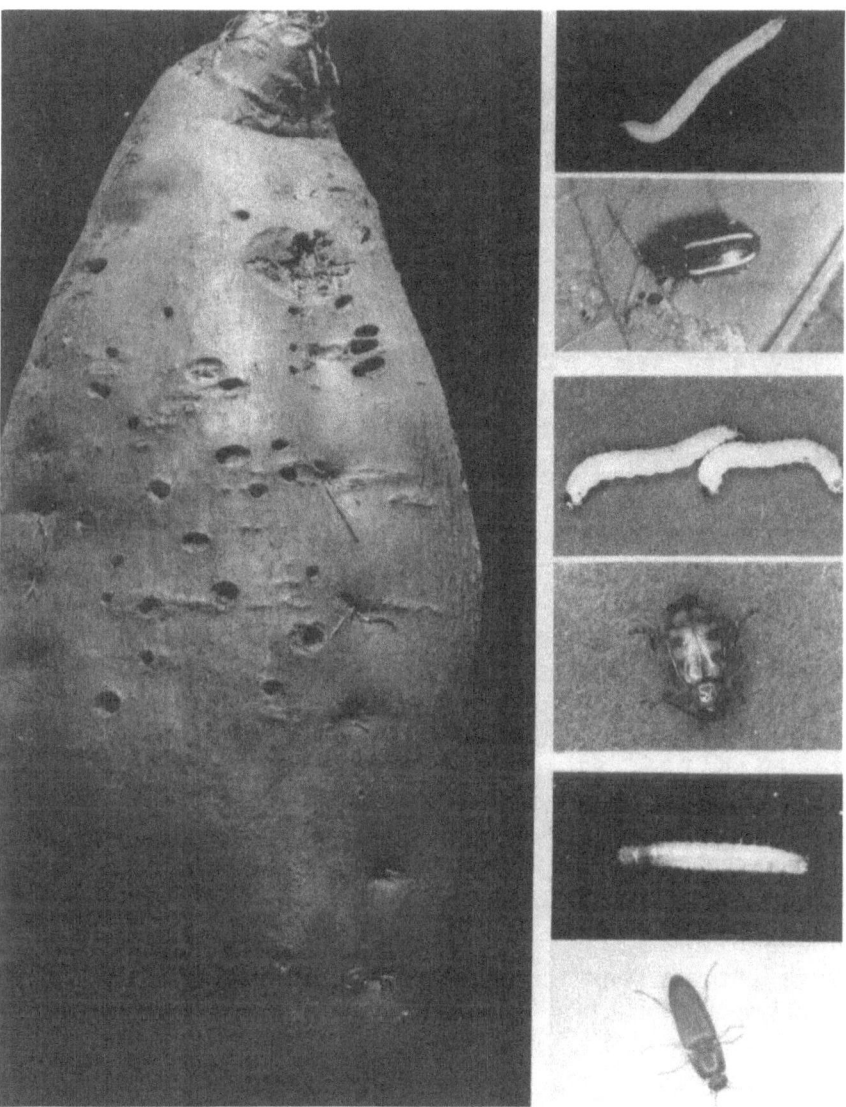

Fig. 15.1. Soil insects of the WDS complex causing similar kinds of injury (left) to sweet potato roots. Pictured top to bottom (right) are larvae and adults of flea beetles, *Systena* spp., cucumber beetles, *Diabrotica* spp., and wireworms, *Conoderus* spp., of the complex.

Flea Beetles

The life cycle and habits of three *Systena* species are similar, and the immature stages look alike. All have a wide range of hosts, including weeds. Larvae eat small holes through the periderm of storage roots and make enlarged cavities and short tunnels just under the periderm. Their damage is similar to that by *Diabrotica* spp.; however, they seldom tunnel into the roots. At harvest, early season damage appears as shallow healed scars, which are elongate or irregularly shaped, in contrast to those by *Diabrotica* spp., which are usually round. Late-season damage by these larvae is referred to as pinhole injury. Adult feeding is restricted to the leaf surface. Adults migrate into sweet potato fields during the spring and summer and oviposit in the soil. After about ten days the eggs hatch and larvae begin feeding. The larval stage lasts 20 to 30 days after which larvae pupate in an earthen cell and emerge in about one week as adults. At least two generations per year occur in the southern United States (Cuthbert 1967, Schalk 1984).

The adult of the sweetpotato flea beetle, *C. confinis*, is an important pest of sweet potato foliage (Smith 1910). The insect utilizes several species in the genus *Convolvulus* spp., including several species of bindweed, as well as sweet potato. This insect is found east of the Rocky Mountains. Adults move into sweet potato fields soon after planting. Eggs are laid in the soil and hatch into larvae within a few days. Larvae resemble those of *Diabrotica* spp. and *Systena* spp., but can be distinguished by the lack of a dark spot or fleshy tubercle on the posterior of the abdomen. The insect pupates in the soil. During the growing season, the flea beetle completes a life cycle in about 30 days. Larvae of this pest make small, winding tunnels just under the skin of sweet potato roots which become more evident as roots grow (Figure 15.2) (Cuthbert 1967, Schalk 1984).

White Grubs

Two species of white grubs, *P. aliena* and *P. ephilida*, are serious pests of sweet potato. Larvae of both species feed mainly on pasture sod, and severe damage to storage roots may result when sweet potato is planted in fields that follow pasture. Larvae of both species gouge out broad, shallow areas on the root which reduces the marketability of the crop (Figure 15.3).

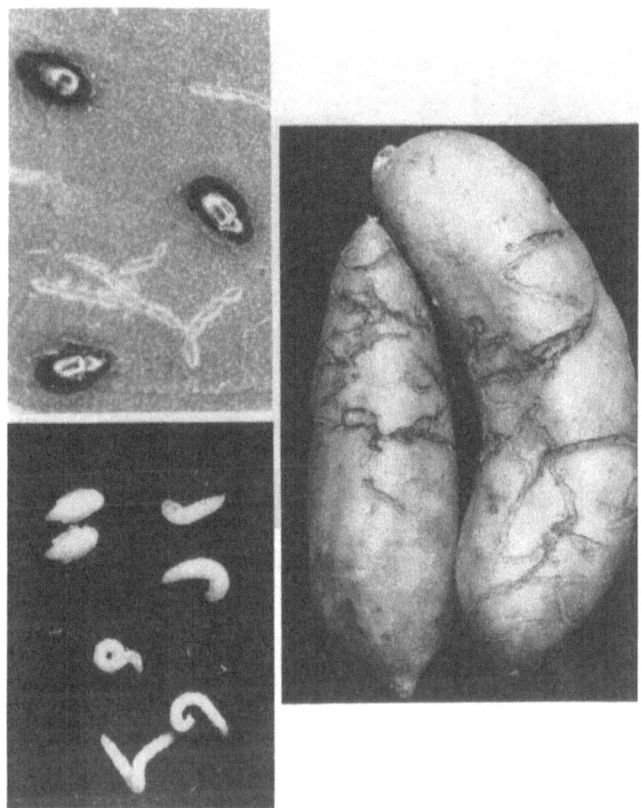

Fig. 15.2. Narrow channels of sweetpotato flea beetle damage (top to bottom: adult, pupae and larvae) to storage roots.

Adults of *P. aliena* live mainly in the soil. *P. aliena* is found in Argentina, Brazil, and Uruguay and was introduced into the United States and Australia about 1900. Adults of *P. aliena* emerge in late spring and mate. Mated females are incapable of feeding, but they can survive for up to 21 days and lay up to 33 eggs. Eggs are laid in the soil during the spring and summer and hatch into C-shaped larvae. Larvae live for one to two years depending upon the environmental conditions. (Jones & Schalk 1985).

Adults of *P. ephilida* are nocturnal and feed on deciduous trees. *Phyllophaga ephilida* is native to the United States and is the only *Phyllophaga* species that attacks sweet potato. Adults of *P. ephilida* emerge in late May; peak flight occurs about the end of June. Females oviposit on bare soil or soil planted to crops. Larvae overwinter in the

288

Fig. 15.3. Grub damage to storage roots by *P.aliena* (top to bottom: adult, pupae, and larvae).

soil; there is one generation per year (Cuthbert 1967, Roberts 1968, Schalk & Jones 1985, L. Rolston, unpubl. data).

White-fringed Beetles

The white-fringed beetles, *Graphognathus* spp., are considered a serious problem on sweet potato in some areas of the United States. The adults are not voracious feeders, but they have been reported on 170 different host plants. Larvae are more destructive and have been reported on 385 plant species, including peanuts, corn, sugarcane, cotton, cowpea, beans, cole crops, chufa, alfalfa, Mexican clover, and sweet potato. Larval injury to sweet potato roots resembles that caused by wireworms, *Diabrotica*, flea beetles and white grubs. However, the predominant injury is usually grub like in appearance (Chalfant *et al.* 1990).

For United States Market Standards, insect injury counts as a defect only when feeding scars exceed an aggregated depth of 3.2cm; on sweet potato, however, any injury that seriously affects the appearance of the product may also lower the grade. Reductions in grade also depend on the availability of the product and market demands. In our research program, we consider a 50% injury level to be realistic and injury up to this level will probably not affect the marketability of a crop.

MANAGEMENT APPROACHES

Breeding for Insect Resistance

Our breeding program emphasizes the development of sweet potato storage roots with high levels of resistance to insects and diseases, as well as good horticultural qualities. We follow quantitative genetic principles using mass selection techniques. True seed are produced in field-grown, open-pollinated nurseries.

Mass Selection. Sweet potato differs from other vegetables in that standard breeding procedures need not be followed. The sweet potato, a hexaploid, has 90 chromosomes with a complex quantitative inheritance. Plants are clonally propagated and Pedigree records are of limited value because it is not possible to reproduce a particular genotype by sexual means. Mass selection procedures which combine rapid generation turn over with high selection pressure provide a sound basis for crop improvement. A breeding program should have both long and short term goals and have one or more mass selection populations to provide new parental types and to assure the development of a wide gene base. The mass selection program moves from one sexual generation to the next by advancing from true seed to true seed each year. In this process, selected plants are not used in the cycle because only their seed is used. Thus, seeds from selected plants can be bulked to start the next cycle (Jones *et al.* 1986) (Figure 15.4).

The number of paternal entries needed to start a mass selection program for long-term objectives (resistance to future pests) is fairly large, and includes as wide a gene base as possible. For example, one of our mass selection populations was started with 350 plant collections from 17 different countries. These plants were open pollinated and

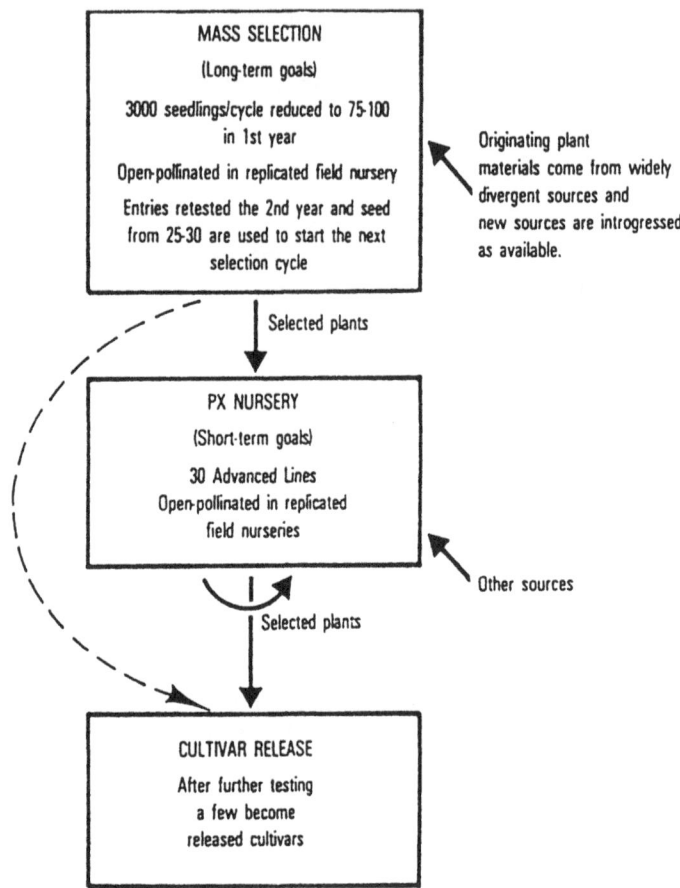

Fig. 15.4. U.S. Department of Agriculture sweet potato breeding program with long and short term goals through the use of mass selection and polycross (PX) breeding techniques.

about 3,000 seedlings were grown from true seed in the greenhouse; 700 of these, which represented all the various sources, were moved to a trellis area (for open pollination by arthropods). Seed from about 200 were then used to start the next cycle. In the third cycle, the number selected as seed parents was reduced to about 100 from the 700 trellised plants. For short-term goals, as few as 6 plants can be used; this will result in a narrow gene base and should provide a better chance for rapid advancement of a particular objective (for example insect resistance) (Jones *et al.* 1986).

After at least three cycles of intercrossing, it becomes necessary to change to a two year cycle to better evaluate yield, sprouting, and storage traits. Seedlings can be evaluated in the greenhouse for disease resistance and horticultural characters; vine cuttings can be transplanted to the field for evaluating insect resistance, and further horticultural selection in nonreplicated plots can be made. About 150 of the best selections can be stored and evaluated for storage quality and palatability. In the spring, the remaining selections may be rated for bedding traits. The best 75 to 100 selections are then planted on trellises in four or five replications. These plants are pollinated by arthropods and seeds are collected and labeled. During the same season, vine cuttings from the plant beds are used to plant replicated field trials. Data are collected on horticultural characteristics and insect susceptibility or resistance, and the best 25 or 30 selections identified. Seed from these plants are then used to start the next cycle of selection (Jones *et al.* 1986). New plant collections can be introgressed into the mass selection population through the seed increase nursery (Jones *et al.* 1986).

Polycross Nursery. This nursery is similar to the mass selection nursery but the number of plant entries is usually limited to no more than 30 advanced lines. These plants are replicated four times in each of two locations each year. These plants are then randomly crossed by a natural population of arthropods to meet our most important short-term objective, the development of insect resistant sweet potatoes. Each time a new line is added to the polycross, one of the previous entries is dropped because of poor insect resistance or other horticultural deficiencies (Jones *et al.* 1986)

Seedlings from the polycross provide a major source of potential resistant cultivars and are screened for the many essential traits as in the mass selection breeding program. Selections made in the first year are tested in the second year seedling trials and in the advanced line trials for two more years. The best selections may be vegetatively increased for submission into regional trials (National Sweet Potato Collaborators Group), where they are evaluated for horticultural qualities and pest resistance (Jones *et al.* 1986).

Using the methods described above, the research team at the U.S. Department of Agriculture, U.S. Vegetable Laboratory, Charleston South Carolina, developed and released ten breeding clones, and six cultivars with multiple resistance to the WDS complex, sweetpotato flea beetle, and grubs. Resistance to white-fringed beetles has not been evaluated (Dukes *et al.* 1987, Hamilton *et al.* 1985 Jones *et al.* 1975, 1980, 1983, 1985, 1987, 1989).

Soil Insect Evaluations and Results. Root damage by the WDS complex is typically rated in two ways. In one method, the percentage of roots injured is determined by dividing the number of roots with injury by the total number of roots. In a second method, a severity index is determined by assigning each root a score between 0 to 4 based on the number of feeding scars (0, no scars; 1, 1 to 5 scars; 2, 6 to 10 scars; 4, > 10 scars) and then calculating the mean of the scores. Damage by sweetpotato flea beetle and grubs is typically recorded as a percentage of roots injured (Jones *et al.* 1979).

Insect populations in experimental sweet potato fields are monitored by taking soil samples (465 cm^2, 15.2 cm in depth, four samples per plot) from around the roots of plants. The samples are then washed with water through a screen (64 mesh) and the numbers of larvae are recorded (Cuthbert & Jones 1978).

Biological Control

Currently, there are no practical biological control agents, such as predators and parasitoids, for cucumber beetles on sweet potato; however, entomopathogens have been found to have potential for suppressing this soil insect pest complex. Lipa (1975) stated that *Beauveria bassiana* (Bals.) Vuill. plays a significant role in the mortality of noxious insects under natural conditions and that mortality of *D. undecimpunctata* from this fungus amounted to 30% in the spring. Bell *et al.* (1972) reported 75 and 100% mortality of *D. balteata* larvae and pupae when exposed to *B. bassiana*, and 0 and 20% mortality of larvae and pupae when exposed to *Metarhizium anisopliae* (Metchnikoff) Sorikin. Virus-like particles have also been reported in *D. undecimpunctata howardi* (Kim 1980); however, their effect on adults was not described. An unidentified microsporidian was found in fat-body tissues of adult western corn rootworm, *D. virgifera* LeConte. This pathogen displayed sublethal effects and reduced rootworm fecundity. Unidentified gregarines (Apicomplexa: Gregarinidia) have been found in feral *Diabrotica* spp. and these organisms reportedly cause insect mortality and reduced fecundity (Schalk 1986).

Two entomophilic nematodes were isolated from *D. balteata* in South Carolina. *Filipjevimermis leipsandra* Poinar and Welch (Nematoda: Mermithidae) and *Heterorhabditis bacteriophora* Poinar (= *H. heliothidis*) (Nematoda: Heterorhabditidae). In cage studies conducted in the field, one application of *F. leipsandra* to the soil resulted in 78% parasitization of *D. balteata* larvae. More multiple infections of insect larvae occurred in soil treated with the highest density of nematode eggs. In a greenhouse pot test and in small field plots, nematodes that were applied to the soil in the egg and preparasitic stages appeared promising for protecting sprouted corn seedlings from damage by *D. balteata* larvae. In a laboratory test, the rate of larval parasitization declined over a 10-day period when *D. balteata* larvae were introduced into the soil inoculated with the nematode (Creighton & Fassuliotis 1983).

Creighton and Fassuliotis (1985) isolated a virulent entompathogenic nematode, *H. bacteriophora* South Carolina strain, from *D. balteata*. In a laboratory test, Schalk (1989) demonstrated that adult emergence and adult weight were reduced 85 and 13%, respectively, when second instars were exposed to densities of 500 infective juveniles per 800 g of soil. When third instars of *D. balteata* were exposed to the same nematode density, 13 and 8% reduction in adult emergence and weight occurred, respectively, indicating that this nematode was more efficacious against younger larvae than older larvae.

Other entomophilic nematodes isolated from *D. undecimpunctata howardi* include *Steinenema* (= *Neoaplectana*) spp. (Nematoda: Steinernematidae), *Howardula* spp. (Nematoda: Allantonematidae), and *Diplogaster* spp. (Nematoda: Diplogasteridae) (Elsey 1977, Fronk 1950, Turco *et al.* 1970).

Several predators of *Diabrotica* spp. have been reported. Lugenbill (1918) isolated a tachinid parasitoid, *Celatoria diabroticae* (Shimer). Egg predators of *D. balteata* have also been reported. These include *Solenopsis germinata* (Fabricius) and *Pheidole* spp. (Risch 1981). Waddill (1978) reported adult predation on *D. balteata* by the earwig *Labidura riparia* (Pallas). Females consumed more *D. balteata* adults than males which was attributed to energy needs for reproduction and nurturing of their progeny.

Sex Pheromone

The discovery of a female-produced sex pheromone of *D. balteata* was reported by Cuthbert and Reed (1964). Chuman *et al.* (1987) later isolated, identified, and synthesized the pheromone, 6,12-dimethylpentadecan-2-one and tested its attractiveness to male *D. balteata* in the field. Male response to traps baited with septa formulated with the synthetic pheromone showed an increase in the numbers of males captured with an increase in pheromone dosage for

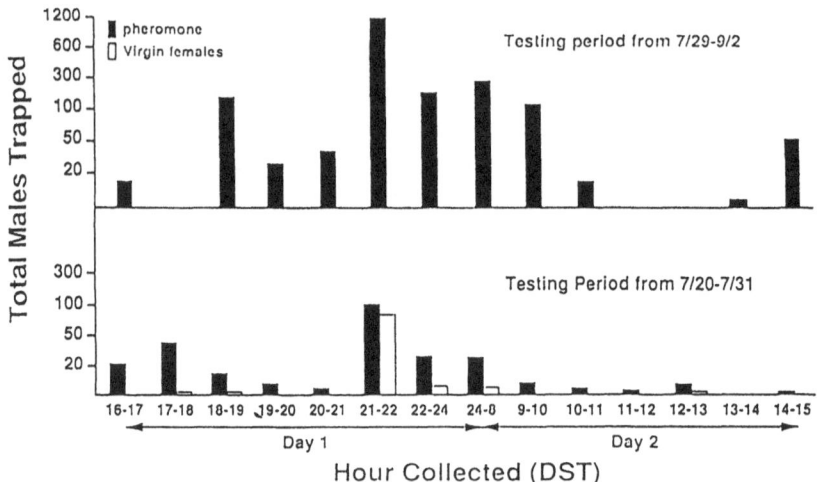

Fig. 15.5. Numbers of male *D. balteata* captured over a 24 hour period in traps baited with septa loaded with 0.5 mg of synthetic sex pheromone (from Schalk *et al.* 1990; reprinted with permission from Fla. Entomol. [Florida Entomological Society, Winter Haven, Florida]).

dosages between 30 and 300 µg of pheromone. The numbers of males captured declined in traps baited with > 300 µg of pheromone. A similar response was observed when pheromone was formulated on filter paper. They indicated that the leveling off and decline of the trap captures at dosages > 300 µg could not be attributed to pheromone release rates. They suggested that males responded positively to one or more of the stereoisomers, and were possibly inhibited by other stereoisomers of the same compound.

Schalk *et al.* (1990) conducted additional field tests and found that trap captures of male *D. balteata* in sex pheromone-baited traps or in virgin female-baited traps peaked at the beginning of the scotophase (2100 to 2200 hours) and declined thereafter (Figure 15.5). Howard (1982) noted similar findings in activity of adult banded cucumber

Table 15.1. Numbers of *D. balteata* males captured during a 24 hour period in traps baited with two virgin females or synthetic sex pheromone (0.5 mg per septa) in 1987.

	Total males captured per day	
Date	Synthetic sex pheromone	Virgin female
13 September	474	194
15 September	326	297
16 September	422	114
Total	1,222	605

beetles using sweep net samples. Virgin females were not as efficacious as the synthetic sex pheromone (0.5 mg per septum) at capturing male *D. balteata* (Table 15.1).

The most efficacious trap type for capturing males was the Sentry wing trap (Table 15.2) (Schalk *et al.* 1990). The Delta, Multi-pher, and 450 ml cup trap types were intermediate in the numbers of males captured, while the Japanese beetle, plastic funnel, and screen-cone boll weevil traps caught the fewest numbers of male *D. balteata*.

The effect of trap height and age of lures on capture of *D. balteata* were also examined (Schalk *et al.* 1990). Male captures did not differ among traps placed 25.4 ($X = 8.6 \pm 12.1$), 50.8 ($X = 10.5 \pm 11.8$) and 101.6 cm ($X = 9.6 \pm 10.6$) above the soil surface. Additional field tests showed that a rubber septum loaded with 0.5 mg of the pheromone was attractive to males for a 12-month period. The highest captures occurred in August and September. Trap captures declined in the cooler season; however, males were collected in traps from October through February (Schalk *et al.* 1990) which indicated that this insect can survive during the winter in Charleston, South Carolina (Figure 15.6). Elsey (1988) noted similar findings and also suggested that Charleston, South Carolina may be the northernmost limit of the range of this beetle.

The sensitivity of male *D. balteata* to the synthetic sex pheromone suggests that the pheromone may be useful in integrated pest management (IPM) programs for this pest. The pheromone can be

Table 15.2. Efficacy of different trap types at capturing adult male *D. balteata*. Each trap was baited with 300 μg of synthetic sex pheromone.

Trap type	Mean no. of males captured per day (± SEM)[a]
Wing	11.3 (5.3)a
Delta	7.6 (4.6)b
Multitrap	5.0 (1.8)b
450 ml cup	4.9 (1.8)b
Japanese beetle	2.1 (0.6)c
Plastic funnel[b]	2.1 (1.1)c
Screen cone boll weevil	1.5 (0.6)c
450 ml cup (nonbaited-blank)	0.0 (0.0)d

[a] Means within column followed by the same letter are not significantly different at the p=0.05 level Duncan's multiple range test.
[b] Sweetpotato weevil trap (see Chapter 6).

employed as a bait in wing traps to detect early infestations, to monitor established populations, and to help time insecticide applications to help prevent population densities of this beetle from causing considerable economic loss. Because the insect is a pest of seasonal crops, the pheromone will probably have little value in mass trapping because there is no long term crop continuity. The pheromone will probably also have little value as a mating disruption tool because adult *D. balteata* live for a long period of time which undoubtedly increases the likelihood that males will locate and mate with females. It should be noted, however, that the potential of these two management approaches for *D. balteata* should be assessed to confirm these beliefs.

OUTLOOK

Registration of new insecticides for use against the soil insect complex on a minor crop, such as sweet potato, is becoming critical. Fewer insecticides are available due to cancellations because of restrictions imposed by the U.S. Environmental Protection Agency. The high costs associated with developing and marketing new insecticides will make availability of new insecticides unlikely. For

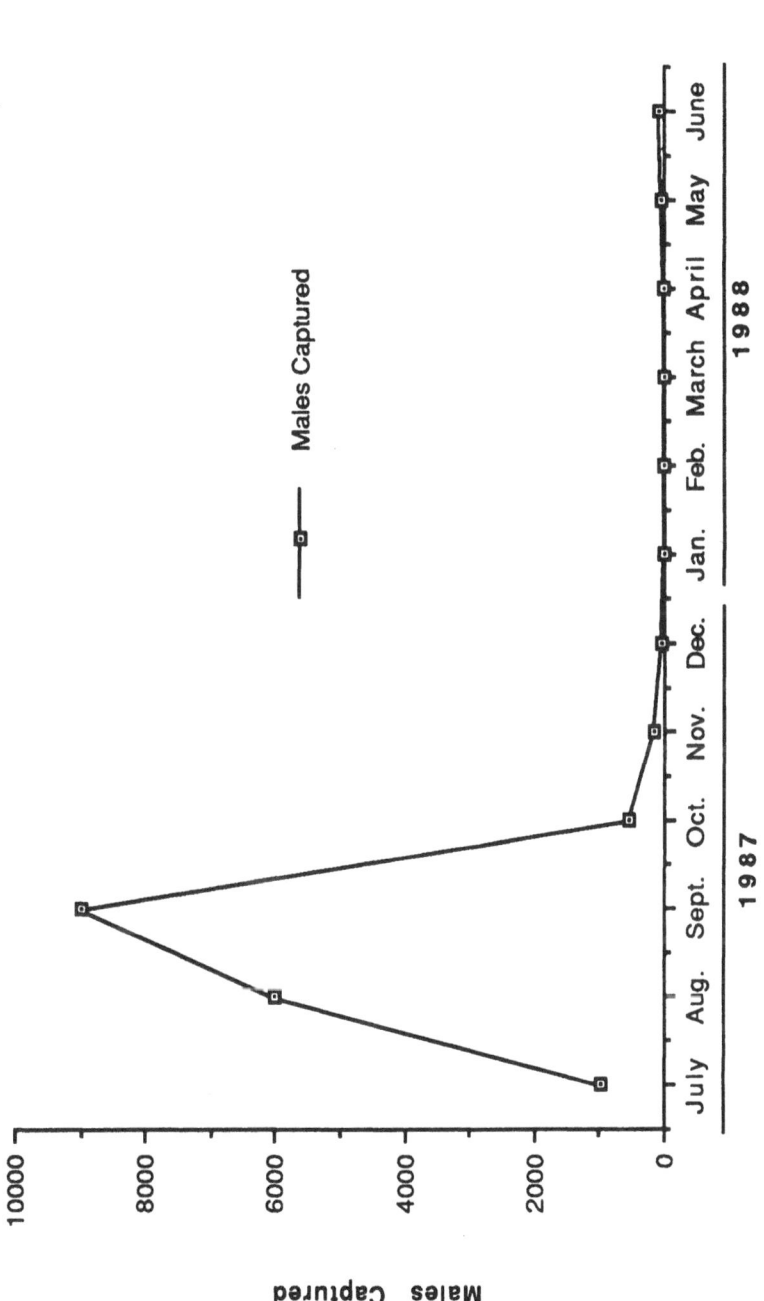

Fig. 15.6. Field evaluation of the sex pheromone (0.5 mg per septa per trap) used continually for a 12 month period to capture males of *D. balteata* in 1987-1988 (from Schalk *et al.* 1990; reprinted with permission from Fla. Entomol. [Florida Entomological Society, Winter Haven, Florida]).

these reasons, alternative management approaches are needed and should be emphasized for managing these pests in the future.

The use of cultivars with resistance to soil insects will permit production of quality sweet potatoes without insecticides. In the southern United States, this could amount to a reduction of almost 136,000 kg AI of insecticides. However, the sweet potato industry is reluctant to change to new cultivars.

Currently, there are no biological control agents that are used commercially for the control of the insects discussed in this chapter, but recent developments in research look promising. Discoveries of new strains of bacteria and fungi, which are effective against a variety of coleopterous pests, have been reported and may have considerable value in IPM programs. Entomophilic nematodes have potential for managing *Diabrotica* spp., but their efficacy against other coleopterous pests is not known. The use of the sex pheromone of *D. balteata* has potential and will help to understand the biology of this pest. This pheromone should ultimately be integrated into IPM programs for this pest on sweet potato. In order for alternative management approaches, such as host plant resistance and the use of biological control agents and sex pheromone, to play a significant role in IPM programs for these pests, more information is needed on insect pest biology, and insect/host plant interactions. Lastly, effective extension programs are needed to disseminate these new technologies to the consumers in the southern United States.

REFERENCES

Bell, J.V., R.J. Hamalle & J. A. Onsager. 1972. Mortality of larvae and pupae of the banded cucumber beetle in soil and sand following topical application of fungus spores. J. Econ. Entomol. 65:605-606.

Chalfant, R.B., R.K. Jansson, D.R. Seal & J. M. Schalk. 1990. Ecology and management of sweet potato insects. Annu. Rev. Entomol. 35:157-80.

Chuman, T., P.L. Guss, R.E. Doolittle, J.R. McLaughlin, J.L. Krysan, J.M. Schalk & J.H. Tumlinson. 1987. Identification of female-produced sex pheromones from the banded cucumber beetle, *Diabrotica balteata* LeConte (Coleoptera:Chrysomelidae). J. Chem. Ecol. 13:1601-1616.

Creighton, C.S. & G. Fassuliotis. 1983. Infectivity and suppression of the banded cucumber beetle (Coleoptera: Chrysomelidae) by the mermithid nematode *Filipjevimermis leipsandra* (Mermithida: Mermithidae). J. Econ. Entomol. 76:615-618.

Creighton, C.S. & G. Fassuliotis. 1985. *Heterorhabditis* sp. (Nematoda: Heterorhabditidae); a nematode parasite from the banded cucumber beetle *Diabrotica balteata*. J. Nematol. 17:150-153.

Cuthbert, F.P., Jr. 1967. Insects affecting sweetpotatoes. U.S.D.A. Agric. Handbook No. 329.

Cuthbert, F.P., Jr. & A. Jones. 1978. Insect resistance as an adjunct or alternative to insecticide for control of sweet potato soil insects. J. Am. Soc. Hort. Sci. 103:443-445.

Cuthbert, F.P., Jr. & W.J. Reed. 1964. Studies of sex attractant of banded cucumber beetle. J. Econ. Entomol. 57:247-250.

Dukes, P.D., M.G. Hamilton, A. Jones & J.M. Schalk. 1987. 'Sumor' a multi-use sweetpotato. HortScience 22:170-171.

Elsey, K.D. 1977. Parasitism of some economically important species of Chrysomelidae by nematodes of the genus *Howardula*. J. Invertebr. Pathol. 29:384-385.

Elsey, K.D. 1988. Cucumber beetle seasonality in coastal South Carolina. Environ. Entomol. 17:496-502.

Fronk, W.D. 1950. Cultural and biological control of the southern corn rootworm in peanuts. J. Econ. Entomol. 43:22-24.

Hamilton, M.G., P.D. Dukes, A. Jones & J.M. Schalk. 1985. 'HiDry' sweet potato. HortScience 20:954-955.

Howard, F.W. 1982. Diurnal rhythm in *Cylas formicarius elegantulus* and some other arthropods in a sweet potato field. Fla. Entomol. 65:194-195.

Jones, A., P.D. Dukes & F.P. Cuthbert, Jr. 1975. W-13 and W-178 sweet potato germplasm. HortScience 10:533.

Jones, A., P.D. Dukes & J.M. Schalk. 1986. Sweetpotato breeding, pp. 1-35. *In* M.J. Bassett (ed.). Breeding vegetable crops. Avi, Westport, Connecticut.

Jones, A., P.D. Dukes & J.M. Schalk. 1989. 'Excel' sweetpotato. HortScience 24:171-172.

Jones, A., J.M. Schalk & P.D. Dukes. 1979. Heritability estimates for resistance in sweetpotato to soil insects. J. Am. Soc. Hort. Sci. 104:424-426.

Jones, A., P.D. Dukes, J.M. Schalk, M.G. Hamilton & R.A. Baumgardner. 1987. 'Southern Delite' sweetpotato. HortScience 22:329-330.

Jones, A., P.D. Dukes, J.M. Schalk, M.A. Mullen, M.G. Hamilton, R. Paterson & T.E. Boswell. 1980. W-71, W-115, W-119, W-125, W-149 and W-154 sweet potato germplasm with multiple insect and disease resistances. HortScience 15:835-836.

Jones, A., P.D. Dukes, J.M. Schalk, M.G. Hamilton, M.A. Mullen, R.A. Baumgardner, D.R. Patterson & T.E. Boswell. 1983. 'Resisto' sweetpotato. HortScience 18:251-252.

Jones, A., P.D. Dukes & J.M. Schalk, M.G. Hamilton, M.A. Mullen, R.A. Baumgardner, D.R. Patterson & T.E. Boswell. 1985. 'Regal' sweetpotato. HortScience 20:781-782.

Kim, K.S. 1980. Cytopathology of spotted cucumber beetle hemocytes containing virus-like particles. J. Invertebr. Pathol. 36:292-301.

Krysan, J.L. & T.A. Miller. 1986. Methods for the study of pest *Diabrotica*. Springer-Verlag, New York.

Lipa, J. 1975. An outline of insect pathology. Foreign Scientific Publication Department of the National Center for Scientific Technical and Economic Information. Warsaw, Poland.

Luginbill, P. 1918. The southern corn rootworm and farm practices to control it. U.S.D.A. Farmers Bull. 950.

Risch, S. 1981. Ants as important predators of rootworm eggs in the neotropics. J. Econ. Entomol. 74:88-90.

Roberts, R.J. 1968. An introduced pasture beetle *Plectris aliena* Chapin (Scarabeidae: Melolonthinae). J. Austr. Entomol. Soc. 7:15.

Schalk, J.M. 1984. Multiple insect resistance in sweet potato, pp. 55-56. *In* M.A. Mullen & K.A. Sorenson (eds.). Proceedings of a sweetpotato weevil workshop. Department of Entomology, North Carolina State University, Raleigh.

Schalk, J.M. 1986. Rearing and handling of *Diabrotica balteata*, pp. 49-56. *In* J.L. Krysan & T.A. Miller (eds.). Methods for the study of pest *Diabrotica*. Springer-Verlag, New York.

Schalk, J.M. & C.S. Creighton. 1989. Influence of sweet potato cultivars in combination with a biological control agent (Nematoda: *Heterorhabditis heliothidis*) on larval development of the banded cucumber beetle (Coleoptera: Chrysomelidae). Environ. Entomol. 18:897-899.

Schalk, J.M. & A. Jones. 1985. Major insect pests, pp. 59-78. *In* J.C. Bouwkamp (eds.). Sweet potato products: a natural resource for the tropics. CRC Press, Boca Raton, Florida.

Schalk, J.M., J.R. McLaughlin & J.H. Tumlinson. 1990. Field response of feral male banded cucumber beetles to the sex pheromone 6,12-dimethylpentadecan-2-one. Fla. Entomol. 73:292-297.

Smith, J.B. 1910. Insects injurious to sweet potatoes in New Jersey. N.J. Agric. Exp. Stn. Bull. 229:3-7.

Turco, C.P., S.H. Hopkins & W.H. Thames, Jr. 1970. Susceptibility of five insect pests to *Neoaplectana glaseri* Steiner, 1929. J. Parasitol. 56:277-280.

Waddill, V.H. 1978. Sexual differences in foraging on corn of adult *Labidura riparia* (Derm.:Labiduridae). Entomophaga 23:339-342.

16. Biology and Management of Wireworms on Sweet Potato

Richard B. Chalfant and Dakshina R. Seal[1]
Department of Entomology
University of Georgia
Coastal Plain Experiment Station
Tifton, Georgia 31793 U.S.A.

The storage root of sweet potato, *Ipomoea batatas* (L.) Lam., is injured by a complex of Coleopterous soil insects. Throughout most of the production areas of the world, the sweet potato weevils are the most important pests (Chalfant *et al.* 1990). In certain areas of the southeastern United States where sweet potato weevils are not a serious threat, wireworms and other larvae of the WDS complex (wireworm, *Systena, Diabrotica*) are the major pests (see Chapter 15). These insect pests are listed as minor pests in other parts of the world (Hill 1983).

Wireworms, larvae of Elateridae, chew small to moderate size, round holes in the storage root, especially in the periderm of the expanding root. On the harvested root, early feeding damage is shallow to deep, healed over scars depending on the depth of initial penetration. Late-season feeding by large larvae causes large, round, ragged holes while smaller larvae produce shallow holes. This damage does not affect biomass or nutritional quality but reduces grade, marketability, and profit. Larvae of cucumber beetles, *Diabrotica* spp., and flea beetles, *Systena* spp., cause similar damage to the storage roots (Cuthbert 1967). It is difficult to distinguish injury to the roots by these species from that by wireworms. Annual losses to the sweet potato industry due to wireworm injury are serious in Georgia ($ 1.6 million on 6,000 acres) (Douce & McPherson 1989). Projected losses

[1] Current address: Tropical Research and Education Center, University of Florida, I.F.A.S., Homestead, Florida 33031 U.S.A.

from wireworm feeding in the southeastern United States, where over 100,000 acres of sweet potato are grown, exceed $25 million annually.

The southern potato wireworm, *Conoderus falli* Lane, had been considered the major wireworm species infesting sweet potato in the southeastern United States (Cuthbert 1967). The tobacco wireworm, *C. vespertinus* (Fabricius), and the gulf coast wireworm, *C. amplicollis* (Gyllenhal) were important in some areas. More recently, *C. scissus* Schaeffer and *C. rudis* Brown were found to be the predominant species in Georgia (Seal 1990). *Conoderus amplicollis* and *C. falli* are also important wireworm species infesting sweet potato fields in Georgia. No recent information about the wireworm complex in other areas has been published.

WIREWORM BIOLOGY

Employment of insect pest management depends upon knowledge of the biology of the pest, and a method for sampling and identifying the various species. This is an onerous task in a soil medium. For this reason, very little information is available on the biologies of the wireworm species infesting sweet potato. Only "the consequences rather than the processes of soil insect behavior have been measured" (Villani & Wright 1990). The following is a review of the general biology of the group; however, there is considerable variation among even closely related species.

The Wireworm Species Complex in Sweet Potato

In Georgia, four species of wireworms are predominant in sweet potato fields (Figure 16.1). *Conoderus scissus* has a life cycle of from two to three years with seven to ten instars. Movement of *C. scissus* larvae is affected by variations in soil moisture. This wireworm infests peanut, cowpea, and corn, in addition to sweet potato. *Conoderus rudis* usually infests sweet potato as an immigrant from nearby fields. The life cycle is short with three to four stadia developing in three to four months. They are found near the soil surface and are prevalent in weedy fields. Populations build up later in the season. The gulf coast wireworm, *C. amplicollis*, is also abundant in Georgia, especially in corn fields. The life cycle lasts from two to three years. Larvae remain at a depth below 10 cm inside smooth, walled cavities. They move to the surface only for feeding and are less sensitive to moisture than *C.*

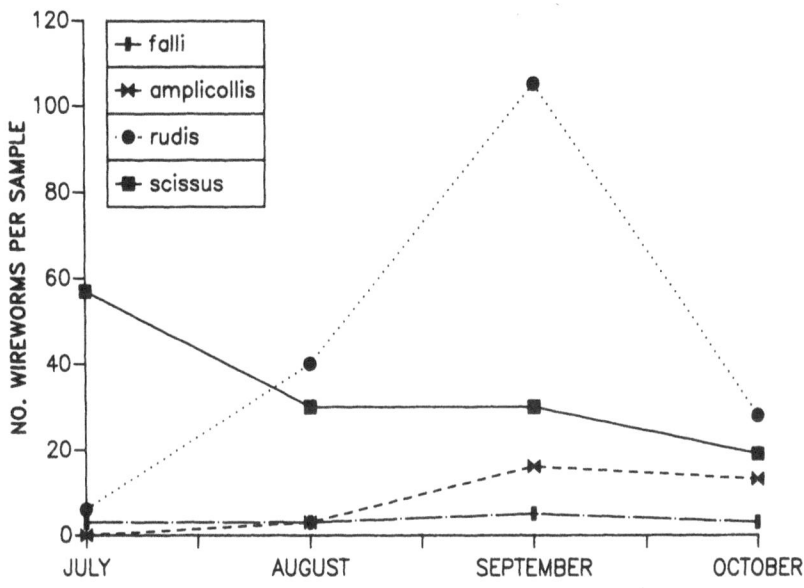

Fig. 16.1. Numbers of wireworms sampled in sweet potatoes during the growing season.

scissus (Seal 1990). The biology of *C. falli* has been studied extensively (Day *et al.* 1971, Dobrovsky 1954, Norris 1957). There are multiple generations. *Conoderus vespertinus,* a minor pest species in sweet potato fields in Georgia, has a life cycle of one year (Conradi & Eagerton 1914, Jewett 1946). During four years of sampling in sweet potato fields in Georgia, only five larvae were collected (Seal 1990). *Glyphonyx* spp. are relatively abundant in old corn fields with high organic residues in the soil and were rarely found in sweet potato fields (Seal 1990). *Conoderus lividus* (Degeer) is also a very minor pest of sweet potato in Georgia (Seal 1990). Seal (1990) developed keys to identify the species affecting sweet potato in Georgia. Keys for other areas and crops are reported (Jewett 1946, Lane 1931, 1953, Rabb 1963, Riley & Keaster 1979, Smith 1975).

Influence of Soil Temperature and Moisture on the Behavior of Wireworms

Variations in crop damage occur during different months of the year. These are caused by changes in wireworm responses to seasonal differences in vegetation cover, soil moisture, temperature, and by the interactions of these factors (Dowdy 1944, Shepard 1973). Soil temperature is an important factor affecting the seasonal vertical distribution of wireworms in Kansas (Bryson 1930) and in the Northwest (Jones & Shirck 1942). In Missouri, *Melanotus* spp. wireworms in corn fields moved to the upper 10 to 15 cm of the soil surface when soil temperatures increased in the lower depths (Fisher *et al.* 1975). Upward movement continued as vegetation began to grow and the soil warmed to 13°C. Falconer (1945) observed no gravitational response when *Agriotes* spp. were exposed to temperatures ranging from 11° to 24°C.

Soil moisture has been recognized as an environmental factor of considerable importance (Thomas 1940). Zacharuk (1962) found that *Ctenicera aeripennis destructor* (Brown) larvae avoided dry soil more at higher than at lower temperatures. Movement of wireworms to baits may have been limited by high soil temperature and low soil moisture (Toba & Turner 1983). Different levels of soil temperatures and moisture evoke behavioral responses in soil inhabiting arthropods that influence their vertical and horizontal distribution within the soil (Shepard 1973, Villani & Wright 1990). Low soil moisture favored intense burrowing activity, whereas high moisture frequently immobilized *Agriotes* spp. Larvae tended to feed when the soil was moderately dry and not when wet (Lees 1943). Jones (1951) found that the Pacific coast wireworm, *Limonius canus* LeConte, and the sugarbeet wireworm, *L. californicus* (Mannerheim), required a moisture-saturated soil atmosphere for survival. Campbell (1937) found that *L. californicus* became conditioned to seasonal changes and had a higher optimum temperature in the summer and fall than in the winter or spring.

Influence of Food and Feeding Behavior of Wireworms

Upward and lateral movements of wireworms were greater than downward movement when food was present. The presence of food also appeared to increase movements more at high temperatures than at low temperatures (Stone & Folley 1955). Differences in larval capture rates within seasons were due to changes in larval activity associated with food searching behavior (Doane 1981).

Various extrinsic and intrinsic factors influence feeding of wireworms. *Agriotes* spp. consumed more wheat at 32°C but fed more regularly at 18°C (Falconer 1945). Wireworm larvae display active and inactive feeding periods under various combinations of temperature and photoperiod (Keaster *et al.* 1975), regardless of substrate or presence of food (Pill *et al.* 1976). Apablaza *et al.* (1977) observed nonfeeding of *M. opacicollis* larvae, probably due to inadequate moisture, food type or ecdysis. Periods of little feeding may follow periods of high feeding activity even during satisfactory environmental conditions (Burrage 1963, Evans 1944, Evans & Gough 1942, Kosmachevskii 1959, Kring 1962, Roberts 1919, Zacharuk 1962).

Wireworms are polyphagous (Guryeva 1969, Kosmachevskii 1959), thus enabling residual populations to exist without preferred hosts. They attack a wide range of crops and weeds (Kosmachevskii 1959, Strickland 1939) and can feed on detritus in the soil (Evans & Gough 1942). Many are predacious and cannibalistic (Bryson 1930, Hyslop 1915, Kabanov 1969, Kring 1959, Negm & Hensley 1969, Seal 1990, Strickland 1939, Thorpe *et al.* 1947, Turnock 1969), including *C. scissus*, *C. amplicollis*, and *C. falli*. Carbon dioxide (CO_2) is a sensory stimulus for several classes of soil animals and may influence the dynamics of crop damage by wireworms. Wireworms use CO_2 gradients to locate respiring plant roots (Doane *et al.* 1975, Klinger 1957, Riley 1979, Stiles & Leach 1960). Diffusion of CO_2 through soil is regulated by soil particle size and moisture content. As the size of soil interstices decreases, CO_2 diffusion decreases; thus, differences in soil type, moisture, and tillage may affect wireworm host searching.

Distribution

An understanding of wireworm larval distribution, including vertical and lateral movements in soil, is useful for effective sampling, for proper placement and timing of insecticides, and for analysis of data. The distribution of wireworm larvae in soil can be characterized horizontally, vertically, and spatially. Seal (1990) characterized the distributions of wireworms in sweet potato fields in Georgia. Horizontal distribution was typically food related and was measured when abiotic factors were favorable within the sweet potato field during the growing season. More wireworms were observed within 10 cm of sweet potato storage roots and numbers of larvae decreased with an increase in the lateral distance away from the roots. Increased lateral movement of wireworms was observed immediately after rain within 0 to 4 cm of the soil profile. Horizontal distribution appears to be limited by physical changes of soil. In sweet potato fields, raised beds limit lateral movement of *Conoderus* spp. larvae. Very few wireworms were observed within the wheel furrows between the beds. Species related horizontal movement was prominent in Georgia sweet potato fields. *Conoderus rudis* were more abundant at the soil surface and showed active lateral movement within a short range of sweet potato plants than other species of *Conoderus*. Earlier stages (up to third instar) of *C. scissus* and *C. amplicollis* showed more active lateral movement than later instars (Seal 1990).

Vertical distribution is also species related and has been observed in wireworms with life cycles greater than one year. Most wireworms are sensitive to soil moisture changes and show seasonal vertical movement. Wireworms change their vertical distribution seasonally. In certain parts of the southern United States, the highest populations occur during July in the upper 10 to 20 cm of the soil profile and within 10 cm laterally to the plant. A downward migration to a depth of 20 to 30 cm typically occurs in November in temperate regions (Burrage 1963, Day *et al.* 1971, Fisher *et al.* 1975, LaFrance 1968, Seal 1990). Size-related vertical distribution was observed in *C. scissus* and *C. amplicollis* during the sweet potato growing season. Late instar *C. scissus* and *C. amplicollis* were mostly found below 10 cm of soil surface and moved upward during feeding and molting. As a result, advanced stage *C. scissus* and *C. amplicollis* remained below the insecticide treated layer and developed a residual population.

Populations of *C. falli* and *C. vespertinus* were usually distributed according to Poisson expectations. However, about half of the samples analyzed also conformed reasonably well to the negative binomial distribution (Onsager & Day 1975). Jones (1937) reported that *Limonius* spp. were distributed according to Poisson expectations, but that conclusion was based on a small fraction of his data. Landell (1938), Yates and Finney (1942), and Finney (1941, 1946) concluded that wireworms did not occur in a Poisson distribution. Seal (1990) determined that the frequency distribution of wireworms in sweet potato fields was a negative binomial early in the sweet potato growing season and approached a random distribution toward harvest. Wireworm damage holes at harvest showed a clumped distribution. In cowpea fields, Nilakhe *et al.* (1982) found that wireworm distributions were adequately described by both a Poisson and negative binomial distribution. They also found that the roots of two cowpea plants were damaged for each wireworm present in the soil. Day *et al.* (1971) established a mathematical relationship between *C. falli* density and damage to sweet potato. An average of 1.6 larvae per ft² resulted in an average of 32% root damage.

Sampling

There is a need for simple and effective techniques to detect soil insect pests, estimate population densities, and to establish damage thresholds. Detection techniques for soil insects have consisted of both absolute (large nonbaited area) and relative (baited smaller area) samplings. Absolute sampling is laborious, time consuming, and impractical for large fields. Relative sampling involves measuring the population in relative units which only allows comparisons to be made in space and time. Corrections are needed to estimate density (Southwood 1966).

Burrage (1963a) used potatoes to estimate seasonal feeding of wireworms economically important in the Canadian prairies. Apablaza *et al.* (1977) assessed the attractiveness of various types of baits in the laboratory for species of wireworms occurring in the midwestern United States. The efficacy of different baiting techniques in the field in Missouri was evaluated by Ward and Keaster (1977).

Relative sampling of wireworms has recently been explored using seed and rolled oat baits (Jansson & Lecrone 1989, Jansson *et al.* 1989). Others have also showed that various food baits, such as wheat seed, wheat:corn seed mixtures, wheat flour, sorghum seed, and oatmeal, were effective at trapping and collecting wireworm larvae in soil (Apablaza *et al.* 1977, Bynum & Archer 1987, Doane 1981, Foster & Ward 1976, Toba & Turner 1983). Kirfman *et al.* (1986) described an efficient bait trap for wireworms in midwest corn fields. Corn and wheat seed are placed in a buried 1.9 liter container filled with vermiculite. Attractants described above (Ivashchenko *et al.* 1986, Ponomarenko *et al.* 1981) could also be used for sampling. Kamm *et al.* (1983) attracted adults using traps baited with geraniol and some of its derivatives.

Several baits and baiting techniques were compared in sweet potato fields in Georgia (Seal 1990). The baits were seeds of various crops. Their attractiveness depended on the wireworm species, soil conditions (Figure 16.2), time of placement, and length of exposure in

310

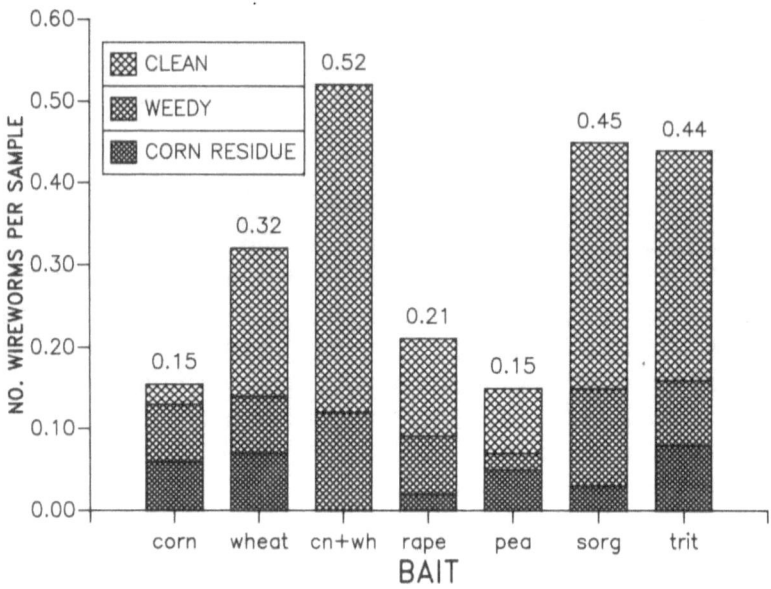

Fig. 16.2. Attraction of various seed baits to wireworms in different field conditions. Numbers on top of bar are the treatment means.

the fields (Table 16.1). A corn:wheat mixture attracted more wireworms than the other seed baits, including corn or wheat alone, cowpea, rape, sorghum, and triticale. Significantly fewer wireworms were found at a depth > 10 cm below the bait (Figure 16.3). Thus, sampling could be confined to the top 10 cm of soil.

PEST MANAGEMENT

Insecticidal Control

Damage tolerances for soil insects on sweet potato in the United States are low. To qualify for U.S. No. 1 grade, storage roots must be virtually void of insect damage. To produce damage-free crops, growers in the United States depend heavily upon soil-applied chemical insecticides that are incorporated into the soil. In the United States, sweet potatoes are produced from transplanted slips or vine

Table 16.1. Effect of age of different seed baits on their attractiveness to wireworm larvae in southern Georgia.

	Mean number of wireworms [a]		
Seed bait	1 Week	2 Weeks	3 Weeks
Corn	2.4b	5.9b	9.1a
Wheat	2.7b	4.1ab	6.9a
Corn:Wheat (1:1)	2.8b	6.3a	5.6a
Rape	1.1a	2.5a	1.4a
Cowpea	1.5a	2.9a	2.1a
Sorghum	1.1b	3.8a	3.5ab
Triticale	1.7c	9.6a	6.7b

[a] Means within a row followed by the same letter are not significantly different ($P = 0.05$) by an LSD mean separation test.

cuttings to raised beds. In the southern United States (e.g., Georgia), growers usually incorporate chlorpyrifos (Lorsban) sprays and granules into the soil for wireworm control before planting using distributors attached to the bedding device. Aldicarb (Temik), usually applied simultaneously to control plant pathogenic nematodes, also aids in wireworm control (Chalfant *et al.* 1987). Some growers broadcast granular or liquid insecticides by air or ground equipment before bed preparation and incorporate chemicals with a disc harrow. The half-life of chlorpyrifos is about six weeks in the loamy sand soils of Georgia (Seal 1990). Sweet potatoes require approximately 20 weeks to harvest. To control the next generation of wireworms and other soil insects, an additional application of insecticide (diazinon or parathion) is usually made at root enlargement, about six weeks after planting. By this time, the foliage interferes with incorporation. Irrigation or a timely rainfall may help. Even with these management practices, growers still sustain losses due to wireworms.

The interrelationship among soil edaphic factors and soil insect biology and management were reviewed by Villani and Wright (1990). They stated that physical factors that cause the movement of target insects by as little as 1 cm into the soil profile may put some insects out of the active zone of some control agents. Management of wireworms and other soil insects has been recently reviewed by Chalfant *et al.* (1990). As discussed above, the soil insect pest species infesting sweet

312

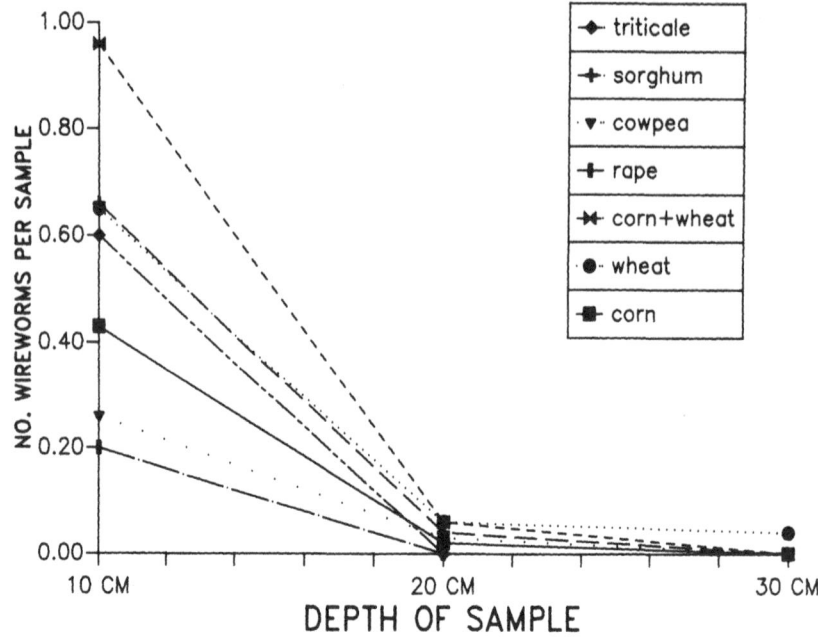

Fig. 16.3. Mean numbers of *Conoderus* spp. wireworm larvae per bait in three soil depths (0 to 10, 10 to 20, and 20 to 30 cm) of different food bait traps in sweet potato fields in southern Georgia.

potato vary considerably in life cycle, distribution within the field, behavior and probably response to insecticides. Most sweet potato crops are planted in fields containing existing wireworm populations. Infestations of *Diabrotica* spp., *Systena* spp., later infestations of second generation wireworms, and other soil insects will cause additional injury (Cuthbert 1967). Long residual, broad-spectrum organochlorines have given the most dependable control in the past (Chalfant *et al.* 1979). Because of insecticide resistance and environmental concerns, these compounds are no longer available. Currently labeled insecticides have relatively short effective residuals and their control does not extend much beyond six weeks (R.B. Chalfant, unpubl. data). Control is erratic (Chalfant *et al.* 1987) and is influenced by formulation, method of incorporation, and soil conditions, including composition, moisture, temperature, and presence of microorganisms (Burkhardt & Fairchild 1967, Getzin 1985, Harris 1966, 1967, Lichtenstein 1966, Villani & Wright 1990, Weidhass *et al.* 1961). Adequate soil moisture is especially critical to activate and incorporate many soil insecticides (Harris 1966, 1967). Insecticides applied to the soil, particularly for prophylactic purposes, are subject to enhanced biodegradation (Felsot 1989). High temperatures that occur on the soil surface in warm areas where sweet potatoes are grown may cause rapid volatilization of insecticides and reduced residual effectiveness.

There are many reports comparing application methods for control of wireworms affecting potato; however, potatoes are planted as cut tubers and those methods do not apply to sweet potato which is planted as vine cuttings and slips. Georgia research from 1983 to 1985 (Chalfant *et al.* 1987) showed that fonofos and chlorpyrifos applied before planting and incorporated (PPI) were effective against wireworms. Granules were more effective than emulsifiable concentrate (EC) formulations. Effectiveness in some tests improved if chlorpyrifos was followed by an additional insecticide (diazinon, ethoprop, or parathion) at root enlargement (RTE) and incorporated by rain or irrigation. Dry soil conditions at RTE rendered the treatment ineffective. Insecticides applied only at RTE were ineffective. Recent research showed that chlorpyrifos was more effective when incorporated as a postplant band (PPB) over newly transplanted vine cuttings (R.B. Chalfant, unpubl. data) (Figure 16.4). Preplant applications of dichloropropene (Telone II), aldicarb, or fenamiphos to control nematodes provided additional control of wireworms (data not shown). Cheshire *et al.* (1988) used a CAT scan to detect the movement of iodine/lead-coated insecticidal granules in soil and their efficacy at controlling soil insects. They found marked differences in the distribution of granules and concomitant wireworm control depending on method of soil preparation and insecticidal incorporation. The best result was obtained when soil was plowed prior to planting followed by broadcasting granular insecticides and then forming beds with subsequent smoothening of the soil. Similar results were obtained when granular insecticides were broadcast before plowing and then harrowing the field, forming the beds, and smoothening the soil.

Various agricultural chemicals can be applied to foliage and soil by injection in center-pivot irrigation systems (= chemigation [Chalfant & Young 1984]). Chemigation of insecticides can be an efficient, cost-effective method for wireworm control on sweet potato (Chalfant *et al.* 1987). Information is needed on the effects of formulations and irrigation rates in relation to depth of penetration in the soil.

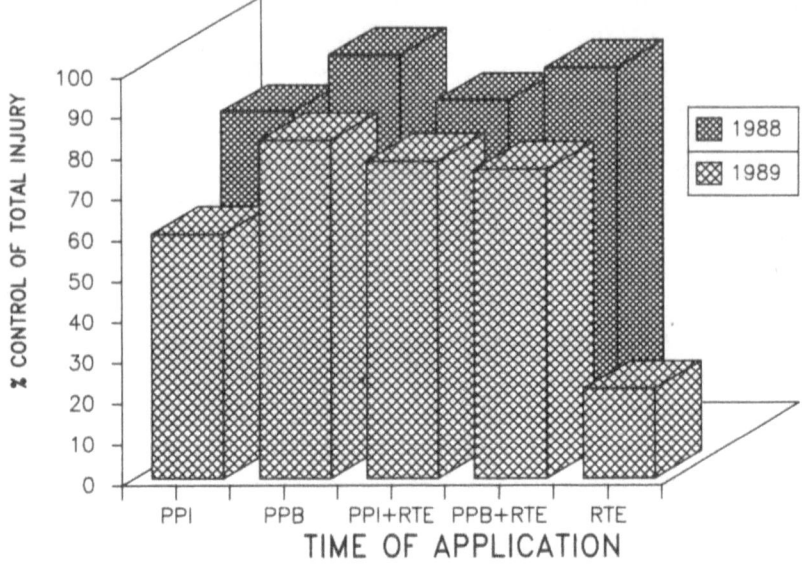

Fig. 16.4. The effect of time and method of application of granular chlorpyrifos on control of wireworms (% injury to storage roots) in sweet potato in southern Georgia (PPI, pre-plant broadcast and incorporated; PPB, post-plant band and incorporated; RTE, broadcast at root enlargement).

Host Plant Resistance

Cultivars resistant to the WDS complex have been developed at the U.S. Department of Agriculture, U.S. Vegetable Laboratory at Charleston, South Carolina (Schalk *et al.* 1986a). This subject was reviewed in Chalfant *et al.* (1990) and by Schalk *et al.* (see Chapters 15, 20). Resistance was related to periderm thickness and materials in the periderm (Schalk *et al.* 1986b). Resistant cultivars can be used as an adjunct or alternative to insecticides according to Cuthbert and Jones (1978). During 1988, several breeding lines and cultivars were evaluated in Georgia with inconclusive results. There were significant differences in wireworm damage among cultivars and breeding lines, but lines classified as 'resistant' were not significantly better than several of the commercial 'susceptible' cultivars evaluated. Differences in the soil insect complexes between Charleston, South Carolina and Tifton, Georgia may account for the differences in results.

Biological Control

There is a dearth of information related to biological control of wireworms in general and wireworms of sweet potato specifically. Day *et al.* (1971) observed that a fungus, *Metarhizium anisopliae* (Metchnikoff) Sorikin and sea gulls caused some larval mortality of *C. falli.* Bell and Hamalle (1971) showed that *C. falli* susceptibility to *M. anisopliae* differed between laboratory and field populations. Carabids, mites, nematodes, and wireworms of the same and different species have been associated with wireworm mortality in laboratory cultures (Seal 1990). In addition to intra- and interspecific predation among wireworms, natural enemies (parasites, predators, and pathogens) cause reductions of wireworm populations in the fields of other crops (Hyslop 1915, Hawkins 1936, Thomas 1940). Potential biological control agents such as *M. anisopliae* and *Beauveria bassiana* (Bals.) Vuill. can be maintained in the soil by moisture management and suitable incorporation. The center-pivot irrigation system is suited for both of these procedures. This form of irrigation is not labor intensive and microbials can be injected into the irrigation water and into the soil (Storey *et al.* 1987). These authors found penetration of propagules up to 20 cm when conidia were formulated in water. When formulated in oil, propagules were restricted to the upper 5 cm of soil.

Cultural Control

Different cropping sequences, winter covers (including weeds), and tillage can affect the occurrence, survival, and population dynamics of pest and beneficial insect species both above and below the ground (All *et al.* 1979, Andow *et al.* 1986, Brust *et al.* 1985, 1986, Rivers *et al.* 1979). Kulash (1943) found that sections of a field that had been in cultivation for a number of years had a higher larval population of *Limonius agonus* (Say) than of *Ludius* spp. or *Melanotus* spp. The wireworm population was much higher in cultivated areas than in grassland. In the plots that had been planted to truck crops or potatoes for five consecutive years, the former had lower wireworm populations than the latter, which was probably due to the fall removal of crop residues, the absence of a winter crop, and the more frequent

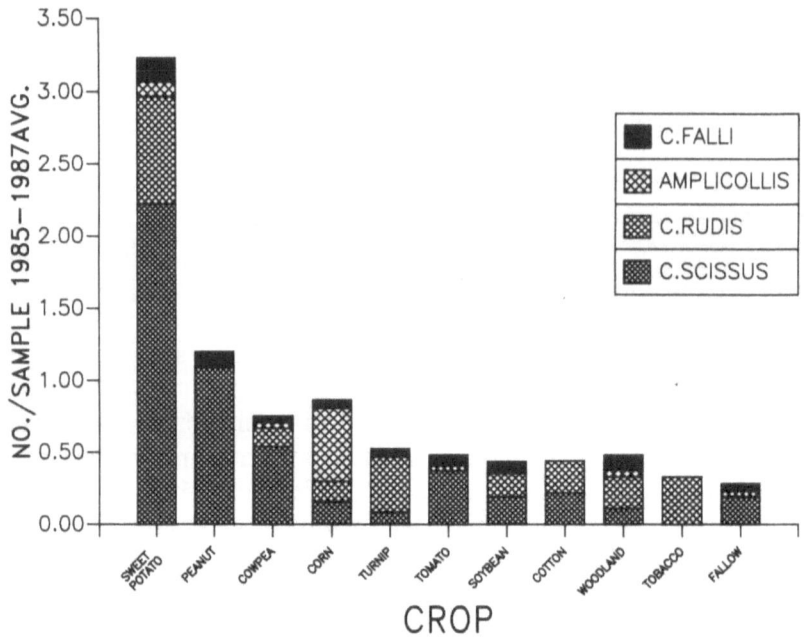

Fig. 16.5. Preference of wireworm species for various crops in the sweet
potato production area.

cultivation of the vegetables. Summer covers can serve as sources of
wireworms and affect crop loss due to wireworm feeding in the
following potato crop (Jansson & Lecrone 1991, McSorley *et al.* 1987).

In Georgia, Seal (1990) found that different wireworm species have
different spectra of crop preference. Most wireworms were sampled
from fields planted with sweet potato. Peanut, corn, and cowpea were
also attractive crops (Figure 16.5). The most prevalent species in
nearly all crops was *C. scissus*, except for *C. amplicollis* in corn and *C.
rudis* in weedy areas (Figure 16.6). Because wireworms overwinter in
the larval or prepupal stage, sweet potatoes rotated with one of these
crops may suffer heavy wireworm damage. For example, *C. scissus* is a
major pest of both sweet potato and peanut (Figure 16.7). Growers
should avoid planting these two crops in sequence and should use non-
hosts in rotations with sweet potato. *Conoderus rudis* is especially
attracted to weedy fields (e.g., Figure 16.6), and good weed control
should help reduce densities of this species. Deep plowing may also
have an adverse affect on wireworm survival by exposing them to

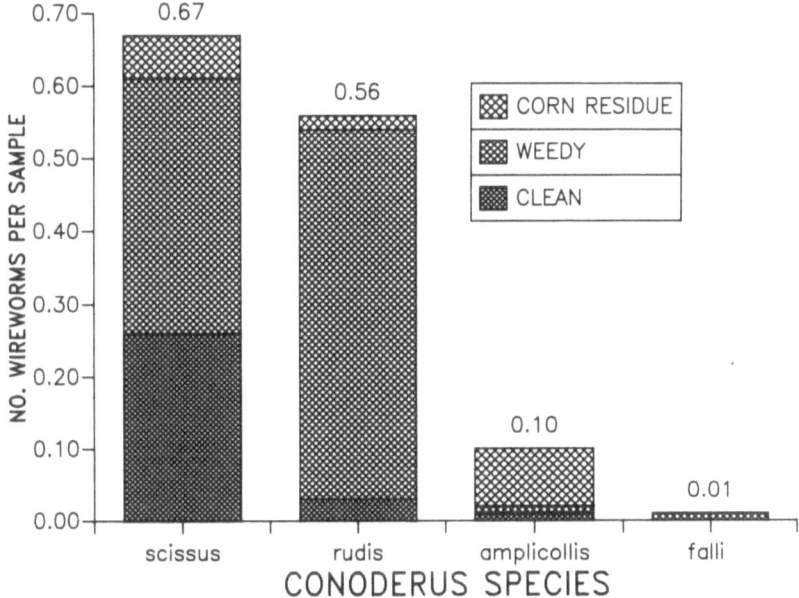

Fig. 16.6. The effect of field conditions on the population density of *Conoderus* wireworm species in southern Georgia.

predators. Day *et al.* (1964) reported that insecticides applied but not incorporated on fall cover crops were effective for control of wireworms in sweet potato planted the following year. Villani and Wright (1990) suggested the use of baits and antifeedants for soil insect management. Solutions containing amino acids (Ponomarenko *et al.* 1981) and a protein-vitamin concentrate (Ivashchenko *et al.* 1986) were found to attract wireworms in the soil and were suggested for use as baits to attract wireworms away from plant parts.

CONCLUSIONS

Management of wireworms and other insect pests of sweet potato for commercial production in the United States is achieved largely by use of insecticides, although growers express interest in resistant varieties (R.B. Chalfant, pers. obser.). There is only one effective soil insecticide (chlorpyrifos) available for control of wireworms on sweet

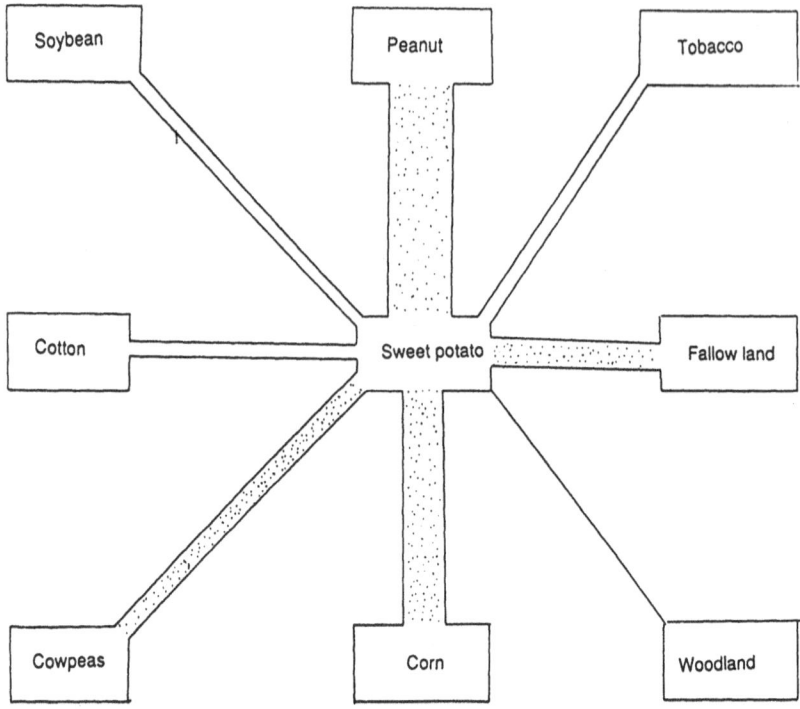

Fig. 16.7. The interrelationships among various crops in the sweet potato production area of Georgia and the life cycle of *C. scissus*. The width of the lines that connect sweet potato to the various crops signifies abundance of *C. scissus* in sweet potato fields that follow after each of the crops.

potato in the southern United States. Alternate insecticides are difficult to obtain because sweet potato is considered a minor crop when compared to other more economically important crops, such as cotton and corn. There is a limited sales potential for insecticides on sweet potato and manufacturers are unwilling to pay for the high costs of labelling. High market standards and low damage tolerances place severe demands on non-insecticidal pest management measures. An integrated approach to management of these pests is long past due. Components should include insecticides (chemical, microbial, antifeedants, growth regulators), cultural control, host plant resistance, and biological control. The data base needed to develop pest management strategies is difficult to obtain. Sweet potatoes are damaged by a complex of pests that is poorly understood because of the difficulty in doing research in a soil medium. Most of the damage is not recognized until well after the pest has gone thus making it difficult to relate cause and effect.

Proper application technique, essential for adequate insecticidal performance, is lacking or poorly understood with soil insecticides. Weak points include improper incorporation and distribution, unsuitable soil conditions, and bad timing. Wireworms migrate in the soil in response to both extrinsic and intrinsic influences as described above. Thus, the target pest may not be in the treated area when the insecticide is applied or may fail to contact an improperly applied toxicant. Poor soil moisture at the time of application may result in a lack of insecticidal activity or early dissipation of the active ingredient (Harris 1966, 1967). Improper use of the soil insecticide can result in rapid degradation of the compound (Felsot 1989).

Resistant cultivars are the most readily available non-insecticidal strategy (Schalk *et al.* 1986a). They have not had widespread acceptance because of the conservatism in the industry and because they have not shown resistance to the soil insect complexes in all areas. When combined with insecticides, control from resistance could be improved and insecticide dosages could be reduced (Cuthbert & Jones 1978).

Many of the insect pests (wireworms, *Diabrotica* spp., and *Systena* spp.) are polyphagous. Growers often rotate sweet potatoes without regard to the previous crop which may serve as a reservoir for wireworms and other soil pests. Thorough knowledge of the effects that previous crops and management practices have on buildup of soil pests would alleviate some problems. The entire agroecosystem rather than the isolated crop should be considered. Use of cover crops that are non-hosts between sweet potato plantings offer promise. Changes in tillage practices could affect both pest and beneficial soil arthropods (Stinner & House 1990). Baits used to sample wireworms could be planted to lure wireworms away from the crop. It should be noted that some of these techniques are incompatible with current production practices.

Use of biological control agents, including fungi, such as *B. bassiana* and *M. anisopliae,* and entomopathogenic nematodes has received little research attention. Limiting factors, in addition to lack of effective agents, include difficulties in reaching the target and in maintaining a suitable soil environment. Proper soil moisture could be maintained and microbials could be applied and incorporated using center-pivot irrigation (Storey *et al.* 1987).

Techniques are available but are not yet in place for improving the management of wireworms and other soil insects affecting sweet potato. Further research on varietal resistance, pest biology as it relates to movement within and between crops, amenable cropping sequences, and efficient insecticidal application techniques offer the most immediate prospects for success. Effective, manageable biological control agents, trapping mechanisms, and changes in tillage systems should also be investigated.

REFERENCES

All, J.N., R. Gallaher & M. Jellum. 1979. Influence of planting date, preplanting weed control, irrigation and conservation tillage practices on efficacy of planting insecticide applications for control of the lesser cornstalk borer in field corn. J. Econ. Entomol. 72:265-268.

Andow, D.A., A.G. Nicholson, H.C. Wien & H.R. Wilson. 1986. Insect populations on cabbage grown with living mulches. Environ. Entomol. 15:293-299.

Apablaza, J.U., A.S. Keaster & R.H. Ward. 1977. Orientation of corn-infesting species to wireworms towards baits in the laboratory. Environ. Entomol. 6:715-718.

Bell, J.V. & R.J. Hamalle. 1971. Comparative mortalities between field-collected and laboratory-reared wireworm larvae. J. Invertebr. Pathol. 18:150.

Brust, G.E., B.R. Stinner & D.A. McCartney. 1985. Tillage and soil insecticide effects on predator-black cutworm interactions in corn agroecosystems. J. Econ. Entomol. 78:1389-1392.

Brust, G.E., B.R. Stinner & D.A. McCartney. 1986. Predator activity and predators in corn agroecosystems. Environ. Entomol. 16:1017-1021.

Bryson, H.R. 1930. A study of field practices as related to wireworm infestations (Coleoptera: Elateridae). J. Econ. Entomol. 23:303-315.

Burkhardt, C.C. & M.L. Fairchild. 1967. Toxicity of insecticides to house crickets and bioassay of treated soils in the laboratory. J. Econ. Entomol. 60:1496-1503.

Burrage, R.H. 1963. Seasonal feeding of larvae of *Ctenicera destructor* and *Hypolithus bicolor* (Coleoptera: Elateridae) on potatoes placed in the field at weekly intervals. Ann. Entomol. Soc. Am. 56:306-313.

Bynum, E.D., Jr. & T.L. Archer. 1987. Wireworm (Coleoptera: Elateridae) sampling for semiarid cropping systems. J. Econ. Entomol. 80:164-168.

Campbell, R.E. 1937. Temperature and moisture preferences of wireworms. Ecology. 18:479-489.

Chalfant, R.B. & J.R. Young. 1984. Management of insect pests of broccoli, cowpeas, spinach, tomatoes, and peanuts with chemigation by insecticides in oil and reduction of watermelon virus 2 by chemigated oil. J. Econ. Entomol. 77:1323-1326.

Chalfant, R.B., M. Hall & D.R. Seal. 1987. Insecticidal control of soil insects of sweet potatoes in the Georgia Coastal Plain. Appl. Agric. Res. 2:152-157.

Chalfant, R.B., S.A. Harmon & A.L. Stacey. 1979. Control of the sweet potato flea beetle and southern potato wireworm on sweet potatoes in Georgia. J. Ga. Entomol. Soc. 14:354-358.

Chalfant, R.B., R.K. Jansson, D.R. Seal & J.M. Schalk. 1990. Ecology and management of sweet potato insects. Annu. Rev. Entomol. 35:157-180.

Cheshire, J.M., Jr., E.W. Tollner, B.P. Verma & W.M. Blum. 1988. Radiographic detection of soil-incorporated granular pesticides and impact of application methods on wireworm management. Am. Soc. Engin. Paper No. 88-1013.

Conradi, A.F. & H.C. Eagerton. 1914. The spotted click beetle (*Monocrepidius vespertinus* Fab.). South Carolina Agric. Exp. Stn. Bull. 179.

Cuthbert, F.P., Jr. 1967. Insects affecting sweet potatoes. U.S. Dept. Agric. Handbook No. 329.

Cuthbert, F.P., Jr. & A. Jones. 1978. Insect resistance as an adjunct or alternative to insecticides for control of sweet potato soil insects. J. Amer. Soc. Hort. Sci. 103:443-445.

Day, A., F.P. Cuthbert, Jr. & W.J. Reid, Jr. 1964. Control of the southern potato wireworm, *Conoderus falli*, on early-crop potatoes. J. Econ. Entomol. 57:469.

Day, A., F.P. Cuthbert, Jr. & W.J. Reid, Jr. 1971. The southern potato wireworm, its biology and economic importance in coastal South Carolina. U.S.D.A. Tech. Bull. No. 1443.

Doane, J.F. 1981. Evaluation of a larval trap and baits for monitoring the seasonal activity of wireworms in Saskatchewan. Environ. Entomol. 10:335-342.

Doane, J.F., Y.H. Lee, J. Klinger & N.D. Westcott. 1975. The orientation and response of *Ctenicera destructor* and other wireworms (Coleoptera: Elateridae) to germinating grain and to carbon dioxide. Can. Entomol. 107:1233-1252.

Dobrovsky, T.M. 1954. Laboratory observations on *Conoderus vagus candeze* (Coleoptera: Elateridae). Fla. Entomol. 37:123-131.

Douce, G.K. & R.M. McPherson. 1989. Summary of losses from insect damage and costs to control in Georgia, 1988. Ga. Agric. Exp. Stn. Special Publ. No. 64.

Dowdy, W.W. 1944. The influence of temperature on vertical migration of invertebrate inhabiting different soil types. Ecology 25:449-460.

Evans, A.C. 1944. Biology and physiology of wireworms of the genus *Agriotes* Esch. Ann. Appl. Biol. 31:235-250.

Evans, A.C. & H.C. Gough. 1942. Observations on some factors influencing growth in wireworms of the genus *Agriotes* Esch. Ann. Appl. Biol. 29:168-175.

Falconer, D.S. 1945. On the behavior of wireworms of the genus *Agriotes* Esch. (Coleoptera: Elateridae) in relation to temperature. J. Exp. Biol. 21:17-32.

Felsot, A.S. 1989. Enhances biodegradation of insecticides in soil: implications for agroecosystems. Annu. Rev. Entomol. 34:453-76.

Finney, D.J. 1946. Field sampling for estimation of wireworm populations. Biomet. Bull. 2:1-7.

Fisher, J.R., A.J. Keaster & M.L. Fairchild. 1975. Seasonal vertical movement of wireworm larvae in Missouri: influence of soil temperature on the genera *Melanotus* Esch. and *Conoderus* Esch. Ann. Entomol. Soc. Am. 68:1071-1073.

Foster, D.G. & C.R. Ward. 1976. Wireworm bait traps in grain sorghum. Texas Agric. Exp. Stn. PR-3425.

Getzin, L.W. 1985. Factors influencing the persistence and effectiveness of chlorpyrifos in the soil. J. Econ. Entomol. 78:412-418.

Guryeva, Y.L. 1969. Some trends in the evolution of click beetles (Coleoptera: Elateridae). Entomol. Rev. 48:154-159.

Harris, C.R. 1966. Influence of soil type on the activity of insecticides in soil. J. Econ. Entomol. 59:1221-1225.

Harris, C.R. 1967. Further studies on the influence of soil moisture on toxicity of insecticides in the soil. J. Econ. Entomol. 60:41-44

Hawkins, J.H. 1936. The bionomics and control of wireworms in Maine. Maine Agric. Exp. Stn. Bull. 381.

Hill, D.S. 1983. Agricultural insect pests of the tropics and their control. Cambridge Univ. Press, London.

Hyslop, J.A. 1915. Wireworms attacking cereal and forage crops. U.S.D.A. Bull. No. 156.

Ivashchenko, I.I. 1986. Feeding attractants in the control of Elaterids. Zachchita Rastenii.(*In* Rev. Appl. Entomol. [Ser. A] 77:2488).

Jansson, R.K. & S.H. Lecrone. 1989. Evaluation of food baits for pre-plant sampling of wireworms (Coleoptera:Elateridae) in potato fields in southern Florida. Fla. Entomol. 72:503-510.

Jansson, R.K. & S.H. Lecrone. 1991. Effects of summer cover crop management on wireworm (Coleoptera: Elateridae) abundance and damage to potato. J. Econ. Entomol. (In press).

Jansson, R.K., S.H. Lecrone & D.R. Seal. 1989. Food baits for pre-plant sampling of wireworms (Coleoptera: Elateridae) in potato fields in southern Florida. Proc. Fla. State Hort. Soc. 102:367-370.

Jewett, H.H. 1946. Identification of some larval Elateridae found in Kentucky. Ky. Agric. Exp. Stn. Bull. No. 489.

Jones, E.W. 1937. Practical field methods of sampling soil for wireworms. J. Agric. Res. 54:123-134.

Jones, E.W. 1951. Laboratory studies on the moisture relations of *Limonius* (Coleoptera:Elateridae). Ecology 32:284-293.

Jones, E.W. & F.H. Shirck. 1942. The seasonal vertical distribution of wireworms in the soil in relation to their control in the Pacific Northwest. J. Agric. Res. 65:125-142.

Kabanov, V.A. 1969. Biology and ecology of *Agriotes ponticus* Stepanov and *Melanotus fusciceps* Gyll. (Coleoptera: Elateridae) in Kransodor territory. Entomol. Rev. 48:307-311.

Kamm, J.A., H.G. Davis & L.M. McDonough. 1983. Attractants for several genera and species of wireworms (Coleoptera:Elateridae. Coleoptcrists Bull. 37:1678.

Keaster, A.J., G.M. Chippendale & B.A. Pill. 1975. Feeding behavior and growth of the wireworm *Melanotus depressus* and *Limonius dubitans*: effect of host plants, temperature, photoperiod and artificial diets. Environ. Entomol. 4:591-595.

Kirfman, G.W., A.J. Keaster & R.N. Story. 1986. An improved wireworm (Coleoptera:Elateridae) sampling technique for midwest cornfields. J. Kansas Entomol. Soc. 59:37-41.

Klinger, J. 1957. Uber die bedeutung des kohlendioxyds fur die orientierung der larven von *Otiorrhyncus sulcatus* F., *Melolontha* und *Agriotes* (Col.) in Boden (Vorlaugife Mitteilung). Mitt. Schweiz. Entomol. Ges. 30:317-322.

Kosmachevskii, A.S. 1959. Biology of *Agriotes litigiosus* var. *tauricus* Heyd. and *Agriotes sputator* L. (Coleoptera: Elateridae). Entomol. Rev. 38:663-672.

Kring, J.B. 1959. Predation and survival of *Limonius agronus* Say (Coleoptera: Elateridae). Ann. Entomol. Soc. Am. 52:534-537.

Kring, J.B. 1962. Feeding and molting in wireworms. Proc. 11th Intern. Cong. Entomol. (Vienna) 3:163-165.

Kulash, W.M. 1943. The ecology and control of wireworms in the Connecticut river valley. J. Econ. Entomol. 36:689-693.

LaFrance, J. 1968. The seasonal movements of wireworms (Coleoptera: Elateridae) in relation to soil moisture and temperature in the organic soils of Southwestern Quebec. Can. Entomol. 100:801-807.

Landell, W.R.S. 1938. Field experiments on the control of wireworms. Ann. Appl. Biol. 25:341-189.

Lane, M.C. 1931. U.S.D.A. Farmers Bull. 1657.

Lane, M.C. 1953. Distribution of wireworm *Conoderus vagus* Cand. U.S.D.A. Coop. Econ. Insect Rpt. 3:536.

Lees, A.D. 1943. On the behavior of wireworms of the genus *Agriotes* Esch. (Coleoptera: Elateridae). II. Reactions to moisture. J. Exp. Biol. 20:54-60.

McSorley, R., J.L. Parrado, R.V. Tyson, V.H. Wadill, M.L. Lamberts & J.S. Reynolds. 1987. Effect of sorghum cropping practices on winter potato production. Nematropica 17:45-60.

Negm, A.A. & S.D. Hensley. 1969. Evaluation of certain biological control agent of sugar cane borer in Louisiana. J. Econ. Entomol. 62:1009-1013.

Nilakhe, S.S., R.B. Chalfant & A. Day. 1982. Soil insect pests in cowpea fields with emphasis on distribution of wireworms. J. Ga. Entomol. Soc. 17:145-149.

Norris, D.M., Jr. 1957. Bionomics of the southern potato wireworm, *Conoderus falli* Lane. 1. Life history in Florida. Proc. Fla. State Hort. Soc. 70:109-111.

Onsager, J.A. & J.A. Day. 1975. Distribution of wireworms (*Conoderus* spp.) in soil samples. J. Ga. Entomol. Soc. 10:9-13.

Pill, B.A., A.J. Keaster & G.M. Chippendale. 1976. Larval survival and pupation of the wireworms *Melanotus depressus* and *Limonius dubitans* in natural substrates. Environ. Entomol. 5:845-848.

Ponomorenko, A.V., V.G. Kalyuzhnyi, P.D. Koktionov & A.A. Kazakaev. 1981. Control of migration of soil-inhabiting Coleoptera. (*In* Rev. Appl. Entomol. [Ser. A] 72:1289).

Rabb, R.L. 1963. Biology of *Conoderus vespertinus* in the Piedmont section of North Carolina (Coleoptera: Elateridae). Ann. Entomol. Soc. Am. 56:669-676.

Riley, T.J. 1979. Wireworms associated with corn. M.S. Thesis. University of Missouri-Columbia, Columbia.

Riley, T.J. & A.J. Keaster. 1979. Wireworms associated with corn: identification of larvae of nine species of *Melanotus* from the north central states. Ann. Entomol. Soc. Am. 72:408-414.

Rivers, R., K.S. Pike & Z.B. Mayo. 1979. Influence of insecticides and corn tillage systems in larval control of *Phyllophaga* J. Econ. Entomol. 70:794-796.

Roberts, A.W.R. 1919. On the life history of wireworms of genus *Agriotes* Esch., with some notes on that of *Athous haemorrhoidalis* F. Ann. Appl. Biol. 6:116-135.

Schalk, J.M., A. Jones & P.D. Dukes. 1986a. Factors associated with resistance in recently developed sweet potato cultivars and germplasm to the banded cucumber beetle, *Diabrotica balteata* LeConte. J. Agric. Entomol. 3:329-334.

Schalk, J.M., J.K. Peterson, A. Jones. P.D. Dukes & W.M. Walker, Jr. 1986b. The anatomy of sweet potato periderm and its relationship to wireworm, *Diabrotica, Systena* resistance. J. Agric. Entomol. 3: 350-356.

Seal, D.R. 1990. Management and biology of wireworms affecting sweet potato in Georgia. Ph.D. Dissertation University of Georgia, Athens.

Shepard, M. 1973. Response of *Melanotus communis* (Coleoptera: Elateridae) larvae to soil temperature and moisture. Can. Entomol. 105:577-580.

Smith, J.W. 1975. The click beetle subfamilies Pyrophorinae and Melanotinae (Coleoptera: Elateridae) in Missouri. Ph.D. Dissertation, University of Missouri, Columbia.

Southwood, T.R.E. 1966. Ecological methods. Methuen and Co. Ltd., London.

Stiles, W. & W. Leach. 1960. Respiration in plants. J. Wiley, New York.

Stinner, B.R. & G.J. House. 1990. Arthropods and other invertebrates in conservation tillage agriculture. Annu. Rev. Entomol. 35:200-218.

Stone, M.W. & F.B. Folley. 1955. Effect of seasons, temperature, and food on the movement of the sugar beet wireworm. Ann. Entomol. Soc. Am. 48:308-312.

Storey, G.K., W.A. Gardner, J.H. Hamm & J.R. Young. 1987. Recovery of *Beauveria bassiana* propagules from soil following application of formulated conidia through an overhead irrigation system. J. Entomol. Sci. 22:355-357.

Strickland, E.H. 1939. Life cycle and food requirements of the northern grain wireworm, *Ludius aeripennis destructor* Brown. J. Econ. Entomol. 32:322-329.

Thomas, C.A. 1940. The biology and control of wireworms. A review of the literature. Pa. Agric. Exp. Stn. Bull. 392.

Thorpe, W.H., A.C. Crombie, R. Hill & J.H. Darrah. 1947. The behavior of wireworms in response to chemical stimulation. J. Exp. Biol. 23:234-266.

Toba, H.H. & J.E. Turner. 1983. Evaluation of baiting techniques for sampling wireworms (Coleoptera: Elateridae) infesting wheat in Washington. J. Econ. Entomol. 76:850-855.

Turnock, W.J. 1969. Predation by larval Elateridae on pupae of the pine looper, *Bupalus piniarius* (L.). Neth. J. Zool. 19:393-416.

Villani, M.G. & R.J. Wright. 1990. Environmental influences on soil microarthropod behavior in agricultural systems. Annu. Rev. Entomol. 35:249-269.

Ward, R.H. & A.J. Keaster. 1977. Wireworm baiting: use solar energy to enhance early detection of *Melanotus depressus*, *M. verberans* and *Aeolus mellillus* in Mid-west corn fields. J. Econ. Entomol. 70:403-406.

Weidhass, D.E., M.C. Bowman & C.E. Schmidt. 1961. Loss of parathion and DDT to soil from aqueous dispersions and vermiculite granules. J. Econ. Entomol. 54:175-177.

Yates, F. & D.J. Finney. 1942. Statistical problems in field sampling for wireworms. Ann. Appl. Biol. 29:157-167.

Zacharuk, R.Y. 1962. Seasonal behavior of larvae of *Ctenicera* spp.. and other wireworms (Coleoptera: Elateridae), in relation to temperature, moisture, food, and gravity. Can. J. Zool. 40:697-718.

17. Vine Borers of Sweet Potato

Narayan S. Talekar
Asian Vegetable Research and Development Center
Shanhua, Tainan, Taiwan, Republic of China

Gene V. Pollard
Department of Plant Science and Biochemistry
Faculty of Agriculture
The University of the West Indies
St. Augustine, Trinidad and Tobago

Among about 300 insect and mite species that feed on sweet potato (*Ipomoea batatas* [L.] Lam.), three species of vine borers, *Omphisa anastomasalis* (Guenee), *Megastes grandalis* Guenee, and *M. pucialis* Snell (Lepidoptera: Pyralidae) are very destructive. They are second in economic importance to sweet potato weevils (*Cylas* spp. and *Euscepes postfasciatus* [Fairmaire]) where the pyralids and the weevils co-occur. The larvae of vine borers develop and feed inside the vines, especially in crowns, and cause significant yield reduction. Because of their cryptic nature, damage from these pests may go unnoticed and farmers often do not undertake any control measures to combat these pests. Yield losses of 30 to 50% due to *O. anastomasalis* have been reported (Ho 1970, Talekar & Cheng 1987). Cowland (1926) reported that *M. grandalis*-infested plants in Trinidad and Tobago failed to produce storage roots. For these reasons, successful management of vine borers is essential.

GEOGRAPHIC DISTRIBUTION AND HOST RANGE

Omphisa anastomasalis occurs mainly in Asia and has been reported in India, Sri Lanka, Malaysia, Indonesia, Taiwan, Hawaii and the Pacific (Zimmerman 1958), Bangladesh (Das & Islam 1985), China, Japan, Philippines, Cambodia, Laos, Vietnam, Burma, and Thailand (Anonymous 1968). Sweet potato is the principal host but

Zimmerman (1958) noted that it also feeds on *Stictocardia campanulata* House. On Penghu island in the Taiwan straits, it also feeds on *Ipomoea pes-caprae* (L.) Sweet. (AVRDC 1989a), and in Bangladesh, this insect attacks *I. aquatica* Forsk. (Das & Islam 1985). Franssen (1935) found them feeding on unidentified convolvulaceous species in Indonesia.

Megastes grandalis occurs in Trinidad and Tobago, Costa Rica, and certain parts of South America including Guyana, Brazil, and Venezuela (Fennah 1947, King & Saunders 1984, Montgomery 1945). In Trinidad, it causes extensive damage to sweet potato (Cowland 1926, Lowe & Wilson 1972). Sweet potato is the only host plant recorded for this pest (Cowland 1926, Fennah 1947), although Montgomery (1945) reported that *M. grandalis* also fed on certain convolvulaceous species in the laboratory; among 13 convolvulaceous species tested, full larval development occurred only on *I. aquatica* and *Merremia glabra*. *Megastes pucialis* occurs sporadically in Brazil (Bondar 1922), but little published information is available on this pest.

LIFE HISTORY

Omphisa anastomalis adults are white with a brownish yellow pattern on the wing (Figure 17.1A). They are active at night with peak mating activity occurring between 0200 and 0400 hours (AVRDC 1989b). They lay ellipsoidal, slightly domed, greenish eggs with a flat base, on the upper and lower surfaces of leaf lamina and on petioles (Figure 17.1B). Eggs are usually laid singly. Egg incubation lasts for about a week. Larvae bore into the stems soon after hatching (Figure 17.1C) and gradually eat their way down the vines. By the time larvae reach the crown, they are half grown. Full-grown larvae are 30 mm long and light purple, although they may also be yellowish white. The head capsule is brown, the ventral surface and legs are white, and the back and lateral sides have yellowish brown grooves (Figure 17.1D). There are six larval instars. The larval period usually lasts 30 to 35 days, but may vary between 21 to 92 days depending upon temperature (Hwa & Chow 1984). Pupation usually takes place in the vine (Figure 17.1E); however, larvae may also bore into storage roots and pupate when roots are close to the soil surface (Figure 17.1F). Pupation in roots is rare. Pupae are light brown and are covered with a white-brown web. Before pupating, the larva makes an exit hole in vines but leaves a papery epidermis on the vine intact. The pupal period lasts

Fig. 17.1. Various stages of *O. anastomasalis*: (A) adult; (B) egg on sweet
potato leaf petiole; (C) larval entry hole in vine; (D) larva feeding inside
sweet potato vine; (E) pupa within stem; (F) pupa within sweet potato
root; and (G) adult exit hole and frass around the crown.

about two weeks. Adults exit through the hole made by the larvae. The
only signs of an *O. anastomasalis* infestation in sweet potato plants are
the presence of frass in the crown and adult exit holes (Figure 17.1G).
Adult females may survive for up to 10 days during which time they
may lay about 300 eggs (Franssen 1935).

The life-history of *M. grandalis* has been studied in Trinidad and
Tobago (Cowland 1926, Fennah 1947, Montgomery 1945, West 1977).
Cowland (1926) indicated that eggs are laid either singly or in groups
of not more than five and never in clusters, whereas West (1977)
observed egg laying in clusters. The egg incubation lasts 6 to 9 days
depending upon temperature. Newly-hatched larvae feed externally on
the plant for a short period after eclosion or may bore immediately
into the stem (Cowland 1926, Dalgarno 1931, Montgomery 1945).

Tunnelling may occur just above the ground level or in stem tips. The site of the initial infestation (internal boring or external feeding) may be important for developing a management program for this pest. There are usually seven larval instars although there may be an eighth (West 1977). The larval period may last for up to nine weeks. The mature larva is typically oligopodous and pinkish with a well developed head capsule, and measures about 30 mm in length. Immature stages have been described by Cowland (1926). The full grown larva spins a silken cocoon inside the stem above ground level; pupation lasts for about two weeks. The adult stage lasts about 4 days (range, 1 to 9 days). The life cycle from egg to egg takes 59 days (range, 41 to 97 days) (West 1977).

NATURE OF DAMAGE

Major damage to sweet potato plants results from the larvae boring into the main stem leading to storage roots. Frass accumulates on the soil surface near the opening of the larval tunnel. Major feeding damage occurs in the crown. The conductive tissues of the vine are damaged by larval feeding which probably decreases the flow of water and photosynthates. Vines with severe tunnelling show weak growth and poor foliage development, which later yellows and wilts (Das & Islam 1985). The distal part of the vine above the damage site may often die. Such plants show poor storage root formation. *Omphisa anastomasalis* larvae may bore directly into storage roots. Chung (1923) reported serious damage by this pest in Hawaii which resulted in death of the plants. The insect reduced yield by about 30% in Malaysia (Ho 1970). Root yield was shown to be negatively correlated with the percentage of plants damaged in Taiwan (Figure 17.2) (Talekar & Cheng 1987).

Megastes grandalis damage results in swelling and splitting of stems and deposit of frass external to the split stem or on the ground; silken threads may also be seen at the split stems. Other symptoms of infestation such as plant stunting and leaf shredding may also be observed, especially during the dry season (COPR 1978, Cowland 1926, Fennah 1947, Montgomery 1945). This may also result in a lack of storage root production (Cowland 1926). Recent studies indicate that the within-plant distribution of larvae, as well as the growth stage at which the plant is attacked, may determine the extent of yield loss. Little or no yield reduction occurs when larvae are confined to the stem or if the plant is attacked late in its development (Lowe 1971). A

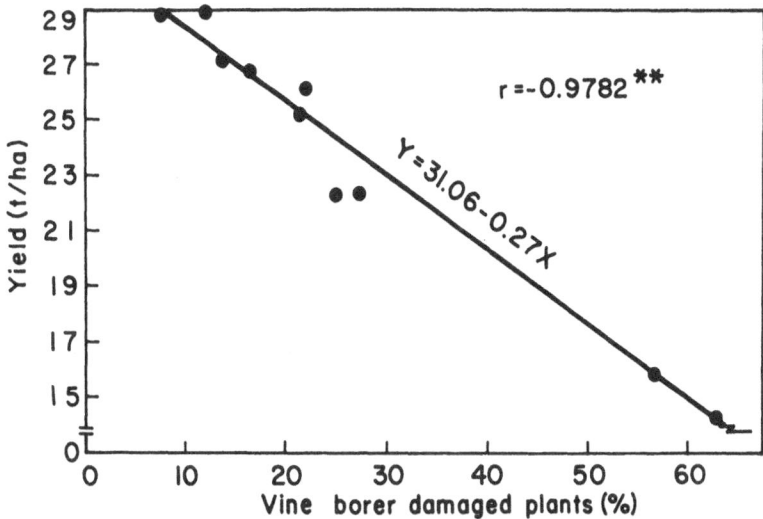

Fig. 17.2. Relationship between sweet potato root yield and percentage damage of storage roots by *O. anastomasalis* (from Talekar & Cheng 1987; reprinted with permission from J. Econ. Entomol. [Entomological Society of America, Lanham, Maryland]).

small percentage (17%) of larvae enter the roots. Larval infestation of the roots is dependent on both the root stalk diameter and the distance between the point of entry of the larvae into the stem and the storage root (West 1977).

One characteristic feature of the insect-host plant relationship is the ability of infested plants to compensate for vine damage. *Megastes grandalis* damage results in loss of almost all, if not all, of medullary phloem and primary and secondary xylem tissue in vines (Duncan 1973, Lowe 1971). However, anomalous secondary cambial growth results in production of new vascular tissue (Duncan 1973), but this is thought to be insufficient to prevent yield loss (West 1977).

MANAGEMENT OF VINE BORERS

Host Plant Resistance

Both vine borer species are rather difficult and costly to control by conventional methods such as chemical insecticides, due to their concealed mode of life history. Their cryptic nature also reduces the

Table 17.1. *Omphisa anastomasalis* damage and root yields of selected sweet potato accessions. [a]

Entry	Identification	% damaged plants [b]	Yield, t/ha [b]
I55	PI 324889	23.5b	7.2b
I92	PI 308208	17.0b	7.6b
AIS35-2	AVRDC selection	83.3a	27.0a

[a] Planting date, 28 April 1982; observation date, 4 October 1982.

[b] Data are means of four replicates. Means in each column followed by the same letter are not significantly different ($P > 0.05$) by Duncan's (1955) multiple range test.

exposure of larvae and pupae to parasitoids and predators. This predicament enhances the use of insect-resistant plants, either alone or in combination with other management approaches, to combat the pests economically. Tomagari and Motoki (1910) screened 22 local sweet potato cultivars for resistance to *O. anastomasalis* in Taiwan. All cultivars were infested and damage ranged from 26 to 94% of the plants. More recently, Talekar and Cheng (1987) screened 383 sweet potato accessions for resistance to *O. anastomasalis* on Penghu island in the Taiwan Straits. Subsequent field screenings of certain selected accessions with the least damage confirmed that resistance to *O. anastomasalis* was present in two accessions, 'I55' and 'I92' (Table 17.1). Significantly less *O. anastomasalis* larvae survived on resistant plants than on susceptible plants in the laboratory when first instars were allowed to feed inside the vines for three weeks (AVRDC 1989b). Resistant accessions have low yield potential and are currently being used to breed insect-resistant sweet potato cultivars at AVRDC in Taiwan.

Lowe and Wilson (1972) found that six West Indies cultivars varied in resistance to *M. grandalis*. One cultivar, '049', was free of *M. grandalis* damage. Other cultivars, '03/62', 'I62', and '28/7' were moderately resistant and had 7, 10, and 13% of all plants damaged, respectively. Nearly half (40%) of all plants were damaged for the most susceptible cultivar, 'C9/9'. West (1977) found no significant difference in *M. grandalis* damage to vines of '049' and three other cultivars, 'AI6/15', 'A28/7', and 'Red Vine' for up to 18 weeks after planting. He did observe, however, that '049' remained free of vine borers for six weeks after planting and was the least acceptable cultivar for oviposition by female *M. grandalis*. Wilson and Lowe (1973) found that '28/7' plants damaged by *M. grandalis* developed extensive secondary phloem to compensate for damage and this enabled uninterrupted translocation of assimilates from foliage to storage roots and concomitantly resulted in adequate storage root growth. Cultivation of such a tolerant cultivar will enable growers to obtain adequate yields in areas where *M. grandalis* is endemic.

Biological Control

Although vine borers spend their larval and pupal stages concealed inside the plant, they are attacked by several hymenopterous parasitoids. Table 17.2 lists all known parasitoids of *O. anastomasalis* and *M. grandalis*. Most adult parasitoids oviposit in vine borer larvae concealed inside the stem, presumably through the transparent membrane that covers adult exit holes. No information exists on the extent of parasitism or its role in reducing vine borer damage to sweet potato.

Cultural Control

No specific cultural practices are undertaken by the farmers in Southeast Asia to control *O. anastomasalis*. Moulding (i.e., earthing up or hilling), which is done by Trinidadan farmers, has been shown to increase storage root formation and may help reduce *M. grandalis* damage. West (1977) investigated the effect of earthing up of plants on blockage of adult exit holes and concomitant control of *M. grandalis*. Adult exit holes are located just above the soil surface. Earthing up at 3 to 6 weeks after planting was not effective; larvae moved upwards in the stem to tunnel exit holes above the soil surface. However, if such earthing up operation was done seven weeks or more after planting, some control was achieved and no new exit holes were made. This was further evidenced by the presence of dead adult moths in the vines.

Chemical Control

Farmers rarely see vine borer damage because the insects develop and feed within the vines and only occasionaly in the storage roots. For

Table 17.2. Known parasitoids of two vine borer species.

Parasitoid species	Location	References
O. anastomasalis		
Hymenoptera		
Chelonus blackburni Cameron	Hawaii	Zimmerman (1958)
Enytus (= *Dioctes*) *chilonis* Cushman	Hawaii	Zimmerman (1958)
Pristomerus hawaiiensis Perkins	Hawaii	Zimmerman (1958)
Tetramorium guineense Fabricius	Taiwan	Sonan (1929)
Xanthopimpla flavolineata Cameron	China	Hwa & Chow (1984)
X. stemmator Thunberg	China, Japan	Hwa & Chow (1984)
	Taiwan	Anonymous (1934)
		Sonan (1929)
X. emaculata Szepligeti	Japan	Anonymous (1934)
M. grandalis		
Hymenoptera		
Apanteles sp.	Trinidad	Montgomery (1945)
A. sesamiae	Trinidad	West (1977)
Brachymeria sp.	Trinidad	West (1977)
Chelonus (as *Chelonius*) *busckiella*	Trinidad	Montgomery (1945)
Eiphosoma (= *Xiphosoma*) *azteca*	Trinidad	Cowland (1926)
E. (as *Eiphosema*) *batatae*	Trinidad	Cowland (1926)
		Montgomery (1945)
Trichogramma minutum Riley [a]	Trinidad	Cowland (1926)
Diptera		
Masicera abdominalis	Trinidad	Cowland (1926)
		Montgomery (1945)
Sarcophaga lambens	Trinidad	Cowland (1926)
		Montgomery (1945)
S. steinodonatis	Trinidad	Cowland (1926)

[a] Egg parasite, all others are larval parasitoids.

this reason, insecticide use for the control of vine borers has been rare. The cryptic nature of the immature stages of these pests also minimizes the contact between residues of conventional chemical insecticides and vine borers. The brief exposure time between egg hatch and the boring into the stem by first instars further reduces the potential for contact between insecticide residues and vine borers, unless chemical insecticides are applied frequently, which may not be economical. Nonetheless, researchers have tested the efficacy of insecticides to obtain control of vine borers at minimum cost.

Ho (1970) controlled *O. anastomasalis* by making six applications of carbaryl or BHC at 14-day intervals starting two weeks after planting in Malaysia. In Taiwan, foliar treatments of deltamethrin (0.025 kg AI/ha) and broadcast applications of carbofuran granules (2 kg AI/ha) over the rows at 14-day intervals resulted in excellent control of *O. anastomasalis* (AVRDC 1981). Yields of plants treated with carbofuran were two times greater than those in nontreated plots. The yield increase was due to *O. anastomasalis* control. Unfortunately, the carbofuran treatment is not economical because the price of sweet potato roots in the local market is usually low. Dipping of stem cuttings in a solution of carbaryl (0.05% AI) for 15 minutes before planting did not control this pest (AVRDC 1981).

Prior to the introduction of modern organic insecticides, Urich (1921) reported high yields were produced when *M. grandalis* was controlled by dipping sweet potato cuttings in Bordeaux mixture and lead arsenate before planting, and then treating fields with two applications of this mixture made once a month for two consecutive months. Various modern insecticides have now been recommended for the control of *M. grandalis* in Trinidad and Tobago (Fennah 1947, Parasram 1969, 1973a, 1973b). Chlorfenvinphos was the standard chemical insecticide recommended following Parasram's (1969) series of experiments in which he obtained a six-fold increase in marketable root yield by the use of this chemical. Lowe and Wilson (1972) obtained partial control of *M. grandalis* when the sweet potato cuttings were dipped in dieldrin (0.05% AI) followed by 14-day interval applications of methomyl (1.4 kg AI/ha) and malathion (1.4 kg AI/ha).

Integrated Control

At present, no single strategy is effective at controlling either species of vine borer. However, several of the control methods outlined above may contribute to the development of an integrated management strategy. For both pest species, several species of natural enemies have been identified. More information, however, is still needed concerning levels of parasitism to develop effective biological control programs. Development of resistant, even moderately resistant, cultivars will contribute considerably to control of both vine borers. Sources of resistance to *O. anastomasalis* have been identified and agronomic materials with resistance to this species will be available from AVRDC in the near future. Resistant cultivars also have potential for controlling *M. grandalis*, considering that the biology and damage caused by this pest are similar to that of *O. anastomasalis*. It may also be possible to develop cultivars that are tolerant to vine borers because both species rarely feed on storage roots. Cultivars that are tolerant to *M. grandalis* already exist in Trinidad and certain AVRDC breeding lines show similar tolerance to *O. anastomasalis*.

Insecticides are rarely used to control *O. anastomasalis* in Southeast Asia, where use of insecticides for control of pests of other crops is widespread. Minimal insecticide use provides an opportunity for biological control agents to proliferate. In Trinidad, however, insecticides are routinely used by farmers which might limit the usefulness of biological control agents.

Research at AVRDC (1989b) showed that females of *O. anastomasalis* produce a sex pheromone that attracts males. Research is currently underway at AVRDC to identify, bioassay, and synthesize the active component(s) of the pheromone. Availability of sex pheromone may facilitate monitoring and help improve control of this vine borer. Lastly, *O. anastomasalis* feeds on various wild *Ipomoea* spp. Removal of these alternate hosts may help reduce the damage by this pest to sweet potato.

In summary, losses due to sweetpotato vine borers can be minimized by integrating various control tactics such as removing alternate, wild *Ipomoea* spp. hosts, using resistant or tolerant sweet potato cultivars, where such germplasm is available, and, when economical, using chemical insecticides judiciously to minimize the exposure of natural enemies to toxicants. The potential sex pheromone may also help improve management programs in the future.

It must be emphasized that the available information on vine borers is very sketchy. For example, no information exists on the seasonality of the pests nor on plant susceptibility in relation to its phenological growth stage. Similarly, information on various mortality factors that govern the population dynamics is not available. Predators and parasitoids may play an important role in the management of vine borers, but data on their ecology and impact on vine borer populations are lacking. Quantitative information on these topics is essential for the development of sustainable, integrated management programs. Unfortunately, due to a decline in production of sweet potato in Asia, which produces 90% of world's total sweet potato, such information may not be available for *O. anastomasalis* in the near future.

REFERENCES

Anonymous. 1934. Insects and diseases of agriculture in Taiwan, No. 2, Field Crop Insect Pests, Government House, Development Office, Taipei, Taiwan (In Japanese).

Anonymous. 1968. Insect pests of quarantine crops 7. Yokohama Plant Quarantine News 337:2-3 (In Japanese).

AVRDC. 1981. Progress report for 1979. Asian Vegetable Research and Development Center, Shanhua, Taiwan, Republic of China.

AVRDC. 1989a. AVRDC Progress Report Summaries 1988. Asian Vegetable Research and Development Center, Shanhua, Taiwan, Republic of China.

AVRDC. 1989b. 1987 Progress Report. Asian Vegetable Research and Development Center, Shanhua, Taiwan, Republic of China.

Bondar, G. 1922. Injurious insects. XXI. *M. pucialis*, a caterpillar infesting the sweet potato. Chacaras Quintaes 25:473-474 (In Portugese) (*In* Rev. Appl. Entomol. 10:472).

COPR. 1978. Pest control in tropical root crops. PANS Manual No. 4, Centre for Overseas Pest Research, London.

Cowland, J.W. 1926. Notes on the sweet potato pyralid moth, *Megastes grandalis* Guen. Bull. Entomol. Res. 17:369-372.

Chung, H.L. 1923. The sweet potato in Hawaii. Hawaii Agric. Exp. Stn. Bull. 50.

Dalgarno, W.I. 1931. Sweet potato growing in Trinidad. D.T.A. Thesis, University of the West Indies, St. Augustine, Trinidad.

Das, G.P. & M.A. Islam. 1985. Sweetpotato vine borer, *Omphisa anastomasalis* Guenee (Pyralidae: Lepidoptera): a newly recorded pest in Bangladesh. Bangladesh J. Agric. 10:65-69.

Duncan, D.B. 1955. Multiple range and multiple F test. Biometrics 11:1-42.

Duncan, E.J. 1973. Anomalous growth in the stem of *Ipomoea batatas* (L.) Lam. infested with *Megastes* larvae. Ann. Bot. 37:981-985.

Fennah, R.G. 1947. Insect pests of food crops in the Lesser Antilles. Department of Agriculture for Windward Islands, St. George's, Grenada, British West Indies.

Franssen, C.J.H. 1935. Insect pests of the sweet potato crop in Java. Korte Meded. Inst., Peziekt., Buitenzorg 10:205-225. (In Dutch) (Translation published. 1986. Asian Vegetable Research and Development Center, Shanhua, Taiwan, Republic of China).

Ho, H.T. 1970. Studies on some major pests of sweet potatoes and their control. Malay. Agric. J. 47:437-452.

Hwa, Y.S. & S.T. Chow. 1984. Sweet potato cultivation in China. Shanghai Scientific Publication Association, Shanghai, China (In Chinese).

King, A.B.S. & J.L. Saunders. 1984. The invertebrate pests of annual food crops in Central America. Overseas Development Administration, London.

Lowe, S.B. 1971. Tuberization in the sweet potato, *Ipomoea batatas* (L.) Lam. M.Sc. Thesis, University of the West Indies, St. Augustine, Trinidad.

Lowe, S.B. & L.A. Wilson. 1972. Preliminary evidence for the existence of differential susceptibility of sweet potato cultivars to *Megastes grandalis* (Guen.) infestation in West Indian sweet potato cultivars. Trop. Agric. 49:361-362.

Montgomery, R.H. 1945. Pests of sweet potato in Trinidad and their control. D.T.A. Thesis, University of the West Indies, St. Augustine, Trinidad.

Parasram, S. 1969. Effects of *Megastes* incidence on yield of sweet potato, p. 102. *In* Annual report of the Department of Crop Science, Section E., Rep. Fac. Agric., University of the West Indies, 1968-1969, St. Augustine, Trinidad.

Parasram, S. 1973a. Control of insect pests of some food crops in the Caribbean. Dept. Agric. Ext., University of the West Indies Ext. Bull. No. 7.

Parasram, S. 1973b. Evaluation of insecticides in *Megastes* control. Annual report of the Department of Crop Science. University of the West Indies, 1971-1972. Department Paper No. 10:36-37.

Talekar, N.S. & K.W. Cheng. 1987. Nature of damage and sources of resistance to sweetpotato vine borers (Lepidoptera: Pyralidae) in sweet potato. J. Econ. Entomol. 80:788-791.

Tomagari, T. & T. Motoki. 1910. Investigation on insect pests in Taiwan, pp 131-134. *In* Taiwan Agricultural Station Special Report No. 4. Taiwan Government Office, Taipei, Taiwan (In Japanese).

Urich, F.W. 1921. Entomologist's reports for 1919 and 1920. Rept. Dept. Agric. Trinidad and Tobago for 1919 and 1920. Port of Spain. (*In* Rev. Appl. Entomol. 10:236).

West, S.A. 1977. Studies on the biology and ecology of the sweet potato stem borer *Megastes grandalis* Guen. in Trinidad. M.Sc. Thesis, University of the West Indies, St. Augustine, Trinidad.

Wilson, L.A. & S.B. Lowe. 1973. Development of supplementary translocatory tissues in intact and *Megastes*-infested 'tuber stalks' of sweet potato (*Ipomoea batatas* [L] Lam). J. Hort. Sci. 43:223-226.

Zimmerman, E.C. 1958. Insects of Hawaii, Vol. 8. University of Hawaii, Honolulu.

18. Management of Insect Vectors of Viruses Infecting Sweet Potato

James W. Moyer and Richard C. Larsen
Department of Plant Pathology
North Carolina State University
Raleigh, North Carolina 27695 U.S.A.

Field observations and assays have revealed virus symptoms and the presence of one or more viruses in virtually all sweet potatoes, *Ipomoea batatas* (L.) Lam., grown from nonvirus-tested material. In many instances the endemic nature of these viruses has facilitated the natural incorporation of high levels of tolerance to regional viruses via selection and propagation of asymptomatic plants. Although resistance is the primary control strategy, it may also be possible to control the pathogen through other strategies directed at the vector.

Accurate virus diagnosis and identification is an important requisite for any management practice. Although tolerance to viruses has improved sweet potato production, it has made diagnosis difficult due to a lack of adequate detection and screening assays for several virus diseases of sweet potato, and in some areas, resulted in a general complacency about the importance of virus diseases in sweet potato. There is, however, justified concern that a virus isolate which is mild or latent in one location on one group of cultivars, may have considerably greater effects, either by itself or in combination with other viruses, when introduced into a new geographic location where local cultivars have a different genetic background. Thus, it is important that necessary precautions be taken to prevent the inadvertent distribution of viruses with germplasm.

A concerted effort is being made in several laboratories to discover the etiology of diseases with symptoms that have been associated with virus infection. Recently, a group of sweet potato virologists developed a list of 14 different viruses or virus-like agents which infect sweet potatoes. A summary of the best characterized members of the list is presented in Table 18.1.

STATUS OF KNOWN VIRUSES AND VIRUS DISEASES

Sweet Potato Feathery Mottle Virus

Many strains of the sweet potato feathery mottle virus (SPFMV) are found nearly everywhere sweet potatoes are grown. Some of the synonyms used for SPFMV isolates include russet crack virus, sweet potato virus A, sweet potato ringspot virus, sweet potato leafspot virus and probably internal cork virus (Cadena-Hinojosa & Campbell 1981, Cali & Moyer 1981, Campbell *et al.* 1974, Loebenstein & Harpaz 1960, Moyer 1986, Sheffield 1957, Yang 1972). Coinfection by SPFMV with an unknown virus is frequently a problem in determining the etiology of disease complexes.

The variability of symptoms associated with SPFMV infection are as much a function of the host genotype and environment as they are the virus strain or isolate (Alconero 1972, Cali & Moyer 1981, Campbell *et al.* 1974, Moyer 1986, Moyer & Cali 1985, Moyer & Kennedy 1978). Symptoms on sweet potato leaves may consist of the classic irregular chlorotic patterns (feathering) associated with the leaf midrib as well as faint or distinct chlorotic spots which have purple pigmented borders in some genotypes. These symptoms are observed predominantly on the older leaves. Veinclearing, veinbanding, and chlorotic spots are the predominant symptoms observed in the indicator host *I. setosa* (Kerr). However, symptoms may be mild and leaves produced after the initial flush may be symptomless. Some strains of SPFMV cause necrotic lesions on the exterior of the roots (russet crack disease) while another strain induces symptoms on the interior of the root (internal cork disease).

SPFMV is the most thoroughly characterized sweet potato virus (Campbell *et al.* 1974, Moyer & Cali 1985, Moyer & Kennedy 1978), and serological procedures have been developed (Cadena-Hinojosa & Campbell 1981, Esbenshade & Moyer 1982). SPFMV has many biological characteristics and cytopathic effects which support its classification as a potyvirus (Cali & Moyer 1981, Campbell *et al.* 1974) even though its biochemical properties such as the capsid protein (Mr 38,000 daltons), RNA (Mr 3.65 X 10^6 daltons) (Moyer & Cali 1985), and virion length (850 nm) (Cali & Moyer 1981, Nome *et al.* 1974) make it an atypically large potyvirus.

Table 18.1. A list of recognized viruses known to infect sweet potato.

Virus	Vector	Distribution	Assay Hosts [a]
CIP-Isolation (2-C6)	?	Unknown	*I. setosa* [b]
Cucumber mosaic virus (CMV)	aphid	Widespread	*N. glutinosa* *Cucumis sativus*
Reo-like	?	Asia	*I. setosa*
Sweet potato caulimo-like virus (SPCLV)	?	Widespread	*I. setosa* *N. megalosiphon*
Sweet potato feathery mottle virus (SPFMV)	aphid	Worldwide	*I. setosa*
Sweet potato latent virus (SPLV)	unknown	Asia	*I. setosa*
Sweet potato leaf curl virus (SPLCV)	*B. tabaci*	Taiwan, Japan, Nigeria	*I. setosa*
Sweet potato mild mottle virus (SPMMV)	*B. tabaci*	East Africa	*I. setosa* *N. tabacum* *N. glutinosa* *N. benthamiana*
Sweet potato mosaic virus (SPMV)		Taiwan	*I. setosa*
Sweet potato ring-spot virus (SPRSV)	unknown	Papua New Guinea	*I. setosa*
Sweet potato vein mosaic virus (SPVMV)	aphid	Argentina	*I. setosa*
Sweet potato yellow dwarf virus (SPYDV)	*B. tabaci*	Taiwan	*I. setosa*
Unknown virus	?	Puerto Rico	*I. setosa*
Whitefly-transmitted component of sweet potato virus disease (SPVD)	*B. tabaci*	Africa, Taiwan	TIB 8 sweet potato infected with SPFMV *I. setosa* [c]

Source: This is a list originally prepared by FAO/IBPGR that has been modified with newly available information.

[a] From FAO/IBPGR technical guidelines for the Safe Movement of Sweet Potato Germplasm.

[b] *Ipomoea setosa* is frequently difficult to inoculate by mechanical transmission; graft transmission from sweet potato is the most reliable means of transmission.

[c] Symptom expression is highly variable in *I. setosa*. This host should only be used after its reliability has been established in environment where test is conducted.

Sweet Potato Vein Mosaic Virus

Sweet potato vein mosaic virus (SPVMV) has been reported only in Argentina (Nome 1973). Direct comparison of the particle morphologies of SPFMV and SPVMV indicated that SPVMV has a modal length of 761 nm, significantly shorter than SPFMV. Sweet potato plants infected with this virus are severely stunted and produce fewer new roots. SPVMV is transmitted nonpersistently by aphids (Nome *et al.* 1974). Antiserum is not yet available to compare this virus to other known potyviruses or to assay sweet potatoes from other countries for the presence of this virus.

Sweet Potato Latent Virus

Sweet potato latent virus (SPLV), formerly designated sweet potato virus N, has been reported only in Taiwan (Chung *et al.* 1986). As the name suggests, infection of many sweet potato cultivars by SPLV does not result in obvious foliar symptoms. The host range of SPLV includes many species of *Convolvulus*, *Chenopodium*, and *Nicotiana*, such as *N. benthamiana* (Domin). Although it induces mild symptoms in *I. setosa*, it can be detected easily in this host by serology.

SPLV also has many characteristics of a potyvirus, including production of characteristic cytoplasmic inclusions. However, all attempts to demonstrate aphid and/or whitefly transmission have been unsuccessful. Thus, definitive classification of this virus awaits further characterization.

Sweet Potato Mild Mottle Virus

Sweet potato mild mottle virus (SPMMV) was isolated in East Africa from sweet potatoes exhibiting leaf mottling, veinal chlorosis, dwarfing, and general stunting of the plant (Hollings *et al.* 1976). SPMMV-infected *I. setosa* exhibit a bright yellow veinal chlorosis in as many as four leaves following inoculation. Subsequent leaves are symptomless. It has been referred to as SPV-T in preliminary reports and may be the same as virus B (Sheffield 1957). Virus B was also isolated from sweet potatoes in East Africa (Sheffield 1957).

Although the morphology of SPMMV and its cytoplasmic inclusions are similar to other potyviruses, its biological characteristics differ greatly from the type member. Most notable among the divergent characteristics is its host range which includes 45 species in 14 plant families. Additionally, SPMMV is vectored by the sweetpotato whitefly, *Bemisia tabaci* Gennadius, and its virions are relatively unstable using purification procedures for other potyviruses (Moyer & Kennedy 1978, Moyer, unpubl. data).

Sweet Potato Yellow Dwarf Virus

Sweet potato yellow dwarf virus (SPYDV), which frequently occurs with SPFMV, was described in Taiwan (Chung *et al.* 1986). The virion morphology and vector of SPYDV are similar to those of SPMMV. Neither virus has been adequately characterized nor has a direct comparison been made to determine the extent of biochemical relationships, but sufficient differences have been reported to continue designating SPYDV as a separate virus.

A Caulimo-like Virus

A virus with some properties similar to those of caulimoviruses was isolated from sweet potato by grafting and has been provisionally designated as sweet potato caulimo-like virus (SPCLV) (Atkey & Brunt 1987). It was first isolated in Puerto Rico and has since been isolated from sweet potatoes grown in Madeira, New Zealand, Papua, New Guinea, and the Solomon Islands (Atkey & Brunt 1987).

Early symptoms on *I. setosa* include chlorotic flecks along the minor veins and interveinal, chlorotic spots. These symptoms may develop into a general chlorosis resulting in wilting and premature death of the leaves. Virions associated with SPCLV were typical of caulimoviruses, but some of the inclusions were similar to the fibrillar ring inclusions induced by geminiviruses.

Other Whitefly Transmitted Agents

Other whitefly-transmitted agents have been isolated from sweet potato in Nigeria, Israel, Taiwan, and the United States (Chung *et al* 1985, Girardeau 1958, Hildebrand 1958, Loebenstein & Harpaz 1960, Schaefers & Terry 1976). These disorders are currently considered as separate agents, however definitive characterization and comparison of these agents has not been conducted. They have properties distinctly different from SPMMV in that they were not mechanically transmitted, they have a narrow host range, and no virions have been identified for these agents. The sweet potato virus disease (SPVD) which was described in Nigeria is one of the most thoroughly investigated disorders (Hahn 1979, Hahn et al. 1981, Schaefers & Terry 1976). This disease is due to the synergistic interaction of a strain of SPFMV and a whitefly-transmitted agent. Diseases similar to SPVD, designated as Georgia mosaic and yellow dwarf, have been reported in the United States (Girardeau 1958, Hildebrand 1958). Sweet potato veinclearing virus reported in Israel also induces symptoms similar to SPVD (Loebenstein & Harpaz 1960). Sweet potato leaf curl disease (SPLC) is another disease whose causal agent has been reported to be transmitted by *B. tabaci* (Chung et al. 1985, Yamashita et al. 1984).

MYCOPLASMALIKE ORGANISMS

Sweet Potato Witches' Broom

A disease of sweet potato causing witches' broom and stunting has been associated with a mycoplasmalike organism (Clark & Moyer 1988, Dabek & Sagar 1978, Jackson & Zettler 1983, Kahn et al. 1972, Summers 1951). The organism is transmitted in a persistent manner by the leafhoppers *Orosius lotophagorum ryukyuensis* and *Nesophrosyne ryukyuensis* (Clark & Moyer 1988).

Pleomorphic bodies were observed in sieve elements of infected plants when thin sections were examined with the electron microscope (Jackson & Zettler 1983). Detection has been restricted to electron microscopy and through the use of leafhopper and graft transmission to alternate hosts.

VECTOR CONTROL

The insect vector control examples used in this chapter are drawn from other virus-vector systems due to the absence of investigations which specifically address sweet potato management systems. In order to develop such strategies for managing vectors, an understanding of virus-vector relationships is essential. Vectors are capable of transmitting a virus for varying time periods which are roughly categorized as persistent, semipersistent, or nonpersistent transmission. Each type of transmission is defined by the time required for the vector to become viruliferous once in contact with a virus-infected plant, and by the length of time that the vector remains viruliferous. For example, if the vector can successfully transmit the pathogen for a minimum of four days, then it is said to be transmitted in a persistent manner. Additionally, these types of viruses are often circulative and occasionally propagative within the insect. Cohen *et al.* (1983) suggested that the unique squash leaf curl virus, transmitted in a persistent manner by *B. tabaci*, may also have deleterious effects on the vector. Whiteflies which were given a 24 hour acquisition access period on infected squash plants had an average lifespan of 19.5 days compared with a lifespan of 25.4 days for those given a 4 hour feeding period.

Semipersistently transmitted viruses are those which may be transmitted for up to 4 days, and viruses transmitted in a nonpersistent manner, often referred to as stylet-borne, are those transmitted for only a brief time by the vector, usually 3 to 4 hours or less (Gibbs & Harrison 1976).

It must be stressed that the time periods for levels of persistence are only approximate, and they may vary according to the type of vector, the dynamics of the vector, host plants, or environmental conditions.

Movement of viruses into a crop by insect vectors can be influenced by several factors. When an insect moves into a field, it seeks a plant on which to feed. During this process it may briefly probe several plants before a suitable host is found. It is during this period of initial probing, that nonpersistent, some semipersistent, and rarely but occasionally, persistent virus transmission occurs. This is referred to as primary virus spread (Kennedy 1976). These primary vectors may have originated from other fields consisting of similar hosts, or from weed host sources in the area. If the vector is not controlled and is allowed to colonize plants, a secondary population and subsequent spread of the virus may occur from plant to plant within the field. In the case of nonpersistently vectored aphid-borne viruses, transient non-colonizing aphids may also be responsible for secondary spread. Secondary infection may appear as small groups of infected plants.

Chemical Control

The use of chemicals and pesticides for control of insects as pests is an important component of sweet potato production. However, the application of pesticides for control of insects as vectors of nonpersistently transmitted viruses is generally not effective. Foliar oils have been used with limited success to control nonpersistently transmitted viruses. The application of insecticides has proven useful, however, for the control of vector populations of persistently transmitted viruses, due to the increased acquisition, latent, and incubation periods necessary for the vector to place the virus within or near the phloem area, and the overall reduction of the vector population within the crop. For example, Jayasena and Randles (1985) studied the movement of the persistently-transmitted subterranean clover red leaf virus (SCRLV) and the semipersistently-transmitted bean yellow mosaic virus (BYMV) on bean plants, *Vicia faba* L., after application of systemic insecticides. Viruliferous aphids were introduced into test plots. Results showed that the insecticide treatment lowered aphid populations and the incidence of SCRLV, but did not lower the incidence of BYMV.

The most effective control strategy to reduce virus spread is through preventive measures. Maintaining low vector populations through good cultural practices and, with the exception of nonpersistently transmitted viruses, through the judicious use of insecticides should significantly reduce virus incidence. The probability, however, that chemical control will reduce the incidence of nonpersistently transmitted viruses is very low.

Host Plant Resistance

There are three types of insect resistance as first proposed by Painter (1951) and later discussed by Kennedy (1976), Horber (1980), and Smith (1989). Antixenosis, formerly nonpreference resistance, alters the feeding or probing behavior of the vector. The host plant has undesirable characteristics for the vector which cause the vector to move to another plant or host. Tolerance encompasses the properties of a plant which supports vector feeding or recovery from feeding damage. Finally, antibiosis exists when the properties of the plant adversely affect the biology of the vector. The above definitions do not assume infection with viruses or other pathogens.

Varying levels of antixenosis may potentially affect both primary and secondary spread of a vector and its associated virus. The host plant must be probed by the vector for at least the inoculation threshold period (ITP) for a virus to be successfully transmitted (Kennedy 1976). If the ITP is not reached, primary virus spread will not occur, and no secondary spread would be expected. Viruses which are transmitted in a nonpersistent manner by a large number of aphid species (McLean 1959, Moyer & Kennedy 1978), such as SPFMV, typically are introduced into a field by use of infected stock or by the vector feeding on susceptible overwintering weed hosts, thereby establishing a primary infection site. Cultivars expressing a high level of resistance to the vector, but less than complete immunity, might effectively increase vector movement within a field by causing vectors to seek more suitable hosts (Kennedy 1976, Kennedy & Kishaba 1977). This results in quick periods of probing by the vector and a concomitant increase in secondary infection. Once secondary infection occurs, virus spread may become more rapid than if the cultivars were more susceptible (Kennedy 1976).

The degree of host susceptibility may also affect viruses which are transmitted in a persistent manner. When the host has a moderate level of resistance to a virus or nonpreference by the vector, persistent viruses which have a relatively short ITP may be successfully transmitted. Kennedy & Kishaba (1977) showed that 51% and 83% of alate melon aphids remained and fed on resistant and susceptible plants, respectively, for 72 hours. They concluded that the resistant plants maintained an "arresting quality" which served to deter aphids from moving.

Secondary spread of viruses is also influenced by transient vectors moving into a field; however, this occurs only when the transient vector is allowed to probe or feed on host plants long enough to acquire the virus. The vector must then probe a noninfected host long enough to complete the inoculation access period (IAP), which is usually 15 seconds or less for nonpersistent viruses. This same principal may be applied to viruses transmitted in a persistent manner with the addition that the virus has a longer IAP and a longer retention time within its vector.

The second category of plant resistance is tolerance, or the level to which a plant can withstand injury by a vector (Kennedy 1976). Under conditions where a vector infestation does not occur, movement of nonpersistent or persistently-transmitted viruses within a field will be relatively slow since probing or feeding patterns will remain relatively unchanged. Should the plants become colonized, yielding higher vector populations, movement of vectors and virus spread will increase. Movement within a field with the accompaniment of transient vectors builds up the secondary inoculum source potential. In this situation, nonpersistent virus transmission is likely, due to the relatively short probing time required for transmission of the virus. Viruses transmitted in a persistent manner will also increase proportionately as vectors feed in phloem cells. Therefore, the concomitant use of systemic insecticides may be effective in controlling high vector populations and further lower the incidence of virus (R.C. Larsen & J.W. Moyer, unpubl. data).

The third type of resistance, antibiosis, can be expressed by the appearance of physical changes, such as pubescence and morphological differences in trichomes, or changes in the biochemistry of the plant (Smith 1989). This type of interference can be partially effective against colonization by whiteflies (R.C. Larsen & J.W. Moyer, unpubl. data). It must be considered, however, that pubescence is not an efficient deterrent and therefore, where probing occurs, antibiosis would have little effect on the spread of nonpersistent viruses. Another form of antibiosis is a chemical defense system. Nielson and Don (1974) showed that resistant alfalfa lines produced phytoalexins that caused spotted alfalfa aphids, *Therioaphis maculata* (Buckton), to withdraw their stylet before entering the phloem. The resulting increase in movement by the vector may increase the spread of nonpersistent viruses but may decrease the incidence of persistent viruses as the retention period expires in its vector.

The deployment of resistant varieties does not assure the grower of a healthy crop, however, with some basic strategies such as utilizing large areas of resistant and tolerant varieties in combination with sound cultural practices, insect vector populations can be maintained at a manageable level.

Barriers

The term "barrier" can be defined as a biological or physical structure designed to inhibit insect movement. These barriers include reflective coarse nettings and interplanting of crops. The use of barriers to study vector movement was first demonstrated by Kunkel (1929). He reported lower leafhopper populations on screened aster plots than on nonscreened plots. Cohen and Marco (1979) described experiments with reflective barriers over potato plants to control aphids vectoring the persistently-transmitted potato leaf roll virus (PLRV). Large mesh (12 x 3 mm), white netting was placed over test plots of both virus-free and seedborne PLRV-infected plants. Virus incidence within plots of non-infected plants was limited to 3% for net-covered plots compared with a maximum of 48% in non-netted plots. Interestingly, the spread of PLRV increased in plots where potato seed was infected with PLRV. It was thought that the high incidence of resident forms of the green peach aphid, *Myzus persicae* (Sulzer), was responsible for spread of the virus in test areas.

Later, Cohen (1981) demonstrated a large reduction in the number of plants infected with cucumber mosaic virus and potato virus Y when several sizes of coarse, white- or gray-colored nets were placed over test plots. The coarse mesh did not significantly reduce light intensity and allowed for normal plant development, and at the same time, it resulted in a 40-fold reduction in alate aphids. He suggested that while yellow colors usually attract aphids, the white or gray colors may either repel aphids, or create an adverse microclimate under the nets. The effects of temperatures created in microclimate areas by mulches and the effects of various colors on whitefly behavior have been reviewed by Cohen (1982). Experiments with plant barriers were also done by Jayasena and Randles (1985). They compared the use of several insecticides and a plant barrier composed of barley on the patterns and rates of infection on bean, *V. faba*, with a potyvirus, bean yellow mosaic virus (BYMV), and a luteovirus, subterranean clover red leaf virus (SCRLV). Plots treated with systemic insecticides reduced aphid populations and subsequently reduced the spread of SCRLV. The spread of BYMV was not controlled by systemic insecticides. The barley barrier had a minimal effect on aphid populations and reduced the spread of BYMV only 1.1 m away from the barley barrier. The effect of the barley barrier on spread of SCRLV was negligible. The advantages of a physical barrier in large cropping situations must first be evaluated in relation to economics and overall practicality before it is implemented. Most studies have shown that there is very little, if any, effect on insect movement once aphids have colonized the plants within barrier plots. Indeed, because most of the known virus diseases of sweet potato are transmitted by aphids in a nonpersistent manner (Clark & Moyer 1988, Moyer & Salazar 1989), or by whiteflies, the utilization of barriers would tend to have little effect on virus movement by these vectors.

Mulches

The use of various types of mulch for the control of aphids and whiteflies has been reported by several workers (Cohen 1982, Cohen & Marco 1973, Dickson & Laird 1966, Kring 1964, Moericke 1954, Smith *et al.* 1964, Smith & Webb 1969, Walker & Randle 1986, Wyman *et al.* 1979). The mechanism of mulches is believed to be determined by color, as the insect may either be repelled (e.g., reflective barriers) or attracted to various light frequencies. Whiteflies and aphids are attracted to the yellow and green spectrum and repelled by varying degrees of white and gray (Cohen 1981, Cohen & Marco 1979). Just as white or gray overhead netting was experimentally used as physical barriers (Cohen 1981, Cohen & Marco 1979), reflective and attractive mulches have also shown some promise for vector management.

Harpaz and Cohen (1965) initially experimented with a straw mulch to control cucumber vein yellowing virus (CVYV), which was transmitted in a nonpersistent manner by *B. tabaci*. Their results showed that CVYV was controlled initially by straw mulch for 10 days but not controlled after 21 days. Wyman *et al.* (1979) attempted to control watermelon mosaic virus II (WMV) with aluminum and white plastic mulch. They found that aluminum and white mulch reduced WMV infection by 96% and 68%, respectively, during the early growing season. Studies by Cohen and Melamed-Madjar (1978) demonstrated that aphids were repelled by aluminum-colored plastic mulch (Smith & Webb 1969), whereas *B. tabaci* was attracted by the mulch. They also showed that there was an even greater attraction to *B. tabaci* when the plastic mulch was yellow. Recent findings indicated that the use of aluminum mulch resulted in fewer individuals of *B. tabaci* alighting on tomato plants (J.B. Kring, pers. comm.). Further, experiments during the fall of 1989 showed that aluminum mulch delayed the onset of a new whitefly-transmitted disease of tomato in Florida by 1.5 to 2 weeks longer than plots where white mulch was incorporated (J.B. Kring, pers. comm.).

The use of colored mulches as an attractant in combination with a sticky substance applied to the mulch could be used effectively to reduce the spread of nonpersistent aphid- and whitefly-transmitted viruses, particularly when used in combination with foliar oil applications. However, the expenses involved in material costs, requirements for specialized machinery, and labor make it impractical for use in most developing countries.

Oils

Bradley *et al.* (1962) demonstrated that foliar applications of oils reduced the spread of nonpersistently-transmitted viruses by aphids, such as potato virus Y. Loebenstein *et al.* (1964) showed that an emulsion of 1% oil was effective at reducing the spread of the aphid-transmitted cucumber mosaic virus. A large percentage of the test aphids lost their ability to transmit to leaves treated with the emulsion. The aphids also lost their ability to acquire the virus when feeding on treated plants. Many of the potentially useful oils have been tested for their effectiveness against semipersistent aphid-transmitted viruses. Results from these experiments showed that mineral oil has the greatest inhibitory properties relative to potato virus Y and other stylet-borne viruses (Bradley *et al.* 1962, De Wijs 1980).

It is easy to be somewhat complacent once the oil has been applied, however, there are additional factors which must be considered. Plants are continuously in a dynamic process during their growth period. The leaves must be inspected closely for adequate oil coverage and further applications are needed as leaves expand. Aphid activity should also be monitored during this period. Infected plants should be carefully rogued to reduce secondary infection. During periods of peak insect population density and activity, thorough coverage of leaves with oil is imperative. The application of oils for control of nonpersistent insect-transmitted viruses of sweet potato may be enhanced by use of reflective mulches or sticky barrier traps; however, research is lacking in this area.

Cultural Practices and Quarantine

A continuing problem facing the sweet potato industry is the production and movement of pathogen tested germplasm on a global basis. Movement of germplasm, which may carry with it introduced genes for vector or virus disease susceptibility, cannot be adequately evaluated due to lack of information currently available on etiology of the diseases (Moyer & Salazar 1989). Meristem-tip culture combined with rigorous virus testing is the recommended strategy for obtaining an initial source of sweet potato stock (Love *et al.* 1989, Moyer *et al.* 1989). A limiting factor is the lack of rapid methods for detecting most sweet potato viruses. Thus, reliable detection is dependent, in most cases, upon grafting into *I. setosa*. This procedure has been recently described (Moyer & Salazar 1989, Moyer *et al.* 1989). The foundations of successful plant disease control are sound agronomic practices. If these basic foundations are not achieved, any additional aids used in suppressing a disease or vector will be of little assistance. This includes maintaining vigorous sweet potato plants through regular fertilization and irrigation. Plants showing unusual growth habits such as stunting, leaf discoloration, or vein patterns, should be rogued because these plants may be infected with a virus disease which may be transmitted to other healthy plants. In addition, removal and destruction of remaining plant material after harvest should be a common cultural practice.

Although the properties and host ranges of many sweet potato viruses or agents remain unknown, frequent and conscientious removal of perennial weed hosts, with special attention given to those belonging to the families Convolvulaceae and Solanaceae, should be efficacious at reducing the overwintering virus inoculum sources.

ACKNOWLEDGMENTS

Portions of this chapter are from a feature article in Plant Disease. 1989. 73:451-455 (Viruses and virus-like diseases of sweet potato by J. W. Moyer & L. F. Salazar).

REFERENCES

Alconero, R. 1972. Effects of plant age, light intensity and leaf pigments on symptomatology of virus-infected sweet potatoes. Plant Disease Reptr. 56:501-504.

Atkey, P.T. & A.A. Brunt. 1987. Electron microscopy of an isometric caulimo-like virus from sweet potato (*Ipomoea batatas*). J. Phytopathol. 118:370-376.

Bradley, R.E.E., C.V. Wade & F.A. Wood. 1962. Aphid transmission of potato virus Y inhibited by oil. Virology 18:327-329.

Cadena-Hinojosa, M.A. & R.N. Campbell. 1981. Serological detection of feathery mottle virus strains in sweet potatoes and *Ipomoea incarnata*. Plant Dis. 65:412-414.

Cali, B.B. & J.W. Moyer. 1981. Purification, serology and particle morphology of two russet crack strains of sweet potato feathery mottle virus. Phytopathol. 71:302-305.

Campbell, R.N., D.H. Hall & N.M. Mielinis. 1974. Etiology of sweet potato russet crack disease. Phytopathol. 64:210-218.

Chung, M.L., Y.H. Hsu, M.J. Chen & R.J. Chiu. 1986. Virus diseases of sweet potato in Taiwan, pp. 84-90. *In*: Plant virus diseases of horticultural crops in the tropics and subtropics, FFTC Book Series No. 33. Food and Fertilizer Technology Center for the Asian and Pacific Region, Taipei, Taiwan, Republic of China.

Chung, M.L., C.H. Liao, M.J. Chen & R.J. Chiu. 1985. The isolation, transmission and host range of sweet potato leaf curl disease agent in Taiwan. Plant Prot. Bull. (Taiwan) 27:333-342.

Clark, C.A. & J.W. Moyer. 1988. Compendium of sweet potato diseases. American Phytopathological Society Press, St. Paul.

Cohen, S. 1981. Reducing the spread of aphid-transmitted viruses in peppers by coarse-net cover. Phytoparasitica 9:69-76.

Cohen, S. 1982. Control of whitefly vectors of viruses by color mulches, pp. 45-56. *In*: K.F. Harris & K. Maramorosch (eds.). Pathogens, vectors and plant diseases: approaches to control. Academic Press, New York.

Cohen, S. & S. Marco. 1973. Reducing the spread of aphid transmitted viruses in peppers by trapping the aphids on sticky yellow polyethylene sheets. Phytopathol. 63:1207-1209.

Cohen, S. & S. Marco. 1979. Reducing virus spread in vegetables and potatoes by net cover. Phytoparasitica 7:40-41.

Cohen, S. & V. Melamed-Madjar. 1978. Prevention by soil mulching of the spread of tomato leaf curl virus transmitted by *Bemisia tabaci* (Gennadius) (Homoptera:Alegoridae) in Israel. Bull. Entomol. Res. 68:465-470.

Cohen, S., J.E. Duffus, R.C. Larsen, H.Y. Liu & R.A. Flock. 1983. Purification, serology and vector relationships of squash leaf curl virus, a whitefly-transmitted geminivirus. Phytopathol. 73:1669-1673.

Dabek, A.J. & C. Sagar. 1978. Witches' broom chlorotic little-leaf of sweet potato in Guadalcanal, Solomon Islands, possibly caused by a mycoplasma-like organism. Phytopathol. Z. 92:1-11.

De Wijs, J.J. 1980. The characteristics of mineral oils in relation to their inhibitory activity on the aphid transmission of potato virus Y. Neth. J. Plant Pathol. 8:291-300.

Dickson, R.C. & E.F. Laird, Jr. 1966. Aluminum foil to protect melons from watermelon mosaic virus. Plant Disease Reptr. 50:305.

Esbenshade, P.R. & J.W. Moyer. 1982. An indexing system for sweet potato feathery mottle virus in sweet potato using the enzyme-linked immunosorbent assay. Plant Dis. 66:911-913.

Gibbs, A. & B. Harrison. 1976. Plant virology: the principles. Edward Arnold Ltd., London.

Girardeau, J.H. 1958. The sweet potato white-fly, *Bemisia inconspicua* Q., as vector of a sweet potato mosaic in South Georgia. Plant. Disease Reptr. 42:819.

Hahn, S.K. 1979. Effects of viruses (SPVD) on growth and yield of sweet potato. Exp. Agric. 15:1-5.

Hahn, S.K., E.R. Terry & K. Leuschner. 1981. Resistance of sweet potato to virus complex (SPVD). HortScience 16:535-537.

Harpaz, I. & S. Cohen. 1965. Semipersistent relationship between cucumber vein yellowing virus (CVYV) and its vector, the tobacco whitefly (*Bemisia tabaci* Gennadius). Phytopathol. Z. 54:240-248.

Hildebrand, E.M. 1958. Two syndromes caused by sweet potato viruses. Science 128:203-204.

Hollings, M., O.M. Stone & K.R. Bock. 1976. Purification and properties of sweet potato mild mottle, a white-fly borne virus from sweet potato (*Ipomoea batatas*) in East Africa. Ann. Appl. Biol. 82:511-528.

Horber, E. 1980. Types and classification of resistance, pp. 15-21. *In*: F.G. Maxwell and P.R. Jennings (eds.). Breeding plants resistant to insects. J. Wiley, New York.

Jackson, G.V.H. & F.W. Zettler. 1983. Sweet potato witches' broom and legume little-leaf diseases in the Solomon Islands. Plant Dis. 67:1141-1144.

Jayasena, K.W. & J.W. Randles. 1985. The effect of insecticides and a plant barrier row on aphid populations and the spread of bean yellow mosaic potyvirus and subterranean clover red leaf luteovirus in *Vicia faba* in South Australia. Ann. Appl. Biol. 107:355-364.

Kahn, R.P., R.H. Lawson, R.L. Monroe & S. Heron. 1972. Sweet potato little-leaf (witches' broom) associated with a mycoplasmalike organism. Phytopathol. 62:903-909.

Kennedy, G.G. 1976. Host plant resistance and the spread of plant viruses. Environ. Entomol. 5:827-832.

Kennedy, G.G. & A.N. Kishaba. 1977. Response of alate melon aphids to resistant and susceptible muskmelon lines. J. Econ. Entomol. 70:407-10.

Kring, J.B. 1964. New ways to repel aphids. 1964. Front. Plant Sci. 17:6-7.

Kunkel, L.O. 1929. Wire screen fences for the control of aster yellows. Phytopathol. 19:100.

Loebenstein, G. & I. Harpaz. 1960. Virus diseases of sweet potatoes in Israel. Phytopathol. 50:100-104.

Loebenstein, G., M. Alper & M. Dewtsch. 1964. Preventing aphid spread of cucumber mosaic virus with oils. Phytopathol. 54:960-962.

Love, S.L., B.B. Rhodes & J.W. Moyer. 1989. Meristem-tip culture and virus indexing of sweet potatoes. 2nd ed. International Board for Plant Genetic Resources, Rome.

McLean, D.L. 1959. Some aphid vector-plant virus relationships of the feathery mottle virus of sweet potato. J. Econ. Entomol. 52:1057-1062.

Moericke, V. 1954. Neue Untersuchungen uber das Farbsehen der Homopteren, pp. 55-69. *In*: Proc. 2nd conference on potato virus diseases. Lisse-Wageningen, Netherlands.

Moyer, J.W. 1986. Variability among strains of sweet potato feathery mottle virus. Phytopathol. 76:1126.

Moyer, J.W. & B.B. Cali. 1985. Properties of sweet potato feathery mottle virus RNA and capsid protein. J. Gen. Virol. 66:1185-1189.

Moyer, J.W. & G.G. Kennedy. 1978. Purification and properties of sweet potato feathery mottle virus. Phytopathol. 68:998-1004.

Moyer, J.W. & L.F. Salazar. 1989. Viruses and viruslike diseases of sweet potato. Plant Dis. 73:451-455.

Moyer, J.W., G.V.H. Jackson & E.A. Frison. (eds.). 1989. FAO/IBPGR technical guidelines for the safe movement of sweet potato germplasm. Food and Agriculture Organization of the United Nations, Rome/International Board for Plant Genetic Resources, Rome.

Nielson, M.W. & H. Don. 1974. Probing behavior of biotypes of the spotted alfalfa aphid on resistant and susceptible alfalfa clones. Entomol. Exp. Appl. 17:477-486.

Nome, S.F. 1973. Sweet potato vein mosaic in Argentina. Phytopathol. Z. 77:44-54.

Nome, S.F., T.A. Shalla & L.J. Petersen. 1974. Comparison of virus particles and intracellular inclusions associated with vein mosaic, feathery mottle, and russet crack disease of sweet potato. Phytopathol. Z. 79:169-178.

Painter, R.H. 1951. Insect resistance in crop plants. MacMillan Co., New York.

Schaefers, G.A. & E.R. Terry. 1976. Insect transmission of sweet potato disease agents in Nigeria. Phytopathol. 66:642-645.

Sheffield, F.M.L. 1957. Virus diseases of sweet potato in East Africa. I. Identification of the viruses and their insect vectors. Phytopathol. 47:582-590.

Smith, C.M. 1989. Plant resistance to insects: a fundamental approach. J. Wiley, New York.

Smith, F.F. & R.E. Webb. 1969. Repelling aphids by reflective surfaces, a new approach to the control of insect-transmitted viruses, pp. 631-639. *In* K. Maramorosch (ed.). Viruses, vectors, and vegetation. J. Wiley, New York.

Smith, F.F., G.V. Johnson, R.P. Kahn & A. Bing. 1964. Repellancy of reflective aluminum to transient aphid virus vectors. Phytopathol. 54:748.

Summers, E.M. 1951. "Ishuko-byo" (dwarf) of sweet potato in Rykyu Islands. Plant Disease Reptr. 35:266-277.

Walker, D.W. & W.M. Randle. 1986. Influence of row covers, mulch, and bedding dates on early production of sweet potato transplant. HortScience. 21:1354-1356.

Wyman, J.A., N.C. Toscano, K. Kido, H. Johnson & K.S. Mayberry. 1979. Effects of mulching on the spread of aphid-transmitted watermelon mosaic virus to summer squash. J. Econ. Entomol. 72:139-143.

Yamashita, S., Y. Doi & K.A. Shin. 1984. Short rod particles in sweet potato leaf curl virus infected plant tissue. Ann. Phytopathol. Soc. Japan 50:438.

Yang, I.L. 1972. Transmission of the feathery mottle complex of sweet potato in Taiwan. Taiwan Agric. Quart. 8:123-134.

19. Biology and Management of Plant-Parasitic Nematodes on Sweet Potato

Parviz Jatala
International Potato Center
P.O. Box 5969, Lima, Peru

One of the major problems in the production of sweet potato, *Ipomoea batatas* (L.) Lam., throughout the world is the damage caused by plant-parasitic nematodes. Although large numbers of these parasites are known to be associated with sweet potatoes, only a few are of economic concern. The degree of damage caused by these nematodes is influenced by the crop, the nematode species, the level of infestation, and other factors such as temperature, soil type, etc. Of the plant-parasitic nematodes associated with sweet potato, the migratory endoparasites constitute the largest number of species, but the extent of their impact on growth and yield has not been sufficiently studied. Stunting and poor plant vigor under field conditions together with yield and quality losses are known to be associated with the presence of a few species of endoparasitic nematodes.

This chapter will focus primarily on the important nematode parasites of sweet potato. The symptomatology associated with infection by these nematodes, their biology and epidemiological aspects, various control alternatives, and the potential of IPM programs for managing of these important nematodes are reviewed.

ROOT-KNOT NEMATODES

Root-knot nematodes, *Meloidogyne* spp., are the most destructive nematode pests of sweet potato. They are widely distributed and are of major concern to sweet potato production in the tropics, subtropics, and warm temperate regions of the world. Several *Meloidogyne* species, including *M. arenaria* (Neal) Chitwood, *M. hapla* Chitwood, *M.*

incognita (Kofoid & White) Chitwood, and *M. javanica* (Treub) Chitwood, infect sweet potato.

On the global scale, *M. incognita* is the most important nematode species attacking sweet potato. It is the only *Meloidogyne* species that is circumglobal in distribution. There are four races of *M. incognita* and their parasitic ability varies according to the susceptibility of the crop. Apparently, all four races of *M. incognita* attack sweet potato to varying degrees; *M. incognita* adversely affects the quality of storage roots in addition to limiting the yield. Infected storage roots crack easily and the cracks provide an avenue for penetration and establishment of many secondary and/or pathogenic organisms, which may subsequently lead to rotting of storage roots. In addition to its direct effects, *M. incognita* interacts with other pathogens to develop disease complexes.

Meloidogyne javanica is often found on sweet potato in the tropical and subtropical regions of the world; however, information on the economic importance of this nematode on sweet potato is not complete. *M. arenaria* and some isolates of *M. javanica*, although inducing necrosis on root systems, do not develop to maturity within roots (Clark & Moyer 1988, Jatala & Bridge 1990). *Meloidogyne hapla* also attacks sweet potato; however, because its distribution is limited to the cooler temperate regions of the world, it is not of economic concern to sweet potato production.

Symptoms and Damage

There are no diagnostic above-ground symptoms associated with *Meloidogyne* infection. Infected plants exhibit general symptoms associated with poor root growth, including stunting and a tendency to wilt during the hot part of the day. The principal and gross morphological changes due to *Meloidogyne* infection are expressed on the root systems. These symptoms consist of necrosis, galling, stunting or pruning, and cracking of storage roots. Infected roots may exhibit different patterns of necrosis. Necrotic flecks and root tip necrosis are principal reactions of resistant plants. In some cases of extreme plant hypersensitivity to nematode infection, there is a pruning effect which results in a lack of vigorous plant growth and subsequent yield loss. General root necrosis is typically associated with susceptibility and its extent varies according to the degree of plant susceptibility. Clark and Moyer (1988) suggested that a large portion of the root system may

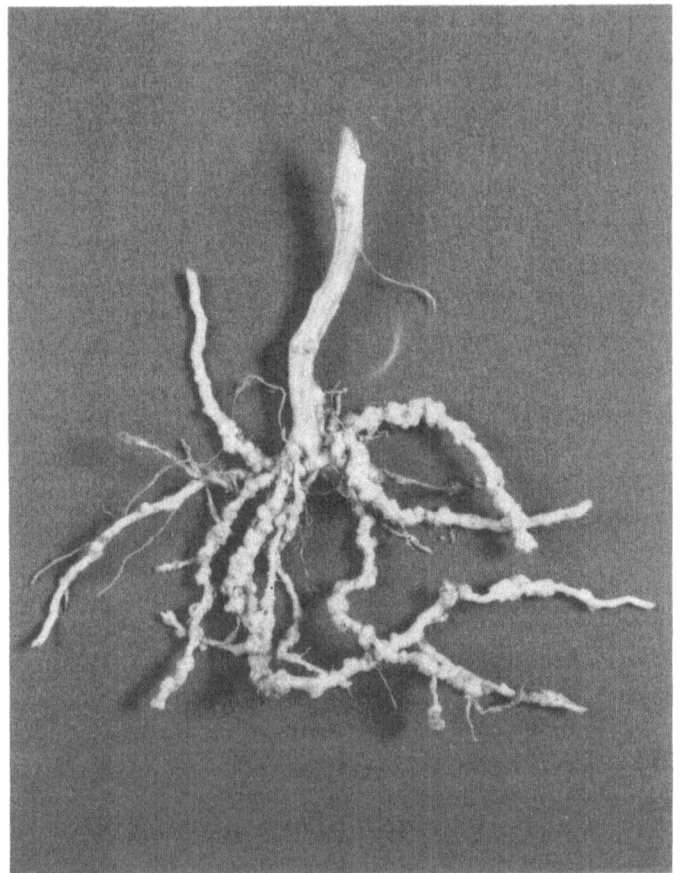

Fig. 19.1. Galls produced by *M. incognita* on roots of a highly susceptible sweet potato.

become necrotic due to the activity of secondary organisms, particularly during the latter stages of the growing season.

Root galling is the most diagnostic symptom of *Meloidogyne* infection (Figure 19.1). It is principally associated with the fibrous roots, but it is also found on the feeder and storage roots. The size of galls varies according to the plant susceptibility and the environmental conditions under which sweet potatoes are grown (Figure 19.1). Extensive and massive root galls are generally associated with plants grown at temperatures between 18-24°C, whereas small galls and more visible egg masses are often observed on plants grown at higher temperatures (Jatala 1965). The gelatinous matrix, which covers the

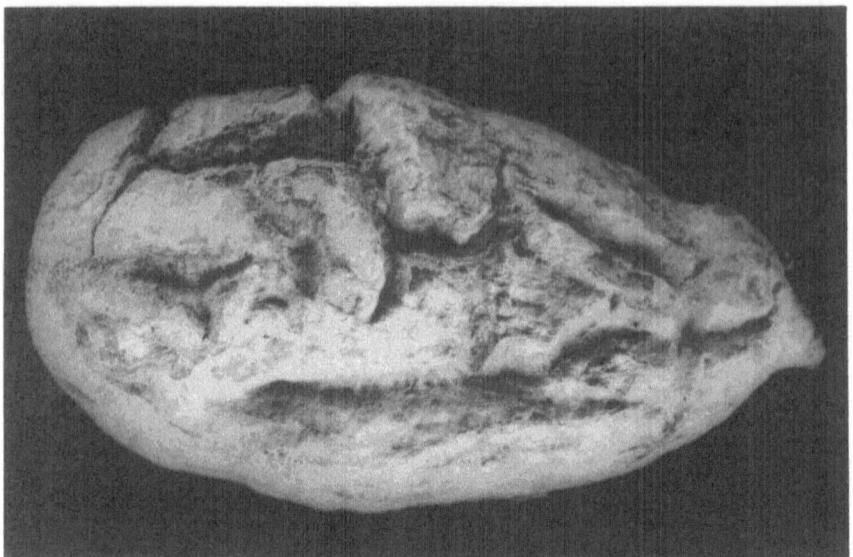

Fig. 19.2. Longitudinal cracking of storage root infected with *M. incognita*.

eggs is translucent white, but may become yellow or orange to dark golden brown during the later stages of plant growth. This color change, may be due, in part, to the effects of edaphic factors in soil. Egg masses usually contain between 500-1,000 eggs at different embryonic developmental stages which hatch readily as they mature. Tyler (1933), however, recorded 2,882 eggs in one egg mass.

Meloidogyne species often cause a pruning effect and stunting of feeder and fibrous roots. Infected feeder roots are usually shorter and have fewer secondary roots and root hairs (Clark & Moyer 1988). Yellowing, reduction of vine growth, transient wilting of foliage (flagging) during the hot part of the day, and abundant flower production are also associated with *Meloidogyne* infection and may result in yield loss.

The most conspicuous and dramatic effect associated with *Meloidogyne* infection is the longitudinal cracking of the storage roots (Figure 19.2). Cracks, which are initiated early in the growing season, are long and deep within the periderm and extend over much of the exposed surface. Those that are initiated in later stages of growth are shorter and shallower and often have necrotic margins as a result of the activity of secondary microorganisms (Clark & Moyer 1988). The difference between the cracking caused by the abiotic factors and that by *Meloidogyne* infection is the presence of nematodes in the fleshy part of the cracked roots infected with these nematodes. Presence of developed females of *Meloidogyne* in the cracked parts, as well as other parts of the storage root, is a consistent characteristic of nematode infection. Such infections often cause darkening and corky tissues in the areas surrounding the nematodes which can be easily detected by dissecting the roots. Nematode infections may also become apparent when roots are prepared for commercial canning (Clark & Moyer 1988).

Ecology and Epidemiology

Root-knot nematodes, *Meloidogyne* spp., are obligate, sedentary endoparasites. Although many species reproduce parthenogenetically, the individual species often exhibit considerable variation in their host range and virulence. Based on their ability to infect the host differentials, four host races have been identified for *M. incognita*; two races of *M. arenaria* are known, and only one race of *M. javanica* and one of *M. hapla* have been identified (Sasser 1979). Apparently, all races of *M. incognita* attack sweet potato. In the United States, however, races 1 and 3 of *M. incognita* are most commonly associated with sweet potato (Clark & Moyer 1988).

The life cycle of *Meloidogyne* spp. starts from the time eggs are deposited into the gelatinous matrix. Shortly after eggs are deposited, differentiation into the first-stage juvenile occurs. After molting within the egg, the first-stage juvenile gives rise to a second-stage juvenile which hatches through a hole at the end of the egg produced by repeated thrusts of the stylet. After hatching, the second-stage juvenile, which is the infective stage, moves out of the egg mass, and into soil until it invades new rootlets near the root tip, just above the root cap where there is intense meristematic activity. They may also enter roots through ruptures caused by the emergence of the lateral fibrous roots from the fleshy roots, as well as through cracks in the fleshy roots (Clark & Moyer 1988). Once inside, they migrate inter- and intracellularly through the tissues until they reach their feeding site. They then attach themselves to the feeding site with the anterior end (head) in contact with the vascular cylinder.

Cells around the nematode head become hypertrophied and multinucleated in susceptible hosts. These specialized nurse cells are called giant cells. Galling may result at this stage. Galls are produced by the introduction of growth regulators from the glands of the second-stage juveniles into plant tissues (Dropkin 1972). As the nematodes grow, they become sedentary, assume a flask shape, and undergo three further molts (Taylor & Sasser 1978). In the last molt, male nematodes pass through a type of metamorphosis and become long and filiform, folded inside the cuticle of the fourth-stage juvenile. Males then escape from the last juvenile cuticle and move freely in the soil. They are often found in the gelatinous matrix where they copulate in the amphimictic species. The females retain the shape of the last larval stage and continue to expand as they mature and become pyriform. They secrete a gelatinous matrix into which they deposit their eggs. The time required to complete the life cycle is dependent upon the prevailing environmental conditions and host susceptibility. Jatala (1965) reported a 28 day life cycle for *M. incognita* on 'Allgold' sweet potato grown under a 16-hour day at 32°C and an 8-hour night at 20°C.

Meloidogyne species tend to thrive in light, friable, sandy loam soil which predominate in the major portion of the world's sweet potato-growing areas (Jatala & Bridge 1990). *Meloidogyne incognita* and *M. javanica* require warm temperatures to complete their life cycle. Extremely cool winter temperatures drastically reduce *M. incognita* populations. Soil moisture may affect nematode densities at planting time and concomitantly affect damage to storage roots. Nematode damage is most severe under moderate drought conditions due to the combined effects of reduced water availability and reduced water uptake capacity of infected roots (Clark & Moyer 1988). Although excess soil moisture reduces nematode populations, the effects of nematodes on the growth and yield of the crop may be minimal when the soil moisture is kept at an adequate level during the growing season. Fluctuation of the soil moisture may induce cracking of the fleshy roots in the absence of nematodes. Although severe nematode infestations do not always cause cracking of the storage roots, nematodes predispose the roots to an increased incidence of growth cracks (Clark & Moyer 1988). The greatest incidence of cracking occurs when moisture is limiting during the early stages and abundant during later stages of the growing season. This flunctuation causes rapid enlargement of the fleshy root and, as a result, cracks develop at weak points in the cortex, particularly at the nematode infection sites (Clark & Moyer 1988).

The initial density of the nematode population determines the level of the damage caused by *Meloidogyne* infection. Factors such as cropping sequences, the degree of susceptibility of the previous crop, and the environmental conditions under which it was grown influence nematode population density and subsequent damage. *Meloidogyne* eggs, juveniles, and females survive in storage roots and can be disseminated in roots, but not in propagative stem material. The use of infected planting material contributes to the initial nematode infection. Irrigation water and unclean farm tools and machinery can aid in the dissemination of the nematodes. Similarly, *Meloidogyne* spp. survive and reproduce on a large number of alternate hosts which can serve as resevoirs of these nematodes.

In addition to intraspecific damage caused by a single species, *Meloidogyne* species also interact with a series of pests and pathogens to develop interspecific disease complexes. Nematodes predispose the fibrous and fleshy roots to decay by secondary invaders, presumably the common soil fungi. *Meloidogyne incognita* interact with *Fusarium* spp. and *Pseudomonas solanacearum* causing severe wilting and premature death (Jatala & Bridge 1990). Although there are several *Fusarium* resistant cultivars, their resistance may be broken in the presence of *M. incognita*. Breeding for double resistance to *M. incognita* and *Fusarium* spp. is an integral part of many breeding programs (Struble *et al.* 1966). *Meloidogyne incognita* also competes with *Rotylenchulus reniformis* Linford & Oliveira on sweet potato (Thomas & Clark 1983). When the initial population of one species is higher than the other, the denser species tends to dominate. Interaction of *Meloidogyne* spp. with other plant-parasitic nematodes on sweet potato has not been studied.

Control

In addition to the use of available resistant cultivars, the major effort for controlling root-knot nematodes has been placed on nematicides. Accurate identification of the nematode population is central to the selection and success of nematode control tactics. Thus, fields should be sampled for nematodes. Sampling should be done immediately after harvesting the previous crop or, in the case of an extended fallow period, about one month before planting.

Use of resistant cultivars is effective at reducing nematode populations, as well as for obtaining high yields and reducing nematode damage to roots. In the United States, the major emphasis to develop *Meloidogyne*-resistant cultivars was initiated in the early 1950's. Similar efforts were made by breeders in Japan and other countries. As a result of these efforts, several cultivars with varying levels of resistance have been released (Duke *et al.* 1978, Giamalva *et al.* 1963, Sasser & Kirby 1979). Results from other studies indicate that several sweet potato clones in the world germplasm collection held at the International Potato Center are resistant to root-knot nematodes (Jatala & Guevara 1988b, 1989). The frequency of resistant clones from Peru and the United States in the germplasm collection is high, with the clones from Peru constituting the larger portion of available resistant germplasm. This high percentage of resistant material is due to the breeding efforts, as well as to the result of selection pressure exerted by farmers throughout the past centuries (Jatala 1989). Jatala (1965) and Jatala & Russell (1972) suggested that a root exudate produced by some resistant cultivars was, in part, the basis for resistance.

Because there is evidence that extensive use of resistant cultivars selects for more aggressive races of *Meloidogyne* (Jatala & Guevara 1988a), use of resistant cultivars should be only a part of an integrated nematode management program. Incorporation of alternate methods of control, such as crop rotation, use of nematode-free propagating material, and the judicious use of nematicides, to develop integrated nematode management programs should reduce the nematode damage below threshold levels.

Selection of nematode- and disease-free propagating material, use of nematode-free seed beds, and cutting slips (planting material) above the soil surface should reduce the danger of nematode dissemination. Use of less suitable or nonhost crops followed by resistant sweet potato cultivars or another nematode-resistant crop in a cropping sequence will help to reduce nematode populations. Hot water treatments also reduce nematode densities on or in seed roots or cuttings (Jatala & Raman 1988, Martin 1970). However, this practice has not been widely adopted because it may cause considerable damage to the planting material. In potato, nematodes within tubers were killed by dipping tubers in nematicides immediately after harvest (Jensen & Jatala 1983). Rodríguez-Kabana *et al.* (1978) used a chemical dip and side dressed the propagating material with oxamyl to control *Meloidogyne*. Such treatments helped to establish the crop by providing early-season protection against nematodes. Studies are needed to determine the effect of dipping sweet potato storage roots in nematicides for control of nematodes on storage roots.

Proper and judicious use of nematicides can effectively reduce nematode populations and their damage. Pre-plant fumigation of seed beds or fields and application of nonfumigants and systemic nematicides are effective approaches for reducing nematode populations. Many organophosphates and carbamates, such as fenamiphos and aldicarb, are effective at controlling *Meloidogyne* species (Clark *et al.* 1980, Gapasin 1981). However, because sweet potato is typically produced in many developing countries that use low input practices, chemical control is usually cost prohibitive (Jatala & Bridge 1990).

Incorporation of the alternative control measures described above should help to reduce the nematode population below the damage threshold. However, it should be noted that IPM strategies are site-specific, and therefore, they should be developed and practiced on a case-by-case basis.

RENIFORM NEMATODES

Rotylenchulus reniformis Linford and Oliveira, the reniform nematode, is a destructive pest of sweet potato and reduces the yield and quality of the crop. Although it is primarily distributed in tropical and subtropical regions of the world, it may also occur in some temperate areas. Its distribution is not as extensive as *Meloidogyne* species, but it occurs in western and northern Africa, India, several Caribbean islands, Central and South America, the southeastern United States, and several countries in the South Pacific, including Fiji, Hawaii, Japan, and the Philippines (Clark & Moyer 1988). Host range and distribution of this nematode have been reported by many investigators (Ayala & Ramírez 1964, Dasgupta *et al.* 1968, Siddiqi 1972). *Rotylenchulus reniformis* is the only species of *Rotylenchulus* known to be of economic importance to agricultural crops. It attacks over 140 plant species in 115 genera of 46 families (Jatala 1990). Although sweet potato is one of the major hosts of this nematode, the extent of its damage to this crop has not always been recognized.

Symptoms and Damage

Although *R. reniformis* has a profound effect on sweet potato, the symptoms caused by this nematode are not as distinctive as those caused by other pathogens. *Rotylenchulus reniformis* feeds on cortical, phloem, and pericycle tissue, and its infection may cause the formation of root necrosis. Stunting of vines, yellowing, and transient wilting of foliage are secondary symptoms that result from damage to fibrous roots. This nematode may cause stunting of the roots when the nematode density is high and may stimulate root growth early in the season (Clark & Moyer 1988). Since there are no distinctive symptoms associated with infection of fibrous roots by this nematode, the only reliable method of diagnosis is identification of the nematodes on the root surface. Because the soil tends to adhere to females and their egg masses on the root system, it is often necessary to clean the roots and stain them to facilitate identification. Large populations of juveniles (e.g., up to 2,000/cc of soil) are often found in soil surrounding the infected roots.

Perhaps the most profound effect of this nematode on sweet potato is the cracking of the fleshy roots. Cracks develop early during enlargement of the roots and deepen and heal over as roots grow. The pattern differs slightly from that caused by root-knot nematodes. Unlike root-knot nematodes, these nematodes are not found within the fleshy roots. However, *R. reniformis* has been observed in small storage roots before enlargement (Clark & Moyer 1988).

Ecology and Epidemiology

The adult female of *R. reniformis* is an obligate, sedentary, semi-endoparasite of roots, whereas the male is nonparasitic. It is bisexual and reproduces by amphimixis. However, it has also been reported to reproduce parthenogenetically. There are two races of this nematode (A and B) which can be identified on three differential hosts (Jatala 1990). The life cycle starts from the first molt which takes place within the eggs; eggs hatch in water without the influence of root exudates. Juveniles develop to the pre-adult stage without feeding and growing and pass through at least three superimposed molts. After infecting the roots, young females pass through the cortex and feed on the endodermis. Females orient perpendicularly to the longitudinal axis of the root with the posterior portion of their body outside the roots. A single cell at the feeding site is converted into a giant cell and the adjacent cells in the pericycle become hypertrophied. Cells of parenchyma and phloem adjacent to the infection site also enlarge. The posterior portion of the females becomes enlarged and they begin laying eggs within 7 to 10 days. Eggs are laid in masses in a gelatinous matrix which may vary from 40 to 100 eggs per mass (Jatala 1990).

This nematode is capable of surviving in air-dried soil for an extended period of time (Birchfield & Martin 1967). Therefore, it can be disseminated on soil adhering to seed roots. It may also be disseminated in the fibrous roots that are attached to seed roots, or in the feeder roots on cuttings or slips produced in infested seed beds.

Damage is severe when the population density of the nematode at planting is high. Therefore, factors affecting the initial population at planting are central to epidemiological and control studies. *Rotylenchulus reniformis* cannot withstand extreme cold winter temperatures and populations are reduced under such conditions. In contrast to root-knot nematodes, *R. reniformis* thrives in heavier and fine-textured soil. However, symptoms such as wilting are enhanced under moderate drought conditions which occur more frequently in sandy soils, a condition favoring *Meloidogyne* species.

In addition to causing direct damage to roots, *R. reniformis* interacts with some important fungal plant pathogens to develop disease complexes. It also competes with *Meloidogyne* on sweet potato (Thomas & Clark 1983). *Rotylenchulus reniformis* produces the same symptoms on sweet potato whether alone or in combination with *Meloidogyne*, and it tends to dominate species of *Meloidogyne* (Clark & Moyer 1988).

Control

Information on the initial nematode population density is critical to determine the control strategies to be used against this nematode. Thus, as with *Meloidogyne* spp., fields should be surveyed to determine the nematode density. Several alternatives for control are available; however, because of a lack of resistant cultivars, control of this nematode is more difficult than that of the root-knot nematode.

An extended fallow period has been recommended for reducing populations of this nematode. Similarly, rotation with nonhost crops for two or more years reduces the nematode population. However, the extensive host range of this nematode limits the choice of rotation crops (Jatala 1990). In general, use of some of the graminaceous crops and resistant cultivars of soybean or cotton followed by a fallow period helps to reduce nematode populations. As with root-knot nematodes, use of clean propagating material, clean seed beds, and vine cuttings or slips that are cut above the soil surface also reduces nematode spread. Hot water treatment of seed roots and infected slips may eliminate nematodes (Martin 1970). However, because such treatments reduce vigor of slips and it is difficult to uniformly treat a large amount of material, this method of control has not been adopted by commercial growers.

Nematicides and soil fumigants are used to control *R. reniformis* populations. Because of a lack of suitable alternatives, this method of control has been used more extensively for *R. reniformis* than for root-knot nematodes. Pre-plant fumigation of the field and application of nonfumigant granular and liquid nematicides have been effective at reducing nematode populations (Birchfield & Martin 1968, Brathwaite 1974, Gapasin 1981). Development of an effective integrated management program is essential to reduce and maintain nematode populations below the damage threshold.

LESION NEMATODES

Lesion nematodes, *Pratylenchus* spp., are associated with sweet potato in some regions of the world. Although this nematode has a cosmopolitan distribution, it is only of major concern to sweet potato production in some countries. *Pratylenchus coffeae* (Zimmerman) Goodey is the most important species attacking sweet potato in Japan, where it causes significant yield loss (Goto 1964, Yoshida 1985). *Pratylenchus brachyurus* (Godfrey) Filipjev & Schuurmans Stekhoven is the species most commonly found on sweet potato in the United States.

Symptoms and Damage

Lesion nematodes cause small to extended necrotic lesions on the fibrous and feeder roots that result in significant yield losses. It also attacks the storage roots and produces brown or black necrotic lesions which affect their marketability (Figure 19.3). As found with other nematodes, necrotic lesions are often invaded by secondary microorganisms, such as fungi and bacteria, which in turn increase the extent of necrosis or decay. Extensive root necrosis leads to stunting of vines and eventual yield loss.

Ecology, Epidemiology, and Control

Lesion nematodes are obligate, migratory endoparasites. Juveniles and adults enter roots in or above the root-hair zone and migrate within the cortex inter- and intracellularly and feed on parenchymatic tissue. Cells which are fed upon turn brown, granular, and necrotic and eventually die. These nematodes usually move out of these necrotic tissues to feed on other tissue or to leave the roots. They mature within the root tissue. Eggs are laid singly or in small masses within the

Fig. 19.3. Storage roots of sweet potato cultivar 'Taihaku' infected with *P. coffeae* (photo by T. Nishizawa).

infected tissue or in soil. Warmer temperatures tend to favor *P. coffeae* because it can complete its life cycle on sweet potato within 30 to 40 days at 25-30°C and within 50 to 60 days at 20°C (Clark & Moyer 1988). These nematodes thrive in sandy, well-fertilized, moist soil. In Japan, it usually completes three generations on sweet potato during the growing season. Necrotic tissue is often invaded by secondary microorganisms. These nematodes also interact with some pathogenic fungi and bacteria to develop disease complexes.

Pratylenchus spp. have a wide host range. Monocots are most susceptible and can harbor large nematode populations. Dicots, such as potato, soybean, etc., are also susceptible. Damage to sweet potato is more severe if it follows crops such as maize, wheat, upland rice, potato, or soybean.

Because of the importance of *P. coffeae* on sweet potato, breeding programs in Japan have screened and selected for resistance to this nematode (T. Nishizawa, pers. comm.). Resistant cultivars are available and these nematodes can be controlled effectively. Some local Peruvian cultivars, such as 'Nemañete' with resistance to *M. incognita*, also exhibit resistance to *P. flakkensis* (M. Canto & P. Jatala, unpubl. data). Soil fumigation and application of granular nematicides are also effective at reducing nematode populations below the damage threshold.

BROWN RING

Brown ring disease is caused by stem and bulb nematodes, *Ditylenchus dipsaci* (Kühn) Filipjev and *D. destructor* Thorne. *Ditylenchus dipsaci* reportedly caused serious crop loss in storage in the United States in the 1930's (Kreis 1937). However, since then, it has not caused serious crop loss to sweet potato. In China, *D. destructor* is reportedly a serious obstacle to sweet potato production and it is widely distributed throughout the major sweet potato producing areas (Anonymous 1987, Yin & Zhang 1983). This nematode causes significant losses where it occurs. Although the initial damage occurs in the field, symptoms are manifested after storage roots are stored over time.

Symptoms and Damage

Infection by the potato rot nematode, *D. destructor*, initially causes sunken areas that are scattered on the storage roots (Figure 19.4). However, the principal symptom is the discoloration of the cortex; in the most advanced stages, the entire flesh turns brown to brownish-black (Figure 19.5). These nematodes are usually confined to the affected tissue. The tissue becomes slightly soft and cork-like in texture. As the disease progresses, the entire storage root becomes decayed. The affected tissue is usually invaded by secondary microorganisms, particularly fungi, which spread throughout the decayed tissue. Infected storage roots fail to sprout. This nematode only affects storage roots and not vines. Because the disorder is principally a storage disease, there are no diagnostic symptoms known for the field. However, older storage roots may exhibit the initial symptoms of the disorder at harvest. Higher temperatures (22-27°C) favor nematode development and symptom expression (Clark & Moyer 1988).

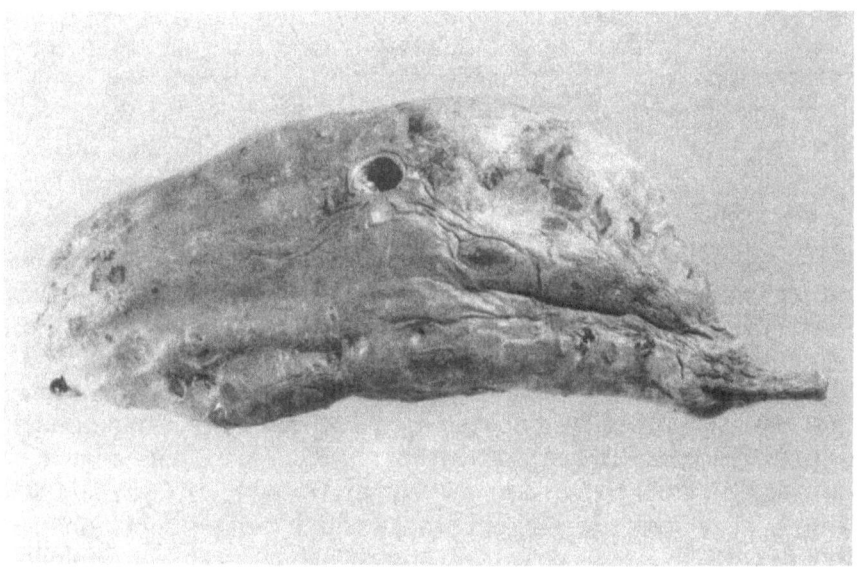

Fig. 19.4. Sunken lesions and crinkling of periderm on storage root of potato infected with *D. destructor.*

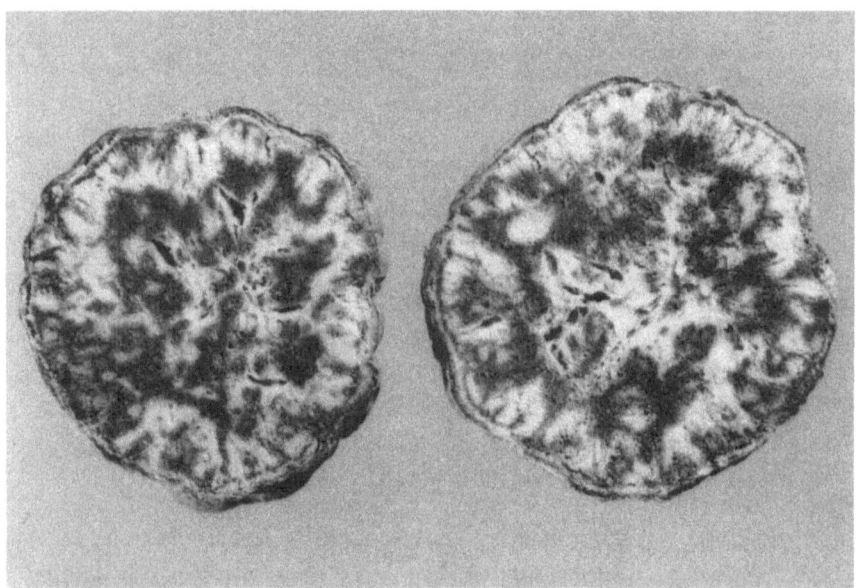

Fig. 19.5. Flesh of storage roots infected with *D. destructor.* Note the discoloration and corky texture of the internal flesh.

Ecology, Epidemiology, and Control

Ditylenchus dipsaci and *D. destructor* are obligate, migratory endoparasites with a rather wide host range. Juveniles and adults attack storage roots of different sizes. They penetrate the roots and move intra- and intercellularly in the cortical tissue and feed on the parenchymatic cells. Parenchyma becomes enlarged and spherical and the intercellular spaces become larger due to the release of pectinase by the nematode during feeding. Cellular collapse may also occur due to nematode feeding. As feeding progresses, brownish lesions form which later darken and increase in size. The periderm over these areas shrinks and crinkles. Eggs are laid within the infected tissue. Cultivars differ in their susceptibility to this nematode and resistant cultivars have been identified (Anonymous 1987). The jersey-type sweet potato is more susceptible (Clark & Moyer 1988). This disease can be reduced by storing roots at cooler temperatures (13-18°C) (Clark & Moyer 1988). Because many weeds may harbor dense populations of this nematode, weed control is essential to reduce nematode populations. Soil fumigation and application of nematicides are also effective at controlling these nematodes.

OTHER NEMATODES OF SWEET POTATO

There are several other genera of plant-parasitic nematodes associated with sweet potato. However, they are either of no economic consequence or their distribution is restricted to small areas. Sting nematode, *Belonolaimus longicaudatus* Rau, severely affects the fibrous root sytems which causes stunting of the plant and significant yield loss. Infection by this nematode stunts the feeder roots which may become swollen behind the root tip. Small, sunken, and discolored lesions may be associated with feeding by this nematode. There is often a proliferation of new roots above the feeding site (Clark & Moyer 1988, Graham & Holdeman 1953).

The stubby-root nematode, *Paratrichodorus minor* (Colbran) Siddiqi, and *Trichodorus* spp. attack sweet potato. They feed near the tips of the fibrous roots causing swelling and stunting of the root growth. High infestations may reduce yield (Roberts & Scheuerman 1984). Although these nematodes are capable of transmitting plant viruses, their role in transmitting sweet potato viruses has not been determined. They have an extensive host range and are generally found in sandy soils.

The spiral nematode, *Helicotylenchus dihystera* (Cobb) Sher, also attacks sweet potato. Although this nematode reproduces on sweet potato, it does not cause significant loss in yield or quality. Low densities of *Helicotylenchus* spp. may even stimulate root growth (Lopez *et al.* 1981).

Soil fumigation and application of nematicides are often effective at reducing population densities of these nematodes. However, because there are no appreciable losses caused by some of these nematodes, control measures are not usually cost effective. Other control alternatives described previously for the important nematodes of sweet potato are also effective at reducing populations of these nematodes.

Recent developments in the areas of biological control and genetic engineering may provide additional control possibilities. Breeding programs should develop cultivars with resistance to the most important nematode pests, and combine it with resistance to other important pests and pathogens.

It is important to note that the application of various integrated control components is generally more successful at reducing nematode populations than the use of only one control component. While eradicating the nematode population is difficult, if not impossible, reducing population levels below the damage threshold can be accomplished by applying a sound and well planned integrated management program.

REFERENCES

Anonymous 1987. Sweet potato research in the People's Republic of China, a CIP/AVRDC/IFPRI study. International Potato Center, Lima, Peru.

Ayala, A. & C.T. Ramírez. 1964. Host range, distribution, and bibliography of the reniform nematode, *Rotylenchulus reniformis*, with special reference to Puerto Rico. J. Agric. Univ. Puerto Rico. 48:140-161.

Birchfield, W. & W.J. Martin. 1967. Reniform nematode survival in air-dried soil. Phytopathol. 57:804.

Birchfield, W. & W.J. Martin. 1968. Evaluation of nematicides for controlling nematodes on sweet potatoes. Plant Disease Reptr. 52:127-131.

Brathwaite, C.W.D. 1974. Effect of DD soil fumigant on nematode population and sweet potato yields in Trinidad. Plant Disease Reptr. 58:1048-1051.

Clark, C.A. & J.W. Moyer. 1988. Compendium of sweet potato diseases. The American Phytopathological Society Press, St. Paul, Minnesota.

Clark, C.A., W. Birchfield, C.P. Yick & D.P. Boniol. 1980. Application rates and methods for Nemacur 3 SC for nematode control on sweet potato, 1979. Fung. & Nemat. Tests 35: 222.

Dasgupta, D.R., D.J. Raski & S.A. Sher. 1968. A revision of the genus *Rotylenchulus* Linford and Oliveira. 1940. (Nematode: Tylenchidae). Proc. Helminthol. Soc. Wash. 35:169-192.

Dropkin, V.H. 1972. Pathology of *Meloidogyne*. Galling, giant cell formation, effect on host physiology. OEPP/EPPO Bull. 6:23-32.

Duke, P.D., A. Jones, F.P. Cuthbert, Jr. & M.G. Hamilton. 1978. W-51 root-knot resistant sweet potato germplasm. HortScience 13:201-202.

Gapasin, R.M. 1981. Control of *Meloidogyne incognita* and *Rotylenchulus reniformis* and its effect on the yield of sweet potato and cassava. Ann. Trop. Res. 3:92-100.

Giamalva, M.J., W.J. Martin & T.P. Hernández. 1963. Sweet potato varietal reaction to species and races of root-knot nematodes. Phytopathol. 53:1187-1189.

Goto, S. 1964. Studies on the control of root rot nematode disease of Goodey, 1951. Miyazaki Agric. Exp. Stn. Bull. No. 5 (In Japanese with English summary).

Graham, T.W. & Q.L. Holdeman. 1953. The sting nematode, *Belonolainius gracitis* Ster.: a parasite on cotton and other crops in South Carolina. Phytopathol. 43:434-439.

Jatala, P. 1965. Nature of resistance and the effect of temperature on *Meloidogyne incognita* in sweet potato. M.S. Thesis, Oklahoma State University, Stillwater.

Jatala, P. 1989. Important nematode parasites of sweet potatoes and their management, pp. 213-218. *In* Improvement of sweet potato (*Ipomoea batata*) in Asia. International Potato Center, Lima, Peru.

Jatala, P. 1990. Reniform and false root-knot nematodes *Rotylenchulus* and *Nacobbus* spp., *In* W.R. Nickle (ed.). Manual of agricultural nematology. Marcel Dekker, Inc., New York (In press).

Jatala, P. & J. Bridge. 1990. Nematode parasites of root and tuber crops, pp. 137-180. *In* M. Luc, R.A. Sikora, & J. Bridge (eds.). Plant parasitic nematodes in subtropical and tropical agriculture. C.A.B. International, Oxen, United Kingdom.

Jatala, P. & E. Guevara. 1988a. Effects of various populations of *Meloidogyne incognita* on the expression of resistance and susceptibility of potatoes generated from true potato seeds. Nematropica 18:12.

Jatala, P. & E. Guevara. 1988b. Reaction of some sweet potato germplasm to *Meloidogyne incognita*. Nematropica 18:12.

Jatala, P. & E. Guevara. 1989. Reaction of Peruvian sweet potatoes to *Meloidogyne incognita*. Nematropica 19:8.

Jatala, P. & K.V. Raman. 1988. Major insect and nematode pests of sweet potatoes and recommendations for transfer of pest-free germplasm, pp. 319-321. *In* Exploration, maintenance, and utilization of sweet potato genetic resources. International Potato Center, Lima, Peru.

Jatala, P. & C.C. Russell. 1972. Nature of sweet potato resistance to *Meloidogyne incognita* and the effects of temperature on parasitism. J. Nematol. 4:1-7.

Jensen, H.J. & P. Jatala. 1983. Promising control measures for elimination or protection of developing tubers from root-knot nematodes, pp. 106-107. *In* W.J. Hooker (ed.). Research for the potato in the year 2000. International Potato Center, Lima, Peru.

Kreis, H.A. 1937. A nematosis of sweet potatoes caused by *Anguillulina dipsaci*, the stem or bulb nema. Phytopathol. 27:667-690.

Lopez, E.A., R.M. Gapasin & M.K. Palomar. 1981. Effects of different levels of *Helicotylenchus* nematode infestations on the growth and yield of sweet potato. Ann. Trop. Res. 3:275-280.

Martin, W.J. 1970. Elimination of root-knot and reniform nematodes and scurf infections from rootlets of sweet potato plants by hot water treatment. Plant Disease Reptr. 54:1056-1058.

Roberts, P.A. & R.W. Scheuerman. 1984. Field evaluations of sweet potato clones for reaction to root-knot and stubby root nematodes in California. HortScience. 19:270-273.

Rodríguez-Kabana, R., J.L. Turner & E.G. Ingram. 1978. Tratamiento de raíces de batata con el nematicida sistémico oxamyl para el control de nematodos fitoparásitos. Nematropica 8:26-31.

Sasser, J.N. 1979. Pathogenecity, host ranges and variability in *Meloidogyne* species, pp. 257-268. *In* F. Lamberti & C.E. Taylor (eds.). Root-knot nematodes (*Meloidogyne* species): systematics, biology and control. Academic Press, New York.

378

Sasser, J.N. & M.F. Kirby. 1979. Crop cultivars resistant to root-knot nematodes, *Meloidogyne* species, p. 24. *In* International *Meloidogyne* project contract AID/ta-C-1234. North Carolina State University, Raleigh.

Siddiqi, M.R. 1972. *Rotylenchulus reniformis*. Plant-parasitic nematodes. Set 1., No. 5. Commonwealth Institute of Helminthology, London.

Struble, F.B., L.S. Morrison & H.B. Cordner. 1966. Inheritance of resistance to stem rot and to root-knot nematodes in sweet potato. Phytopathol. 56:1217-1219.

Taylor, A.L. & J.N. Sasser. 1978. Biology, identification and control of root-knot nematodes (*Meloidogyne* species). North Carolina State University Graphics, Raleigh.

Thomas, R.J. & C.A. Clark. 1983. Population dynamics of *Meloidogyne incognita* and *Rotylenchulus reniformis* alone and in combination, and their effect on sweet potatoes. J. Nematol. 15:204-211.

Tyler, J. 1933. Reproduction without males in aseptic root culture of the root-knot nematode. Hilgardia 7:373-388.

Yin, G. & Y. Zhang. 1983. A revision on the pathogenic nematode of the stem nematode disease of sweet potato. Acta Sci. Nat. Univ. 4:118-127. (In Chinese with English summary).

Yoshida, T. 1985. Correlation between successive yield tests for agronomic characters in sweet potato. Japan J. Breed. 35:204-208.

20. Breeding Sweet Potato for Insect Resistance: A Global Overview

Wanda W. Collins
Department of Horticulture
North Carolina State University
Raleigh, North Carolina 27695 U.S.A.

Alfred Jones
U.S. Department of Agriculture, Agricultural Research Service
U.S. Vegetable Laboratory
Charleston, South Carolina 29414 U.S.A.

Michael A. Mullen
U.S. Department of Agriculture, Agricultural Research Service
Stored-Product Insects Research and Development Laboratory
Savannah, Georgia 31406 U.S.A.

Narayan S. Talekar
Asian Vegetable Research and Development Center
P.O. Box 42
Shanhua, Tainan, Taiwan, Republic of China

Franklin W. Martin [1]
Lehigh, Florida 33639 U.S.A.

Sweet potato, *Ipomoea batatas* (L.) Lam., is a member of the Convolvulaceae family. It is a hexaploid with 90 chromosomes and exhibits complex patterns of inheritance for most characters. On a global basis, soil insects are the major biotic factors that damage storage roots and subsequently reduce yield of sweet potato. Cuthbert (1967) reported at least 19 species of soil-inhabiting insects that damage the storage roots of sweet potato. Resistance to most of those pests was identified and, in some cases, resistance to a number of

---

[1] Previous address: U.S. Department of Agriculture, Agricultural Research Service, Tropical Agriculture Research Station, Mayaguez, Puerto Rico 00708 U.S.A.

species seemed to be conditioned by similar genetic factors (Cuthbert & Davis 1970).

The most serious soil insect pests are sweet potato weevils, *Cylas* spp. and *Euscepes postfasciatus* (Fairmaire). Cockerham and Deen (1947) first studied the resistance of sweet potato to these pests. Since then, several attempts have been made to develop resistant clones in breeding programs around the world because of the high level of economic damage caused by these pests.

In general, breeding sweet potato for insect resistance has received much less attention than breeding for other agronomic traits. Relative to other crops, such as grain crops which are considered more important food crops on a global scale, sweet potato has lacked allocated attention and resources. In addition, breeding for insect resistance is a difficult, long-term process which requires high resource inputs. Few sweet potato breeding programs exist with the physical, financial, and human resources necessary to conduct effective breeding and genetics studies on plant resistance to insects. Three international agricultural research centers, Asian Vegetable Research and Development Center (AVRDC), International Institute for Tropical Agriculture (IITA), and the International Potato Center (CIP) have been involved with genetics and breeding of sweet potato, especially as they relate to insect resistance. In addition, the United States Vegetable Laboratory (Charleston, South Carolina) and the Tropical Agriculture Research Station (Mayaguez, Puerto Rico) have actively bred sweet potato for insect resistance. The focus of the United States Vegetable Laboratory program has been on weevil, *Cylas formicarius* (Fabricius), resistance, as well as on resistance to a complex of soil insects particularly important in the southeastern United States. These constitute the major research programs in the world that are breeding sweet potato for resistance to insect and disease pests.

Because of the severe yield reductions due to insects in tropical growing areas, many programs have screened and evaluated local germplasm for resistance. For example, results of some of these evaluations have been published for Asia (Bong & Saad 1987, Pillai & Kamalan 1977, Pillai & Nair 1981), the South Pacific (MacFarlane 1984, Pole 1988), Africa (Anota & Odebiyi 1984, Munthali 1988), the Caribbean (Alleyne 1982), and the United States (Waddill & Conover 1978). This chapter reviews efforts and progress in developing resistance in sweet potato to insect pests in Taiwan at AVRDC, in Nigeria at IITA, in the Caribbean, and in the United States.

BREEDING FOR RESISTANCE TO THE WDS COMPLEX

Studies involving resistance of sweet potato to insects have been conducted at the U.S. Department of Agriculture (USDA) Vegetable Laboratory in Charleston, South Carolina since 1961 (Cuthbert 1967, Cuthbert & Davis 1970). Nineteen different species of insects were found to attack storage roots of sweet potato (Cuthbert 1967). The complex of soil insects that consistently causes considerable damage to sweet potato storage roots is composed wireworms, *Conoderus* spp., rootworms (cucumber beetles), *Diabrotica* spp., and flea beetles, *Systena* spp.. This complex has been referred to as the WDS pest group (see Chapters 15 and 16). Collectively, the WDS complex includes the the southern potato wireworm, *Conoderus falli* Lane; tobacco wireworm, *C. vespertinus* Fabricius; banded cucumber beetle, *Diabrotica balteata* LeConte; spotted cucumber beetle, *D. undecimpunctata howardi* Barber; elongate flea beetle, *Systena elongata* Fabricius; pale-striped flea beetle, *S. blanda* Melsheimer; and another flea beetle, *S. frontalis* Fabricius. In addition, the sweetpotato flea beetle, *Chaetocnema confinis* Crotch, and white grubs, *Plectris aliena* Chapin and *Phyllophaga ephilida* Say, have been studied.

Damage by *C. confinis*, *P. aliena*, and *P. ephilida* is distinct from that by the WDS complex. Storage root injury caused by *C. confinis* (narrow channels or grooves just under the skin) and *P. aliena* and *P. ephilida* (broad shallow channels gouged in the root) are easily recognized and can be rated separately (Jones *et al.* 1987b). Injury caused by other members in this complex is characterized by small round feeding holes or scars which may become larger or more distorted as the root grows. The WDS type of injury is given a single rating for the entire complex (see Chapter 15).

Heritability estimates were made of injury caused by the WDS complex and by *C. confinis* (Jones *et al.* 1979). The percentage of roots with injury was found to be a good, reliable indicator of pest injury index. Heritability based on the percentage of roots injured was approximately 45% for WDS, 40% for *C. confinis*, and 51% for all insects, regardless of type. These results indicated that heritability was sufficiently high to suggest that further progress could be made in this population for higher levels of insect resistance.

Resistance was not shown to be genetically linked to any undesirable characters so there did not appear to be a major barrier to development of insect-resistant cultivars with good horticultural characters comparable to those in commercial use (Jones & Cuthbert 1972).

Recurrent selection was used to increase resistance to WDS (Cuthbert & Jones 1972). As shown in Figure 20.1, both frequencies and levels of resistance to the WDS complex were increased during four cycles of selection.

A mass selection program was then initiated to combine this WDS resistance with other insect, nematode, and disease resistance, and with other desirable production and market qualities (Jones *et al.* 1976). Early (second) and late (sixth) generations of the new recurrent selection population were compared to further evaluate genetic progress for WDS resistance. Results showed that although WDS populations were sufficient to measure change between the generations in resistance, no progress could be demonstrated. The original level of resistance to WDS was retained but not increased; by the sixth generation, an unfavorable association of WDS with root size and yield had developed. However, intensified selection pressure and more precise evaluation techniques could overcome these restrictions on genetic progress.

Insect injury to storage roots by the WDS complex, *C. confinis*, and all species combined without identification was compared among cultivars with insect resistance developed by recurrent selection and with insecticide-treated (fonofos; Dyfonate 10 G) plants (Cuthbert & Jones 1978). Results showed that some clones (e.g., W-13) had high levels of resistance to soil insects and could be grown without soil insecticides to prevent sustained economic damage (Table 20.1) (Jones *et al.* 1986). Other clones could be grown without economic loss when supplemented with insecticide applications. Thus, using resistance as a component of a pest management system could replace or enhance the effectiveness of chemical controls. They also concluded that resistant cultivars would probably not reduce population levels of some of the more common pests. The abundance of WDS insects did not differ among clones; however, WDS insects fed on resistant clones less than on other clones. Fewer larvae of *C. confinis* were present on resistant clones than on other clones, suggesting that resistance to this insect might reduce populations of this flea beetle on sweet potato.

Six sweet potato clones with moderate levels of resistance to *C. formicarius* plus resistance to other soil insects and diseases have been released from the United States Vegetable Laboratory breeding program (Jones *et al.* 1980). Injury to these clones was no more than 50% of that to a susceptible cultivar, Centennial, when averaged over six to thirteen laboratory or field studies.

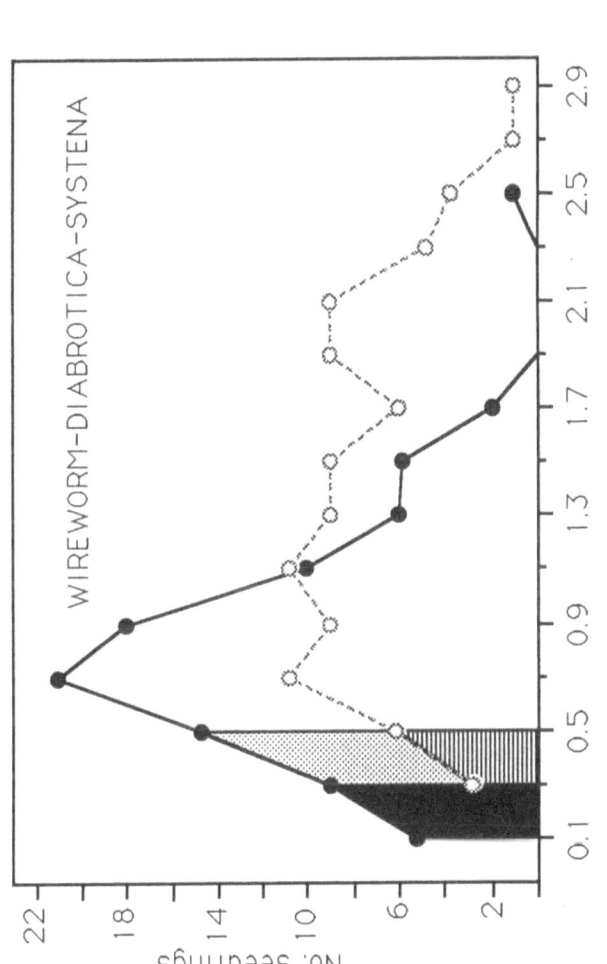

Fig. 20.1. Frequency polygon for WDS injury to selected (solid line) and nonselected (broken line) sweet potato populations. The cross-hatched area shows the portion of the nonselected population with an acceptable level of resistance; the lightly shaded area shows the increased frequency of the same level of resistance in the selected population; and the dark-shaded area shows the portion of the selected population with higher levels of resistance than found in the nonselected population. Thus, through four cycles of mass selection, both frequency and level of resistance were increased (Cuthbert & Jones 1972).

Table 20.1. Estimates of genetic and insecticide control of soil insects. [a]

Selection or cultivar	Roots damaged, %			Control by source, % [b]		
	Nontreated	Treated	Mean	Genetic	Insecticide	Both
WDS at Charleston, South Carolina, 1975						
W-13	35	19	27	61	18	79
W-3	65	34	50	28	34	62
Goldrush	90	57	74	--	37	--
C. confinis at Charleston, South Carolina, 1975						
W-13	3	2	3	89	4	93
W-3	4	3	4	85	4	89
Goldrush	27	17	22	--	37	--
All insects including the white grub, *P. aliena*, at Charleston, South Carolina, 1975						
W-13	27	20	24	70	8	78
W-3	65	39	52	29	29	58
Goldrush	91	66	79	--	27	--
White grub, *P. ephilida*, at Sunset, Louisiana, 1978						
L#-64	8	0	4	87	12	100
W-94	12	3	7	81	14	96
W-99	16	5	11	75	17	92
L4-89	31	10	20	52	32	84
SC 1149-19	46	23	34	28	36	64
Centennial	64	46	55	--	28	--

[a] Adapted from Cuthbert and Jones (1978) and from Rolston *et al.* (1981).
[b] Control estimates based on injury to the nontreated susceptible cultivar.

Three additional cultivars have been released with high levels of resistance to WDS insects (Jones *et al.* 1983, 1985, 1987a). Two of these cultivars, Southern Delite and Regal, were compared with two more susceptible and widely grown cultivars, Jewel and Centennial. Each of the four cultivars was compared with an extremely susceptible experimental selection, SC1149-19. 'Regal' and 'Southern Delite' provided 79.2 and 81.0% control, respectively, of all insect injuries which was comparable to the level of control by insecticides. Rolston *et al.* (1981) showed similar levels of white grub control by resistant clones.

BREEDING FOR RESISTANCE TO VINE BORER

Sweet potato vine borer, *Omphisa anastomasalis* (Guenee), is an insect pest with a limited distribution, but it is very destructive in the areas of the world where it occurs (Talekar & Cheng 1987, also see Chapter 17). It occurs in tropical and subtropical Asia and the Pacific. Because insecticidal control is limited and difficult, AVRDC initiated studies to determine if differences in resistance to this insect existed in sweet potato germplasm. Results of two years of research identified two accessions that were significantly and consistently less damaged by the vine borer than susceptible cultivars. The progeny of one of these accessions was evaluated and found to have substantially less damage overall than the progenies of nonselected agronomic clones. Work is continuing to select clones as breeding lines to further increase the level of resistance to sweet potato vine borer (see Chapter 17).

BREEDING FOR RESISTANCE TO *Cylas* SPP.

In tropical growing areas, three weevil species, *C. formicarius*, *C. puncticollis* (Boheman), and *Euscepes postfasciatus* (Fairmaire), are the most destructive of nearly 300 arthropod species that feed on sweet potato. At least one weevil species occurs wherever sweet potato is grown in the tropics. These weevils attack sweet potato roots in the field and in storage. In the field, the adult female lays eggs in the stem or in developing storage roots. Legless, white grubs, which are responsible for most of the damage, tunnel through stems and roots. Weevil feeding induces terpenoid production which makes even slightly damaged roots unpalatable (Akazawa *et al.* 1960, Uritani *et al.* 1975). Because of the nature of their feeding habit, weevils are difficult to control with insecticides once an infestation is present. Various cultural control methods are used worldwide; however, their use and effectiveness differs from region to region. The use of weevil-resistant cultivars is a practical and economical approach for managing these weevils.

Several studies have been conducted during the past 50 years to identify resistance to *Cylas* spp. and to introduce it into commercial cultivars. The earliest attempts to identify cultivars with resistance to *C. formicarius* were made by a USDA research group in Baton Rouge, Louisiana (Cockerham & Deen 1947, Cockerham & Harrison 1952). Other research groups in the United States have been involved intermittently since that time (Mullen *et al.* 1980a,b, 1982, Rolston *et al.* 1979, Waddill & Conover 1978), and a number of clones with varying levels of resistance have been reported. Breeding programs for weevil resistance have also been conducted at AVRDC in Taiwan and at IITA in Nigeria.

In addition to these breeding programs, many researchers have screened local germplasm for resistance to these weevils (e.g., Anota & Odebiyi 1984, Bong & Saad 1987, Jayaramaiah 1975, MacFarlane 1984, Munthali 1988, Pillai & Nair 1981, Pole 1988). Plant traits which have been identified as important in weevil resistance include fleshy root density (Martin 1984), dry matter and starch content (Cockerham & Deen 1947, Hahn & Leuschner 1981), vine thickness (Pole 1988), neck length (Pillai & Kamalan 1977), root depth (Burdeos & Gapasin 1980, Jayaramaiah 1975, Pillai & Kamalan 1977), crown hardness (Cockerham & Deen 1947, Jayaramaiah 1975) and fleshy root surface chemistry (Nottingham *et al.* 1989, Wilson *et al.* 1988, also see Chapter 11). Sutherland (1986) reviewed plant traits associated with resistance and their relationship to screening of plants.

The Approach at AVRDC, Taiwan

Screening sweet potato germplasm for resistance to *C. formicarius* was initiated at AVRDC in 1974 (AVRDC 1975), and since then, it has been a continuous research activity. Because weevil damage is a production constraint in the field, sweet potato germplasm has been screened at AVRDC under relatively uniform weevil population pressure maintained in the field. Sweet potato plants were planted in 1-m wide parallel rows spaced 2 m apart throughout experimental fields. These plants were infested at about 8 weeks after planting by releasing laboratory-reared weevils (Talekar 1982a). These infested sweet potato plants served as a source of weevil pressure. The materials to be screened were then planted about 10 to 12 weeks after the weevil-infested rows were planted. Each entry was planted in single-row plots (5 m long x 1 m wide) between two weevil-infested rows. This planting arrangement enabled each test entry to be bordered on two sides by weevil-infested plants to obtain a uniform infestation and minimize plant escapes from weevils. Yield in these plants was adequate (over 1 kg) and provided information on weevil infestation. During the season, normal cultural practices, such as weed control, fertilization, and irrigation, were used; however, no insecticides were applied.

At harvest, roots were cut into thin slices, the weevil-damaged portion was removed, and the number of larvae, pupae, and adults were recorded. In earlier studies, the weights of weevil-damaged and non-damaged slices were recorded and the percentages of roots damaged were calculated. However, because an earlier study found significant positive correlation between the total numbers of weevils (larvae+pupae+adults) per unit root weight and percentage of roots damaged (Talekar 1982b), only the numbers of weevils per unit root weight was used in more recent evaluations.

In preliminary studies in which the number of entries was large and the number of replicates small, the mean (x) and standard deviation (sd) of the number of weevils found per unit root weight was used to classify resistance levels (Talekar 1987a). Accessions that had weevil densities less than x-$2sd$ were considered highly resistant (HR); between x-$2sd$ and x-$1sd$, moderately resistant (MR); between x-$1sd$ and x, low resistance (LR); between x and x+$2sd$ susceptible (S); and more than x+$2sd$, highly susceptible (HS). In the absence of a reliable standard resistant cultivar for comparison, this procedure proved to be very useful. It was also flexible enough to be useful under varying levels of insect population pressure. After each evaluation, the susceptible accessions were discarded and those with at least some resistance were advanced to the next screening phase which determined the stability of resistance. In advanced tests with few entries and a large number of replicates, the insect count data were analyzed by analysis of variance and the least damaged entries were retested until a stable resistance level was achieved.

By the end of 1988, the entire sweet potato germplasm at AVRDC, which consisted of 1,300 accessions, had been evaluated. Wide variations in weevil damage to a particular accession were observed among test locations, seasons, replicates within a single trial, and even among plants within a single plot, which made resistance ratings difficult to assess. For example, two experimental clones, I123 and I152, were used for breeding weevil resistance into agronomic cultivars of sweet potato based on their performance in several early tests over a number of locations and seasons. However, all progenies derived from crosses with these two clones were susceptible to weevils. In later tests, the apparently resistant parents also proved to be susceptible.

This type of wide variation in the characters of sweet potato under varying environmental conditions is not confined to weevil resistance. Environmental variations as well as genotype x environment interactions have been reported for carotenoid content (Ezell & Wilcox 1958), alcohol soluble solids (Austin et al. 1970), protein content (AVRDC 1975, Collins & Walter 1981, Collins et al. 1987), yield (Collins et al. 1987, Jong & Park 1975), and other quality factors such as dry matter, intercellular space, and baking quality (Collins et al. 1987). These types of variations make it difficult to breed for weevil resistance because of the necessity of testing over many environments (locations and years) to reduce specific sources of variation. Collins et al. (1987) suggested that by testing in four locations during two years with 4 replications per test for yield and internal root quality traits, variances attributable to environments and genotype x environment interactions would be reduced and result in acceptable and reliable data. Knowledge of the magnitude of these sources of variation for weevil resistance would provide a valuable step in developing sweet potato clones with stable levels of weevil resistance.

Talekar (1987b) suggested that adequate sources of resistance to sweetpotato weevil in storage roots may not exist in sweet potato germplasm and the past approach of AVRDC to develop resistant cultivars by conventional breeding methods may be inappropriate. Consequently, AVRDC has adopted a new nonconventional approach to identify a stable source of resistance. Crosses have been made between *I. trifida*, a close relative of sweet potato, and *I. batatas* (Iwanaga et al. 1987). Both cross and backcross progenies were tested for resistance to weevils at AVRDC (AVRDC 1987, 1988, 1989). Certain progeny showed considerable levels of resistance. The resistance was thought to be due to the presence of thin stems in the progenies. Under natural conditions in sweet potato plantings before storage roots enlarge (about seven to ten weeks after planting depending upon the cultivar), weevils are typically found in vines, especially in and near the crown. Infestations of crowns last from shortly after planting until harvest. This source of weevils serves as a reservoir for later infestations in storage roots. The narrow stems of the *I. trifida* x *I. batatas* progeny help to reduce infestations of vines and subsequently reduce infestations of storage roots. Thus, most of the research on weevil resistance at AVRDC is now focused on identifying genotypes with reduced vine infestations. Preliminary screening of germplasm at AVRDC has identified several thin stem accessions with reduced susceptibility to weevil damage. Cultivars with resistance to weevils in vines alone, however, will probably not control these weevils adequately because of late-season damage that occurs when roots enlarge and subsequently become more accessible to weevils.

The Approach in the United States

Cylas formicarius is also the most destructive insect pest of sweet potato in the United States. The search for weevil resistant germplasm in the United States was reported by Cockerham (1943) and Cockerham and Deen (1947). Cockerham and Harrison (1952) found that two breeding clones in Louisiana had a lower weevil infestation than other clones in the field. They concluded that there was no direct relationship between crown and root infestation, and that deeply set roots were relatively free of weevils. Despite their premise that there was potential to breed sweet potato for resistance to weevils, little work was done until the 1970's. Waddill and Conover (1978) evaluated several white-flesh sweet potato cultivars in southern Florida and found considerable variation in weevil resistance within the germplasm. They identified the mechanisms of resistance (antixenosis [nonpreference], antibiosis, and tolerance) that existed in the germplasm that was tested.

The identification of resistant genotypes is difficult because resistance occurs as a continuous gradient and is probably inherited in a quantitative fashion (Rolston *et al.* 1979). Also, field tests to evaluate germplasm for resistance are expensive, time consuming, and often inconclusive. For these reasons, a laboratory technique was needed to evaluate resistance and reduce the number of genotypes to be tested in the field.

The method developed by Mullen *et al.* (1980a) facilitated the rapid selection of sweet potato clones for resistance to weevils. This method used a plexiglass chamber (17.9 x 17.9 x 5.7 cm high) containing a plexiglass platform into which 25 evenly spaced holes (1.8 cm diam.) were drilled. The holes were arranged in a 5 x 5 Latin square and the platform was raised 3 cm from the bottom of the chamber. Chambers were fitted with a top covered with a fine nylon screen.

Sweet potato storage root plugs were cut with a cork borer (no. 10) and pushed (periderm out) into a vial (1.8 cm diam.). Each test consisted of 22 or 23 test clones and two or three standard entries (susceptible checks). Fifteen pairs of adult weevils were placed in each chamber and allowed to feed in total darkness for 48 hours. The weevils were then removed and the feeding punctures were counted. Clones were replicated four to six times. Twenty-four clones with the lowest levels of feeding (≤ 10.6 feeding punctures) were considered resistant to feeding by the weevils.

Promising sweet potato clones were also field tested (Mullen *et al.* 1980b, 1982, 1985). Each clone was replicated eight times using ten plants per plot. Sweet potato slips, grown in greenhouse beds, were planted 30 cm apart in plots 90 cm apart and rows 1 m apart in randomized complete block designs with eight replications. The field was bordered by two buffer rows of 'Jewel' sweet potato. Standard horticultural practices were followed. Approximately 12,000 *C. formicarius* adults were released into each field from four waterproof wooden shelters to ensure an adequate weevil population. Each shelter was filled with sweet potato storage roots (approximately 12 kg). About 3,000 weevils were released into each shelter 60 days after planting. This method ensured that the developing plants would be exposed to high weevil pressure. At harvest, the roots were examined for weevil damage and evaluated by several criteria.

Roots were rated for weevil damage with an index from 1 (no visible damage) to 5 (high visible damage). Roots were also examined for feeding and oviposition holes, as well as for adult emergence holes. The percentage of crowns that were damaged was calculated. Also damage to the crowns was rated with an index similar to that used for the roots. The numbers of plants with infested roots were recorded. All storage roots produced in each plot were weighed. The numbers of weevils emerging from infested roots and the percentage of infested crowns were also recorded periodically; however, the information from these data were not adequate to justify the labor and time involved. Of the 24 clones shown to be resistant by the rapid laboratory evaluation method, 18 were also found to be resistant in field trials.

Several clones have been developed with moderate levels of weevil resistance. Two clones, W-125 and W-152, which were selected as having some resistance, were released as 'Resisto' and 'Regal' (Jones *et al.* 1983, 1985). However, Jansson *et al.* (1987) showed that these new cultivars were not resistant to weevils in southern Florida, where weevil population pressure is much higher than in more northern regions (Georgia and South Carolina) where the clones were selected. Therefore, the actual value of this resistance is probably minimal in areas where high weevil densities consistently occur.

In the USDA sweet potato breeding program at the Tropical Agriculture Research Station in Mayaguez, Puerto Rico, early efforts failed to identify weevil-resistant germplasm, although differences in cultivar susceptibility to weevils were evident (Bravo et al. 1983, Martin 1983, 1984). It was observed that harvesting early resulted in less weevil damage because lower weevil populations occurred early in the growing season. For this reason, this breeding program emphasized the development of short-season cultivars with increased yield and quality that could be harvested three months after planting. Three clones which showed acceptable yield and root quality combined with lower weevil damage after a three month growing season were identified before this program ended.

The Approach at IITA, Nigeria

While studies at AVRDC and in the United States focused mainly on *C. formicarius*, efforts at IITA were directed toward resistance to the African sweet potato weevil, *C. puncticollis*. Since the program began in 1971, a number of promising clones have been selected and resistance characteristics have been incorporated into the breeding populations (Hahn & Leuschner 1981, IITA 1976, 1979, 1985). These were the first reports of resistance of sweet potato to the African sweet potato weevil. In addition, broad-sense heritabilities of resistance were reported as $H=0.84+0.00$ for roots and $H=0.79+0.03$ for shoots (Hahn & Leuschner 1981). These heritability estimates are extremely high, but they do not indicate whether additive or dominance effects are more important. Significant environment and genotype x environment interactions were shown to exist, in the same study, but the cultivar resistance effect was much larger than the interaction effects.

Breeding sweet potato for resistance to weevils has been discontinued at IITA; however, materials previously selected remain available. It is not known whether the resistance to *C. puncticollis* is related to resistance to *C. formicarius*. Testing of the IITA material at AVRDC showed that it was not resistant to *C. formicarius* in Taiwan (N.S. Talekar, unpubl. data). However, some cross-resistance was demonstrated in the United States (Rolston et al. 1979).

BREEDING FOR RESISTANCE TO *Euscepes postfasciatus*

Euscepes postfasciatus is a serious pest of sweet potato storage roots in many areas of the Caribbean, Pacific Islands, and Central and South America. Searches have been made to identify resistant clones (Pole 1988, Raman 1988, also see Chapter 14). Alleyne (1982) evaluated clones in Barbados in an effort to use varietal resistance as a control strategy. This is discussed in Chapter 14. However, no organized breeding programs have been conducted to develop weevil-resistant germplasm until recently. CIP began work on the genetics and breeding for resistance to this weevil in 1987. Screening for resistance in the germplasm collection at CIP resulted in the identification of 15 clones that were resistant to this insect (CIP 1988, Raman 1988). Efforts are continuing at CIP to develop higher levels of resistance in the sweet potato germplasm (see Chapter 14).

SUMMARY

This chapter reviews the efforts that have been made to identify and increase genetic resistance of sweet potato to insect pests. Progress has been made in development of screening techniques, selection procedures, identification of resistance, and increasing resistance through breeding. Increased efforts in all these areas should result in more useful levels of resistance which can be used as components of integrated pest management programs to control sweet potato insects (see Chapter 21 for future outlook).

REFERENCES

Akazawa, T., L. Uritani & H. Kubota. 1960. Isolation of ipomeamarone and two coumarin derivatives from sweet potato roots injured by the weevil, *Cylas formicarius elegantulus*. Arch. Biochem. Biophys. 88:150-156.

Alleyne, E.M. 1982. Varietal resistance as a control strategy against West Indian sweet potato weevil, *Euscepes postfasciatus* (Fairmaire) on sweet potatoes in Barbados. Proc. Caribb. Food Crops Soc. 18:54-62.

Anota, T. & J.A. Odebiyi. 1984. Resistance in sweet potato to *Cylas puncticollis* (Coleoptera: Curculionidae). Biol. Afr. 1:21-30.

Austin, M.E., L.H. Aung & B. Graves. 1970. The use of carbohydrate contents as an index of sweet potato maturity, pp 42-44. *In* D.L. Plucknett (ed.). Tropical root and tuber crops tomorrow. vol. 1. College of Tropical Agriculture, University of Hawaii, Honolulu.

AVRDC. 1975. Annual report for 1974. Asian Vegetable Research and Development Center, Shanhua, Taiwan, Republic of China.

AVRDC. 1987. AVRDC progress report. Summaries 1986. Asian Vegetable Research and Development Center, Shanhua, Taiwan, Republic of China.

AVRDC. 1988. AVRDC progress report. Summaries 1987. Asian Vegetable Research and Development Center, Shanhua, Taiwan, Republic of China.

AVRDC. 1989. AVRDC progress report summaries. Asian Vegetable Research and Development Center, Shanhua, Taiwan, Republic of China.

Bong, C.F.J. & M.S. Saad. 1987. Preliminary screening for resistance to *Cylas formicarius* (Fab.) in sweet potato in east Malaysia. Malay. Appl. Biol. 16:297- 302.

Bravo, R., J.A. Santiago & F.W. Martin. 1983. Techniques for developing resistance to weevil, *Cylas formicarius elegantulus*, in sweet potato. Proc. Am. Soc. Hort. Sci., Trop. Region 27(B):93-94.

Burdeos, A.T. & D.P. Gapasin. 1980. The effect of soil depth on the degree of sweet potato weevil infestation. Ann. Trop. Res. 2:224-231.

CIP. 1988. Annual Report CIP 1988. International Potato Center, Lima, Peru.

Cockerham, K.L. 1943. The host preference of the sweet potato weevil. J. Econ. Entomol. 36:471-472.

Cockerham, K.L. & O.T. Deen. 1947. Resistance of new sweet potato seedlings and varieties to attack by the sweetpotato weevil. J. Econ. Entomol. 40:439-441.

Cockerham, K.L. & P.K. Harrison. 1952. New sweet potato seedlings that appear resistant to sweetpotato weevil attack. J. Econ. Entomol. 45:132.

Collins, W.W., L.G. Wilson, S. Arrendell & L.F. Dickey. 1987. Genotype x environment interactions in sweet potato yield and quality factors. J. Am. Soc. Hort. Sci. 112:579-583.

Collins, W.W. & W.M. Walter, Jr. 1981. Potential for increasing nutritional value of sweet potatoes, pp. 355-364. *In* R.L. Villareal & T.D. Griggs (eds.). Sweet potato. Proceedings of the 1st international symposium. Asian Vegetable Research and Development Center, Shanhua, Taiwan, Republic of China.

Cuthbert, F.P., Jr. 1967. Insects affecting sweet potatoes. U.S.D.A. Agric. Handb. 329.

Cuthbert, F.P., Jr. & B.W. Davis. 1970. Resistance in sweet potatoes to damage by soil insects. J. Econ. Entomol. 63:360-361.

Cuthbert, F.P., Jr. & A. Jones. 1972. Resistance in sweet potatoes to Coleoptera increased by recurrent selection. J. Econ. Entomol. 65:1655-1658.

Cuthbert, F.P., Jr. & A. Jones. 1978. Insect resistance as an adjunct or alternative to insecticides for control of sweet potato soil insects. J. Am. Soc. Hort. Sci. 103:443-445.

Ezell, B.D. & M.S. Wilcox. 1958. Variation in carotene content of sweet potatoes. J. Agric. Food Chem. 6:61-65.

Hahn, S.K. & K. Leuschner. 1981. Resistance of sweet potato cultivars to African sweetpotato weevil. Crop Sci. 21:499-503.

IITA. 1976. Annual report 1975. International Institute of Tropical Agriculture, Ibadan, Nigeria.

IITA. 1979. Annual report 1978. International Institute of Tropical Agriculture, Ibadan, Nigeria.

IITA. 1985. Annual report 1984. International Institute of Tropical Agriculture, Ibadan, Nigeria.

Iwanaga, M., J.Y. Yoon, N. S. Talekar & Y. Umemura. 1987. Evaluation of the breeding value of 5X interspecific hybrids between sweet potato cultivars and 4X *I. trifida*. *In* Proceedings of an international sweet potato seminar-workshop. Visayas State College of Agriculture, Baybay, Philippines (In press).

Jansson, R.K., H.H. Bryan & K.A. Sorensen. 1987. Within-vine distribution and damage of sweetpotato weevil, *Cylas formicarius elegantulus* (Coleoptera: Curculionidae), on four cultivars of sweet potato in southern Florida. Fla. Entomol. 70:523-526.

Jayaramaiah, M. 1975. Reaction of sweet potato varieties to the damage of the weevil, *Cylas formicarius* (Fab.) (Coleoptera: Curculionidae) and on the possibility of picking up of infestation by weevil. Mysore J. Agric. Sci. 9:418-421.

Jones, A. & F.P. Cuthbert, Jr. 1972. Associated effects of mass selection for soil insect resistances in sweet potato. J. Am. Soc. Hort. Sci. 98:480-482.

Jones, A., P.D. Dukes & F.P. Cuthbert, Jr. 1976. Mass selection in sweet potato: breeding for resistance to insects and diseases and for horticultural characteristics. J. Am. Soc. Hort. Sci. 101:701-704.

Jones, A., P.D. Dukes & J.M. Schalk. 1986. Sweet potato breeding, pp. 1-35. *In* M.J. Bassett (ed.). Breeding vegetable crops. Avi, Westport, Connecticut.

Jones, A., J.M. Schalk & P.D. Dukes. 1979. Heritability estimates for resistance in sweet potato to soil insects. J. Am. Soc. Hort. Sci. 104:424-426.

Jones, A., J.M. Schalk & P.D. Dukes. 1987b. Control of soil insect injury by resistance in sweet potato. J. Am. Soc. Hort. Sci. 112:195-197.

Jones, A., P.D. Dukes, J.M. Schalk, M.G. Hamilton & R.A. Baumgardner. 1987a. 'Southern Delite' sweet potato. HortScience 22:329.

Jones, A., P.D. Dukes, J.M. Schalk, M.A. Mullen, M.G. Hamilton, R. Paterson & T.E. Boswell. 1980. W-71, W-115, W-119, W-125, W-149 and W-154 sweet potato germplasm with multiple insect and disease resistances. HortScience 15:835-836.

Jones, A., P.D. Dukes, J.M. Schalk, M.G. Hamilton, M.A. Mullen, R.A. Baumgardner, D.R. Paterson & T.E. Boswell. 1983. 'Resisto' sweet potato. HortScience 18:251-252.

Jones, A., P.D. Dukes, J.M. Schalk, M.G. Hamilton, M.A. Mullen, R.A. Baumgardner, D.R. Paterson & T.E. Boswell. 1985. 'Regal' sweet potato. HortScience 20:781-782.

Jong, S.K. & K.Y. Park. 1975. Variety x environmental interaction in sweet potato (*Ipomoea batatas* Lam.) tests in Korea. The Research Reports of the Office of Rural Development. 17(C):125-130.

MacFarlane, R. 1984. Solomon Islands Report of the entomologist 1981 and 1982, pp. 10-16. *In* Min. Home Affairs Natl. Dev., Honiara, Solomon Islands.

Martin, F.W. 1983. Goals for breeding the sweet potato for the Caribbean and Latin America. Proc. Am. Soc. Hort. Sci., Trop. Region 27(B):61-71.

Martin, F.W. 1984. Development of resistance to weevil *Cylas formicarius* in sweet potato. Proc. Caribb. Food Crops Soc. 18:272-276.

Mullen, M.A, A. Jones, R. Davis & G.C. Pearman. 1980a. Rapid selection of sweet potato lines resistant to the sweetpotato weevil. HortScience 15:70-71.

Mullen, M.A, A. Jones, D.R. Paterson & T.E. Boswell. 1982. Resistance of sweet potato lines to the sweetpotato weevil. HortScience 17:931-932.

Mullen, M.A, A. Jones, D.R. Paterson & T.E. Boswell. 1985. Resistance in sweet potatoes to the sweet potato weevil, *Cylas formicarius elegantulus* (Summers). J. Entomol. Sci. 20:345-350.

Mullen, M.A., A. Jones, R.T. Arbogast, J.M. Schalk, D.R. Paterson, T.L. Boswell & D.R. Earhart. 1980b. Field selection of sweet potato lines and cultivars for resistance to the sweetpotato weevil. J. Econ. Entomol. 73:288-290.

Munthali, D.C. 1988. Susceptibility of five locally grown sweet potato (*Ipomoea batatas*) cultivars to *Cylas puncticollis* infestation in the field in Zomba (Malawi). Unpubl. manuscript.

Nottingham, S.F., K.-C. Son, D.D. Wilson, R.F. Severson & S.J. Kays. 1989. Feeding and oviposition preferences of sweet potato weevil, *Cylas formicarius elegantulus* (Summers), on storage roots of sweet potato cultivars with differing surface chemistries. J. Chem. Ecol. 15:895-903.

Pillai, K.S. & P. Kamalan. 1977. Screening sweet potato germplasm for weevil resistance. J. Root Crops 3:65-67.

Pillai, K.S. & S.G. Nair. 1981. Field performance of some pre-released sweet potato hybrids to weevil incidence. J. Root Crops 7:37-39.

Pole, F.S. 1988. Vine thickness in sweet potato (*Ipomoea batatas*): its inheritance and relationship to weevil damage. M.A. Thesis, University of the South Pacific, Western Samoa.

Raman, K.V. 1988. Major sweet potato insect pests and selection for resistance to sweet potato weevil, *Euscepes postfasciatus* (Fairmaire), pp. 83-89. *In* Improvement of sweet potato in East Africa. International Potato Center, Lima, Peru.

Rolston, L.H., T. Barlow, A. Jones & T. Hernandez. 1981. Potential of host plant resistance in sweet potato for control of a white grub, *Phyllophaga ephilida* Say (Coleoptera: Scarabaeidae). J. Kansas Entomol. Soc. 54:378-380.

Rolston, L.H., T. Barlow, T. Hernandez, S. Nilakhe & A. Jones. 1979. Field evaluation of breeding lines and cultivars of sweet potato for resistance to the sweetpotato weevil. HortScience 14:634-635.

Sutherland, J.A. 1986. A review of the biology and control of the sweetpotato weevil *Cylas formicarius* (Fab). Trop. Pest Manage. 32:304-315.

Talekar, N.S. 1982a. A search for sources of resistance to sweetpotato weevil, pp. 147-156. *In* R.L. Villareal & T.D. Griggs (eds.). Sweet potato. Proceedings of the 1st international symposium. Asian Vegetable Research and Development Center, Shanhua, Taiwan, Republic of China.

Talekar, N.S. 1982b. Effects of sweetpotato weevil infestation on sweet potato root yields. J. Econ. Entomol. 75:1042-1045.

Talekar, N.S. 1987a. Resistance in sweet potato to sweetpotato weevil. Insect Sci. Applic. 8:819-823.

Talekar, N.S. 1987b. Feasibility of the use of resistant cultivar in sweetpotato weevil control. Insect Sci. Applic. 8:815-817.

Talekar, N.S. & K.W. Cheng. 1987. Nature of damage and sources of resistance to sweet potato vine borer (Lepidoptera: Pyralidae) in sweet potato. J. Econ. Entomol. 80:788-791.

Uritani, I., T. Saito, H. Honda & W.K. Kim. 1975. Induction of furano-terpenoids in sweet potato roots by the larval components of the sweet potato weevils. Agric. Biol. Chem. 37:1857-1862.

Waddill, V.H. & R.A. Conover. 1978. Resistance of white-fleshed sweet potato cultivars to the sweetpotato weevil. HortScience 13:476-477.

Wilson, D.D., R.F. Severson, K.-C. Son & S.J. Kays. 1988. Oviposition stimulant in sweet potato periderm for sweet potato weevil, *Cylas formicarius elegantulus* (Coleoptera: Curculionidae). Environ. Entomol. 17:691-693.

21. Breeding Sweet Potato For Weevil Resistance: Future Outlook

Wanda W. Collins
Department of Horticulture
North Carolina State University
Raleigh, North Carolina 27695 U.S.A.

Humberto A. Mendoza
International Potato Center
P.O. Box 5969
Lima, Peru

Sweetpotato weevil, *Cylas formicarius* (Fabricius), is the most destructive insect pest of sweet potato, *Ipomoea batatas* (L.) Lam., in the tropical and subtropical growing areas of the world and is also a serious pest in many temperate growing areas, such as the United States. The fact that this insect attacks stems, crowns, and enlarged storage roots makes it a difficult pest to control. Biological, chemical, and cultural methods that minimize damage have been described in other chapters of this book (see Chapters 6, 7, and 9); however, weevil control is still difficult to achieve. Considering the magnitude of damage caused by this weevil on a worldwide basis, cultivars with any level of sustainable resistance would be of value in reducing losses. While *C. formicarius* may be the most important sweet potato weevil worldwide, *C. puncticollis* (Boheman), *C. brunneus* (Fabricius), and *Euscepes postfasciatus* (Fairmaire) also cause serious damage to sweet potato storage roots. Resistance to these weevils would also be valuable. This chapter will explore the future potential for breeding sweet potato for resistance to weevils. It should be noted that the breeding techniques discussed in the present chapter also have potential for increasing levels of resistance to other insect pests.

PROSPECTS FOR IMPROVING RESISTANCE THROUGH CONVENTIONAL PLANT BREEDING TECHNIQUES

Sweet potato is a hexaploid with 90 chromosomes. While there is still uncertainty about the mode of origin and the ancestral parents of sweet potato, it is clear that this species acts as a functional diploid and that most traits of commercial importance are quantitatively inherited with additive gene effects usually being more important than dominance effects (Jones *et al.* 1976). This makes sweet potato particularly well-suited to mass selection which is inexpensive and does not require extensive pedigree records compared to other selection techniques. However, the most critical aspects of a mass-selection improvement program are still the use of good parental material with high levels of the desired character and the ability to precisely evaluate the trait in question.

Previous studies have shown that resistance to certain insects exists in sweet potato germplasm. Jones *et al.* (1980) developed eight sweet potato clones and released two cultivars with resistance to southern potato wireworm, *Conoderus falli* Lane, banded cucumber beetle, *Diabrotica balteata* LeConte, spotted cucumber beetle, *D. undecimpunctata howardi* Barber, elongate flea beetle, *Systena elongata* Fabricius, white grub, *Plectris aliena* Chapin, and sweetpotato flea beetle, *Chaetocnema confinis* Crotch. Heritability of resistance was estimated as 0.45 for the complex of insects composed of wireworms, *Diabrotica* spp., and *Systena* spp. (WDS) and 0.40 for flea beetle (Jones *et al.* 1979). Mean resistance to insects was improved by 25% for the WDS complex, by 21% for sweetpotato flea beetle, and by 57% for white grub during only four cycles of selection (Cuthbert & Jones 1972). It was shown that both frequencies of resistant clones and levels of resistance can be increased by the use of mass selection techniques combined with careful and precise screening techniques.

Several researchers have demonstrated that differential levels of resistance to sweetpotato weevil exist in sweet potato germplasm (Cuthbert & Jones 1972, Hahn & Leuschner 1981, Mullen *et al.* 1981, Pillai & Nair 1981, Rolston *et al.* 1979, Waddill & Conover 1978). Preference, nonpreference (antixenosis), and antibiosis have been suggested as mechanisms of resistance in sweet potato. The levels of resistance which have been reported are low and do not stand up under high weevil population pressure. Jansson *et al.* (1987) stated that cultivars reportedly resistant in Georgia and South Carolina were not resistant in Florida. However, it is likely that the level of resistance expressed by those cultivars was not detectable under the higher population pressure in southern Florida. Even so, there is potential to increase weevil resistance in sweet potato through conventional plant breeding in a mass selection improvement program because of the differential levels of resistance that have been consistently documented in many screening programs and because resistance to other insects has been demonstrated and increased through mass selection.

The Asian Vegetable Research and Development Center (AVRDC) and the International Potato Center (CIP) are investigating other methods of developing resistance to weevil through the use of wild relatives of *I. batatas*. At AVRDC, hybridizations between cultivated sweet potato (6X) and *Ipomoea trifida* (4X) relatives resulted in hybrids which showed little to no weevil damage (Takagi & Opena 1988). Further crosses and evaluations are being conducted to determine the usefulness of these hybrids. Scientists at CIP are currently working with similar germplasm of *I. trifida*.

Other researchers have approached the problem of identifying resistance by studying biochemical factors which might be associated with cultivars showing higher levels of resistance. In studies of the surface chemistry of sweet potato roots, differences in certain chemical compounds which affect oviposition of adult weevils have been identified (Nottingham *et al.* 1987). This line of research is continuing and could lead to biochemical parameters for identifying resistance levels. The value of these procedures would be in the ability to identify resistance levels in the absence of the insect, and they could reduce problems which currently arise with environmental sources of variation in resistance responses (see Chapter 11).

PROSPECTS FOR IMPROVING RESISTANCE THROUGH NONCONVENTIONAL PLANT BREEDING TECHNIQUES

Nonconventional plant breeding techniques have had limited use in sweet potato. Among those which might have an impact on the future development of weevil resistance in sweet potato are a) transformation and b) quantitative trait loci identification through restriction fragment length polymorphism (RFLP) analysis.

Transformation

Sweet potato has been successfully transformed through *Agrobacterium*-mediated DNA transfer (Jaynes *et al.* 1986, also see

Chapter 10) with the incorporation of a high protein gene. This opens the possibility that a transformation system could be used to transfer genes directly to *I. batatas*, should single genes for resistance to weevil be located in related sweet potato germplasm that is now difficult to use because of hybridization problems.

Use of RFLP Analysis for Identification of Quantitative Trait Loci (QTL's)

Based on previous studies and experience, resistance to sweetpotato weevil is a quantitatively inherited trait and may have low heritability. This type of trait is usually very difficult to evaluate in a selection and breeding program. RFLP's are differences which can be observed in different genotypes after DNA is digested by specific restriction endonucleases (REN's). Specific RFLP patterns have been linked to genes which control quantitative traits (QTL's) in certain crops. They function just as normal conventional markers but they have a heritability of 1.0 because they are a product of DNA digestion and are not affected by temperature, plant age, or other environmental factors which may affect the gene or genes with which the RFLP is linked. Large numbers of RFLP markers can be generated using different REN's and could serve as indirect selection tools for the QTL's controlling weevil resistance. RFLP markers have been associated with QTL's for insect resistance in other crops (Nienhuis *et al.* 1987). Direct selection for RFLP marker loci associated with resistance QTL's may result in a correlated response for increase of the frequency of favorable alleles for resistance.

FUTURE OUTLOOK

Low levels of resistance to weevil exist in different germplasm pools of sweet potato. Differential reactions have been well documented in the literature. The correct and efficient use of these low levels of resistance is crucial to the development of economically usable and consistent weevil resistance. Higher levels of resistance are desirable for direct control of the insect; however, lower levels of resistance may also be important because they may influence the population dynamics of the weevil itself and help to increase the efficiency of other control methods. Resistance should be viewed as one component of an integrated pest management program.

There are several breeding strategies which should be implemented at this point in order to progress towards higher levels of weevil resistance in sweet potato. It is apparent, after hundreds of clones from many different germplasm pools have been evaluated, that higher levels of resistance might not be found within *I. batatas* by further searching, and must be "created" by a logical, efficient, and well-directed program of mass selection which will combine resistance alleles. Therefore, the first strategy should be the implementation of mass selection programs to increase frequency and levels of resistance through a recurrent selection program. Several factors are crucial to the success of such a program:

1) Non-preference and antibiosis must be treated as separate mechanisms of resistance if they, in fact, do exist. More than likely, different underlying genetic mechanisms exist for each. These genetic mechanisms must be identified in order to determine how they will respond to mass selection and to predict genetic gain. Likewise, other resistance mechanisms, such as presence or absence of specific chemical stimuli, must be separately identified and mass selection must be practiced for each character. Selection for each identified resistance trait will be more effective than selection for a group of traits which may have different mechanisms and may respond differently to a specific selection procedure.

2) Precise measurement techniques should be used which do not depend on natural infestations for identifying resistance. Mullen *et al.* (1980) developed a measurement technique which was quick and efficient. Although they stated that intermediate classification was difficult and correlations between tests were poorest in the intermediate group, the identification of individuals in the intermediate classification is a perpetual problem in almost all evaluations of quantitative traits of sweet potato and does not preclude rapid progress in a selection program. It will be necessary to test genetic populations for resistance under controlled weevil population pressure because the levels of resistance that currently exist are likely to be overcome by high population pressure. Population pressure must be adjusted so that resistance is detectable in the early generations of a breeding program. Population pressure can then be increased as additive genes are accumulated.

3) It is essential to know the heritabilities of each of the resistance mechanisms in question as well as the environmental and genotype x environmental interaction effects on those traits. Heritability can be increased in several ways (e.g., increase in replications, locations, years) to give more rapid progress once these effects are known.

With the levels of resistance which have been described, the possibility of increasing resistance to sweetpotato weevil exists. It is impossible to determine what level of resistance is attainable; however, the prospect of developing an economically important level which may be used as a component in an integrated control system seems realistic.

The second strategy which must receive more attention is the elucidation and transfer of resistance, it it exists, from wild relatives of sweet potato. This could circumvent years of work in a mass selection program but, at present, is more uncertain in its applicability than a standard mass selection program within *I. batatas*. Breeding methods, both conventional and nonconventional, must be developed to transfer any resistance that is identified. Studies such as those currently underway at AVRDC and CIP on methods of utilizing wild relatives, and those on gene transfer through transformation of sweet potato, should be supported and continued in order to provide a mechanism for using resistance found in other *Ipomoea* species.

A third breeding strategy which could be important and which deserves attention is RFLP mapping of quantitative trait loci. This is a more risky venture than the first two strategies but preliminary successes in other crops indicate that it could provide results that would make the search for resistance both extremely reliable and efficient.

The major impediments to the implementation of these breeding strategies are lack of personnel working in actual sweet potato improvement programs in both developed and developing countries, and lack of proper levels of funding to undertake a large multi-faceted program of improvement. A joint effort, directed by a central coordinating organization mobilizing and utilizing the various capabilities which exist in many developing countries, as well as developed countries, is necessary. Research expertise for various components of such an effort exists in many national programs of developing countries, such as India, China, and the Philippines. However, to be effective, such an effort would have to be initiated and coordinated by an international center such as CIP or AVRDC or some other central organizing agency which could most effectively mobilize world sweet potato resources to achieve that goal.

REFERENCES

Cuthbert, F.P., Jr. & A. Jones. 1972. Resistance in sweet potatoes to Coleoptera increased by recurrent selection. J. Econ. Entomol. 65:1655-1658.

Hahn, S.K. & K. Leuschner. 1981. Resistance of sweet potato cultivars to African sweet potato weevil. Crop Sci. 21:499-503.

Jansson, R.K., H.H. Bryan & K.A. Sorenson. 1987. Within-vine distribution and damage of sweetpotato weevil, *Cylas formicarius elegantulus* (Coleoptera: Curculionidae) on four cultivars of sweet potato in southern Florida. Fla. Entomol. 70:523-526.

Jaynes, J.M., M.S. Yang, N. Espinoza & J.H. Dodds. 1986. Plant protein improvement by genetic engineering:use of synthetic genes. Trends Biotechnol. 4:314-320.

Jones, A., P.D. Dukes & F.P. Cuthbert, Jr. 1976. Mass selection in sweet potato: Breeding for resistance to insects and diseases and for horticultural characteristics. J. Amer. Soc. Hort. Sci. 101:701-704.

Jones, A., J.M. Schalk & P.D. Dukes. 1979. Heritability estimates for resistance in sweet potato to soil insects. J. Amer. Soc. Hort. Sci. 104:424-426.

Jones, A., P.D. Dukes, J.M. Schalk, M.A. Mullen, M.G. Hamilton, D.R. Paterson & T.E. Boswell. 1980. W-71, W-115, W-119, W-125, W-149 and W-154 sweet potato germplasm with multiple insect and disease resistances. HortScience 15:835-36.

Mullen, M.A., A. Jones, R. Davis & G.C. Pearman. 1980. Rapid selection of sweet potato lines resistant to the sweetpotato weevil. HortScience 15:70-71.

Mullen, M.A., A. Jones, R.T. Abrogast, D.R. Paterson & T.E. Boswell. 1981. Resistance of sweet potato lines to infestations of sweet potato weevils, *Cylas formicarius elegantulus* (Summers). HortScience 16:539-540.

Nienhuis, J., T. Helentjaris, M. Slocum, B. Riggero & A. Schafer. 1987. Restriction fragment length polymorphism analysis of loci associated with insect resistance in tomato. Crop Sci. 27:797-803.

Nottingham, S.F., D.D. Wilson, R.F. Severson & S.J. Kays. 1987. Feeding and oviposition preferences of the sweet potato weevil *Cylas formicarius elegantulus* on the outer periderm and exposed inner core of storage roots of selected sweet potato cultivars. Entomol. Exp. Appl. 45:271-275.

Pillai, K.S. & S.G. Nair. 1981. Field performance of some pre-released sweet potato hybrids to weevil incidence. J. Root Crops 7:37-39.

Rolston, L.H., T. Barlow, T. Hernandez, S.S. Nilakhe & A. Jones. 1979. Field evaluation of breeding lines and cultivars of sweet potato for resistance to the sweet potato weevil. HortScience 14:634-635.

Takagi, H. & R.T. Opena. 1988. Sweet potato breeding at AVRDC to overcome production constraints and use in Asia, pp. 233-245. *In* Exploration, maintenance and utilization of sweet potato genetic resources. International Potato Center, Lima, Peru.

Waddill, V.H. & R.A. Conover. 1978. Resistance of white-fleshed sweet potato cultivars to the sweet potato weevil. HortScience 13:476-477.

22. Sweet Potato Pest Management: A Social Science Perspective

Douglas E. Horton [1]
International Potato Center
P.O. Box 5969
Lima, Peru

Peter T. Ewell
International Potato Center
P.O. Box 25171
Nairobi, Kenya

The sweet potato, *Ipomoea batatas* (L.) Lam., is one of the world's principal food crops. It is a rustic, low-input crop which is widely grown in Africa, Asia, and Latin America, where it has significant untapped potential to meet the increasing demands for food, livestock feed, and processed products (Horton *et al.* 1989, TAC 1987). Historically, the sweet potato has received little research attention in relation to its actual and potential value. In the mid-1980's, the Consultative Group on International Agricultural Research (CGIAR) increased the priority for sweet potato improvement, and the International Potato Center (CIP) began research and training on the crop (Gregory *et al.* 1990).

Sweet potato improvement faces many challenges, of which the scarcity of funding for sweet potato improvement and the lack of scientific information on the major constraints to sweet potato production and use are most important (Gregory *et al.* 1990). Most sweet potatoes are grown in poor countries, which, aside from China, invest relatively little finances into agricultural research. Within these countries, sweet potato tends to be grown in isolated areas by low-

---

[1] Current Address: International Service for National Agricultural Research, P.O. Box 93375, 2509 AJ, The Hague, The Netherlands.

income families who have little political influence on their countries' research agendas. As a result, little domestic funding goes for sweet potato improvement. The sweet potato is also not an economically important crop in industrial countries and in foreign trade, and both international agencies and donor organizations tend to overlook this crop. As a result, the global budget for sweet potato improvement is quite limited.

The scarcity of funding for sweet potato improvement necessitates that researchers focus on only the most important constraints to production and use, and on the most promising avenues for expanding production and use. Consequently, little scientific information has been generated on technical and particularly socioeconomic aspects of the crop, its producers, and its consumers. Even the most basic information, such as where and how the crop is grown, marketed, and utilized, is absent from the scientific literature. Also, little is known about the pests and diseases of sweet potato, and the varieties that are most commonly grown by farmers or preferred by consumers.

The dearth of information on basic aspects of the crop discourages many scientists from working on sweet potato. It also deprives managers of the facts and figures needed to set appropriate priorities for sweet potato improvement and to justify needed programs to ministries of finance and to external donors.

In this paper, we contribute to the body of information that researchers and managers can use for planning and evaluating sweet potato improvement programs and for justifing additional support for them. We have divided this subject into three sections.

The first section, based on FAO statistics and early results of CIP-sponsored field studies, highlights salient features of global sweet potato production and notes some of the crop's valuable features in the context of food systems in developing countries. The second section presents a framework for assessing constraints to sweet potato production and use and outlines results of an international survey of researchers. We conclude this section by noting the urgent need for field studies. To illustrate the potential value of such studies, the third section describes how on-farm research in Peru and Tunisia has contributed to IPM work with another important crop, the potato.

THE SWEET POTATO IN DEVELOPING COUNTRIES

Knowledge of the current status of global sweet potato production and utilization is limited. The best source of data on global production is the FAO's Basic Data Unit. Unfortunately, the FAO data are not very reliable, particularly in developing countries where most sweet potatoes are grown for on-farm use in isolated areas. On the basis of existing publications, recent field studies, and personal observations, it seems likely that FAO statistics generally underestimate sweet potato production and bear little relation to actual utilization. Despite these limitations, the FAO statistics presented below provide a useful starting point for characterizing global sweet potato production patterns and trends. More detail on this subject can be found in Horton *et al.* (1989).

Production Patterns and Trends

The sweet potato, which has its origin in the tropical Americas (see Chapter 3), has spread to most of the world's tropical, sub-tropical, and warmer temperate regions. According to the FAO, the sweet potato is grown in 111 countries, of which 101 are classified by the United Nations as "developing." Among the world's root and tuber crops, the sweet potato ranks second only to the potato in economic importance (Horton 1988a).

Developing countries produce and consume nearly all of the world's sweet potatoes (Horton 1988b). Roughly 90% of all sweet potatoes are grown in Asia, just under 5% in Africa and about 5% in the rest of the world (Table 22.1). A small percentage, 2%, of the crop is grown in industrialized countries, mainly in North America and Japan. With an annual harvest of nearly 100 million tons, China is the world's largest producer of sweet potato. China also has much higher yields and production per ha than most other countries. Indonesia, Uganda, and Vietnam, which follow China in production, each harvest about 2 million tons of sweet potato roots annually. Many smaller countries, including the Solomon Islands, Tonga, Papua New Guinea, Rwanda, and some of the Caribbean Islands have higher levels of per capita production, and sweet potato has an important economic and dietary role in these countries.

According to FAO estimates, world sweet potato production increased by 40% from 1960 to about 1975 and then declined to a level about 15% higher than that of the early 1960's. Over the last quarter century, production has fallen sharply in Japan and in the United States. In Latin America, sweet potato production rose in the 1960's and then fell to just below its initial level. In Asia, production followed a similar but less pronounced trend; production now stands at a level

Table 22.1. World sweet potato production and changes since the early 1960's.[a]

Geographic region	1983-1985 average				
	Change in production 1961/63-1983/85, %	Production per capita, kg	Production, 000 t	Yield, t/ha	Harvested area, 000 ha
World	13	24	114,185	14	7,998
Asia	12	38	104,603	16	6,413
China	23	91	93,550	18	5,067
Africa	78	11	6,100	6	1,094
North and Central America	10	4	1,442	7	213
South America	-37	5	1,371	9	153
Oceania	52	23	560	5	116
Europe	-44	0	108	11	10
Developing countries	20	32	111,979	14	7,867
Developed countries	-70	2	2,206	17	131

[a] Derived from FAO Basic Data Unit.

of approximately 25% above the early 1960 level. Africa is the only region in which sweet potato production has increased throughout the entire period; it is now nearly double the 1960 level (Horton 1988b).

Energy and Protein Production

The sweet potato is often, but wrongly, thought to be a source of starch and little else in the human diet. Sweet potato roots are a good source of food energy and while their protein content is relatively low, the protein quality is extraordinarily high. In fact, among major food crops, the sweet potato has the highest recorded net protein utilization.

Sweet potato crops produce large amounts of edible energy, protein, and vitamins per unit of land and time. At present average yields in developing countries, the sweet potato ranks first among major food crops in production of edible energy per ha and third in production of edible protein per ha (Table 22.2). Even though it has a

Table 22.2. The ten food crops with the highest production value per hectare in developing countries.[a]

Crop	Production value, U.S. $/ha	Dry matter, t/ha	Edible energy, million kcal/ha	Edible protein, kg/ha
Tomato	3,159	1.1	3.1	157
Cabbage	3,026	1.3	3.2	175
Potato	1,633	2.3	7.1	196
Yam	1,581	2.6	8.4	175
Sweet potato	1,210	4.0	12.6	187
Cassava	595	3.4	7.3	32
Cocoyam	554	1.2	3.7	72
Rice, paddy	493	2.6	7.1	130
Banana	492	1.5	3.9	36
Groundnuts in shell	297	0.9	4.1	190

[a] Derived from FAO (1984) and U.S. Department of Agriculture (1975). FAO estimates are 1981-1983 averages; price estimates are for 1977.

long growing season, sweet potato has the highest rate of production of edible energy per ha per day of any major crop. Furthermore, the production of edible protein per day is superior to that of rice and ten times greater than that of cassava.

The high yield of sweet potato in developing countries is due largely to China, where successful breeding and sweet potato improvement programs have had a major impact at the farm level. China's success may auger well for sweet potato improvement in other areas, which are just now beginning to organize the needed research and extension efforts.

Nutritional Value

As detailed in Woolfe (in press), sweet potato produces two useful food types from the same plant: fleshy storage roots and green tops. Both can be used as a nutritious food for humans and animals, and in fact, the tops are a better source of high quality protein than the roots.

Tops contain twice the level of protein on a fresh weight basis. Because the tops are often among the cheapest vegetables available in local markets, they are an inexpensive source of dietary protein in some countries. This gives sweet potato an added advantage for feeding families from home gardens. Tops are also a highly nutritious animal feed that is used extensively in some countries.

Sweet potato provides a number of dietary nutrients which are valuable for combating certain severe and widespread nutritional problems in the developing world. The foremost of these characteristics is extremely high provitamin A carotenoid content in the roots of certain clones and in the leaves. Vitamin A deficiency (xerophthalmia) is one of the major public health problems confronting some developing countries at the present time. Both roots and tops are important sources of other vitamins, especially ascorbic acid. This makes them a valuable complement to cereal-based diets which are often limited in this amino acid. The sweet potato is also a good source of thiamin, iron, and calcium in the human diet, and it supplies modest amounts of several B group vitamins, including niacin, pyridoxine, folic acid, and riboflavin. Riboflavin is a B vitamin that is generally deficient in rice-based diets in Asia.

Sweet Potato in Food Systems

The sweet potato has a number of attributes which auger well for its future role in combating food shortages and malnutrition in poor areas where population and land pressures are rising. As fertile arable land per person decreases, there will be a growing need to open up marginal areas and to intensify existing cropping patterns. Often, neither the farmer nor the national economy can afford to increase inputs such as agricultural chemicals. These circumstances will highlight the value of sweet potato because it produces high yields with few inputs under marginal conditions, and it can adapt to many cropping systems.

Despite the wide geographical distribution and value of the sweet potato crop, socioeconomic studies of sweet potato production and use are rare. A number of anthropological studies describe certain aspects of sweet potato production or use as part of more general accounts of social or cultural organization, especially in Oceania. However, few studies have documented production, distribution, and utilization patterns in a systematic manner, and virtually none have analyzed production constraints or the potential benefits of sweet potato research and development.

The few published studies and early results of field studies being carried out in a number of countries in association with CIP indicate that sweet potato is adapted to a wide range of agroclimatic conditions and has been successfully incorporated into diverse food systems.

The sweet potato has a broad ecological and agroeconomic adaptability, which is related to the genetic diversity within the cultivated species. The crop is grown from 35/ N to 35/ S and from sea-level to almost 3,000 m in elevation. In South America, it is grown in the Andes mountains, in the Amazonian jungle, on the great sub-tropical and temperate plains of the southern cone, and, under irrigation, in the desert on the Pacific coast. In the Caribbean and the Pacific, it is grown on small tropical islands; in Africa, it is grown at mid-elevations and in parts of the tropical lowlands; and in Asia, it is grown at a wide range of altitudes from temperate to tropical zones. Many of the environments in which the crop is grown have poor, degraded soils that support few other crops.

Sweet potato has many roles in diverse food systems across the world. Perhaps the most typical system involves small-scale sweet potato production, primarily for household consumption and secondarily for livestock feed and sale. Intercropping is a common feature. While globally, this may be the most common system, in many areas, sweet potato is important as a cash crop, a livestock feed, or as an industrial input.

Field studies in Asia, Africa, and Latin America show a diversity of sweet potato production practices and final use in different kinds of food systems. For example, on the island of Java, sweet potato is intensively cultivated as a cash crop by farmers who are linked to a well-organized marketing chain supplying the major cities for fresh consumption or food processing into snacks. In the neighboring island of Sumatra, sweet potato is grown as an off-season staple food between rice crops.

In Rwanda, where sweet potato is one of the major staples, the intensity of production and the importance of marketing vary according to topography and altitude. In Kenya, where there are very large urban populations, large-scale commercial sweet potato production occurs. One example is the coastal strip that supplies Mombasa, where both fresh roots and processed sweet potato snacks are consumed.

On the coast of Peru, where desert conditions restrict agriculture to irrigated valleys with little or no pasture land, sweet potato foliage is the major source of animal feed. Varieties are selected by farmers for their two qualities: good root production and abundant foliage.

On the southern plains of South America, in Argentina and Uruguay, sweet potato is cultivated both in small "kitchen" gardens and in fields of 200 ha or more. Most production is either for household consumption or for sale to the large cities. Buenos Aires annually consumes around 75,000 tons of fresh sweet potato. Despite high levels of production, only a relatively small percentage of production in Argentina and Uruguay is processed for making a sweet cake eaten for desert. The most sophisticated and diversified system of post-harvest utilization is in China, where sweet potato is grown for its fresh roots or greens for human consumption, for animal feed, and for production of noodles.

CONSTRAINTS TO SWEET POTATO PRODUCTION AND USE

A knowledge of the main constraints to sweet potato production and use in different production zones is essential for effective research planning and for targeting regional efforts. When CIP began to work on sweet potato in the mid 1980's, it was confronted with a lack of both published information and in-house experience with the crop. In order to quickly assemble basic information on production zones and constraints, we undertook an international survey of sweet potato researchers. A previous study conducted a similar survey in Asia (Horton 1989).

The first step involved meeting with scientists in each of the disciplinary departments of CIP to develop a questionnaire that listed the most likely constraints to sweet potato production and use. A list of hypothetical constraints was developed that included 50 separate items grouped under 10 headings: varieties, planting material, fungal and bacterial diseases, virus diseases, nematodes, insects, environmental problems, storage, marketing, and demand.

The complete list of constraints was included in a questionnaire that solicited the opinions of researchers on the relative importance of each constraint in the major sweet potato growing regions of their country. In filling out the questionnaire, researchers first listed the major sweet potato producing regions and estimated the amount of land planted to sweet potato and the yield in each region of their country. Researchers then indicated the relative importance of each constraint in each region using a numerical score from 0 to 3 as follows: 0, not present; 1, of little practical importance; 2, somewhat important; and 3, very important. Questionnaires were also mailed to sweet potato researchers throughout Africa, Asia, and Latin America. They were also distributed to participants at several regional workshops organized by CIP to plan collaborative sweet potato research and training activities with national agricultural research organizations. In total, 61 national sweet potato researchers filled out and returned questionnaires that provided information on constraints in 171 production sites in 35 countries.

Although a random sampling technique was not used to select informants, a broad representation of African, Asian and Latin American countries was obtained in the CIP-sponsored workshops. As a result, survey informants provided information on regions that account for approximately 60% of the sweet potato grown in all developing countries. Because of the great importance of China as a producer of sweet potato, a special effort was made to obtain representative data on that country. In a national workshop, data on constraints in 42 locations were provided by 11 Chinese scientists.

Many constraints to crop production are location-specific, and must be analyzed in relation to the agro-ecological conditions of different regions. Ideally, results of the constraints survey would be presented for specific sweet potato production zones. However, this is not possible because basic information on the crop and its ecological niches is not yet available, and the agroecological mapping of sweet potato production is still in an exploratory stage.

In this paper, we use the Koppen classification of climates, as presented in Trewartha and Horn (1980), as a proxy for classifying sweet potato production zones. This approach has been applied in earlier work on potato (Midmore & Rhoades 1987).

In Africa, Asia, and Latin America, sweet potato is grown in areas with four broad types of climate: tropical rainy climates (A climates in the Koppen system), semi-arid climates (B climates), mild temperate rainy climates (C climates), and cold climates (D climates).

In the following paragraphs, we relate salient results of the constraints survey to each of these climates and to the geographical regions in which they occur.

Production Zones

About one-half of the sites covered by the survey have mild, temperate, rainy climates (C climates). Based on estimates of the reseachers who filled out the questionnaire, in developing areas, about 80% of all sweet potato is grown under temperate conditions. Temperate climates account for nearly all the sweet potatoes grown in China and half of those grown in the rest of Asia and in South America. Although the estimates based on this survey are subject to error and should be considered as preliminary, it seems safe to conclude that a mild, temperate, rainy climate is the most typical climate for sweet potato production in developing countries.

According to the survey, the second-most typical climate for sweet potato production is tropical and rainy (A climate). Tropical areas account for nearly 40% of the sites surveyed and an estimated 20% of the area planted to sweet potato in all developing countries. In Central America, in Sub-Saharan Africa, and in many parts of Asia, most sweet potato is grown in tropical areas. Countries with relatively large sweet potato production areas (hectareage) in tropical areas include Brazil, Bangladesh, Haiti, India, Uganda, and Vietnam.

The survey indicated that relatively few sweet potatoes are grown in semi-arid areas (B climates) or in cold areas (D climates). Cold-climate sweet potato production is primarily restricted to China. Semi-arid production occurs in several countries, including Argentina, Ethiopia, Peru, Thailand, and Venezuela. While the semi-arid and colder growing areas are relatively small on a global scale, the sweet potato plays an important role in such areas in several countries.

Results of the Survey

Most of the researchers who filled out the questionnaire were production specialists (in contrast to post-harvest specialists or social scientists), and it was feared that their responses might be biased in favor of the technical production problems with which they are most familiar. However, this does not appear to have been the case. Irrespective of their discipline or work experience, most respondents indicated that post-harvest and marketing problems were the most important constraints (Figure 22.1).

When the data were pooled to include all sites, the top-ranked constraints both related to post-harvest problems: unstable sweet potato supplies and prices and the lack of suitable processed products. The survey indicated that the leading production constraints were: low soil fertility, drought, and the sweetpotato weevil, *Cylas formicarius* (Fabricius) (Table 22.3).

417

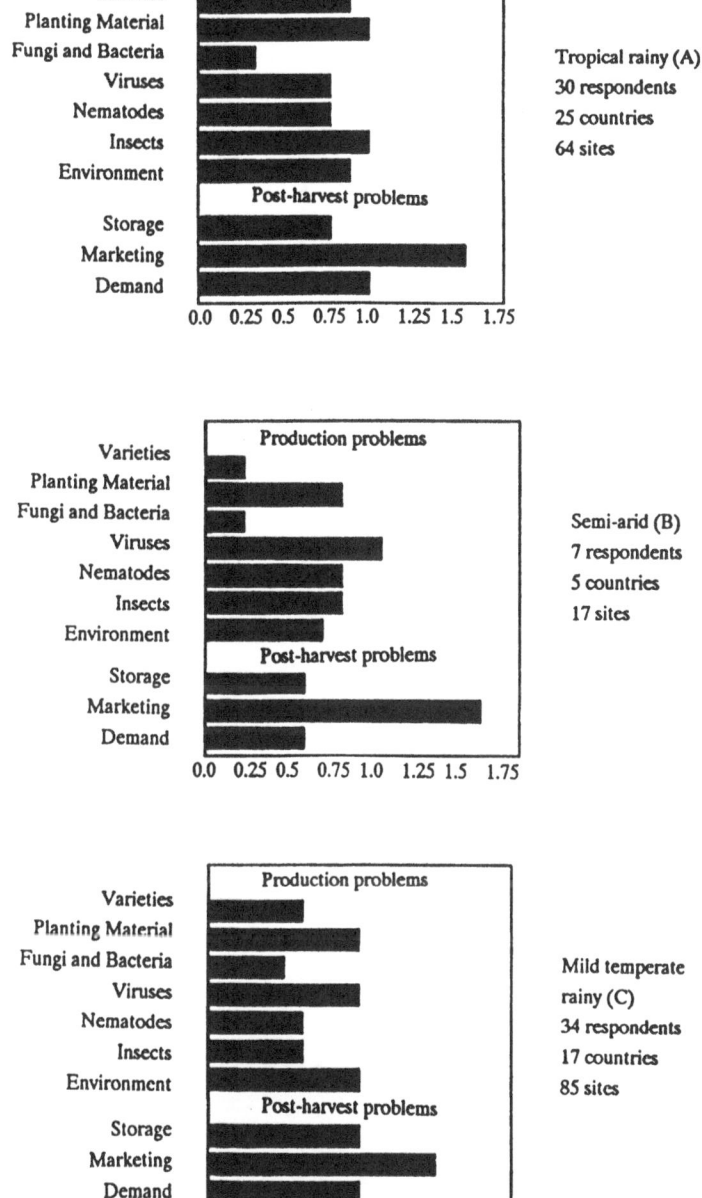

Fig. 22.1. Mean scores for importance of various constraints to sweet potato production and use in three different climatological zones.

Table 22.3. Data profile of scientists that were surveyed and mean scores of constraints to sweet potato production and use by climate.

Data profile,	Tropical, rainy A	Semi- arid B	Mild, temperate, rainy C	Cold D
		Climate		
No. of respondents	30	7	34	3
No. of countries	26	6	17	2
No. of sites	64	17	85	5
Mean yield, t/ha	5	7	15	6

Production constraint	Mean score			
Varieties				
Too late maturing	0.9	0.6	0.6	1.2
Poor market consumer acceptance	1.1	0.2	0.6	0.0
Not suited for processing	0.7	0.0	0.3	0.6
Not suited for livestock feeding	0.4	0.0	0.3	0.0
Planting material				
Poor health	1.3	1.2	1.0	1.0
Poor physiological condition	0.6	0.5	0.7	0.2
Scarcity	1.2	0.8	0.8	0.4
High cost	0.3	0.1	0.5	0.0
Fungal and bacterial disease				
Scab	0.5	0.1	0.4	0.0
Fusarium wilt	0.3	0.2	0.5	0.0
Erwinia spp.	0.1	0.2	0.2	0.0
Virus disease	0.7	0.9	0.8	1.0
Plant-parasitic nematodes				
Meloidogyne incognita	0.7	0.7	0.5	1.0
Insects				
Cylas spp.	2.2	0.9	0.7	0.0
Euscepes postfasciatus	0.4	0.9	0.7	0.0
Vine borers	0.7	0.5	0.4	0.0
Diabrotica spp.	0.3	0.4	0.4	0.0
Environment				
Cold, frost, or hail	0.1	0.4	0.9	0.6
Heat	0.4	0.2	0.3	0.0
Drought	1.7	0.9	1.5	2.0
Excess soil moisture	0.8	0.1	0.7	0.0
Low light energy	0.4	0.2	0.5	0.0
Weeds	1.3	1.7	1.1	1.0
Low soil fertility	1.5	1.1	1.7	1.8
Soil salinity	0.5	0.8	0.5	1.6

Production constraint	Climate			
	Tropical, rainy A	Semi-arid B	Mild, temperate, rainy C	Cold D
	Mean score			
Environment continued				
Low soil pH	0.6	0.3	0.4	0.4
Aluminum toxicity	0.4	0.2	0.2	0.0
Poor soil structure	0.9	0.6	0.7	0.2
Storage				
Diseases (rots)	1.2	1.1	1.5	0.8
Insects	1.4	1.2	0.5	0.6
Rodents	1.2	0.4	0.7	0.2
Sprouting	0.9	0.5	0.4	0.2
Shriveling	1.0	0.8	0.5	0.2
Poor store design	0.9	0.3	0.8	0.6
Insufficient storage capacity	0.4	0.6	0.7	0.4
High storage costs	0.3	0.6	0.7	0.4
Marketing				
Transportation problems	1.6	0.8	1.5	1.6
Unstable supplies and prices	1.7	1.8	1.5	1.4
High marketing costs	1.1	1.4	1.1	0.6
High handling losses	1.0	1.5	0.8	1.0
Demand				
Consumer acceptance/taste	0.7	0.4	0.5	0.6
Lack of variety with good eating qualities	0.9	0.2	1.0	1.6
Not in traditional diet	0.7	0.9	0.7	0.4
Sweet potato not available	0.6	0.0	0.5	0.0
Price too expensive	0.5	0.8	0.6	0.6
Lack of processed products	1.9	0.9	1.5	1.0
Lack of good processing qualites	1.0	0.2	1.0	0.8

Post-harvest and marketing problems were considered to be most significant, relative to production problems, in mild temperate areas that have relatively high yields (e.g., in China). Production problems, in general, and pest problems, in particular, appear to be most severe in the tropical rainy areas, where yields are lowest.

The researchers who filled out the questionnaire considered the sweetpotato weevil, *C. formicarius*, to be the single most important constraint to production and use of sweet potato in the tropics. In these areas, researchers also noted that insect damage in storage was severe. Unfortunately, the survey did not provide information on the type and cause of this damage.

The survey results indicated that the highest global priority should be placed on overcoming post-harvest and marketing constraints to sweet potato production and use. In this context, integrated pest management (IPM) specialists should consider the impact of insect damage on storage, marketing, and demand. The survey also indicated a clear role for IPM specialists in relation to the sweetpotato weevil, which was considered the single most important production constraint in tropical areas. It is in these areas where people are poorest and where the sweet potato could play the most crucial role in meeting the rapidly expanding demands for food.

Results of the constraints survey should be viewed as working hypotheses, not firm conclusions, because they are based on the opinions of researchers, and not on results of careful field studies. Field studies are needed to confirm or reject these hypotheses. The following section illustrates how field studies have contributed to the development and transfer of IPM technologies in potato in developing countries.

LESSONS FROM ON-FARM RESEARCH WITH POTATOES

Potato production is expanding rapidly in many developing countries, and chemical pesticides are being used more widely. Entomologists at CIP and in many national agricultural research systems are working to develop effective pest control strategies that reduce dependence on chemical pesticides. The examples from Peru and Tunisia presented in this section are based on Ewell *et al.* (1990) and on von Arx *et al.* (1988).

Highland Peru

Potatoes were domesticated in the highlands of Peru, where the risks of damage and loss from natural hazards, such as pests and diseases, are high. Surveys were carried out in two highland regions, Cuzco and the Mantaro Valley, where potato is the major crop. As the first part of each interview, we asked farmers to identify the insects which affected their potato crop. A box of insect specimens with pinned adults and larvae in alcohol proved a very useful survey tool.

We soon learned that older people, especially women, were the best informants.

We asked farmers to assess the importance of the insects and other problems which affected their crop in the most recent growing season. Informants scored the various risks from 0 to 3, where 0 = unimportant and 3 = a serious problem.

Most informants cited natural hazards (frost, hail, and drought) and uncertain prices as the most important production problems. Andean potato weevils, *Premnotrypes* spp., wireworms, fungal diseases, potato tuber moth, *Phthorimaea operculella* (Zeller), and flea beetles were considered by farmers to be the most problematic insect pests.

Farmers were found to use various strategies to manage insect pests. One common practice was to place aromatic plants on or near stored tubers. These plants served as both both a barrier and a repellent. About one-third of the farmers in our sample reported the use of the local weed Muña (*Minthostachys* sp.) in their storage areas; a smaller number were using leaves of *Eucalyptus* sp., hot peppers, and other plants.

Since the late 1940's, chemical pesticides have been used widely on potato crops in the Peruvian highlands. Our surveys indicated that chemical pesticides were the most commonly purchased input used in potato production. Approximately 95% of the farmers interviewed used a large number of chemical pesticides. The 241 farmers interviewed used 46 different insecticides and 18 different fungicides.

The most important source of information about new pesticides, cited by nearly one-third of the farmers, was radio advertising by the chemical companies. The second and third most important sources were recommendations of merchants in commercial outlets and the advice of neighbors. By and large, farmers did not consider the National Research Institute or the extension agents to be useful sources of information on pest management.

The Andean weevils, *Premnotrypes* spp., are generally considered to be the most serious insect pests of potato in highland Peru. However, field studies indicated that field damage was spotty and farmers accepted a certain amount of loss as inevitable. Monitoring throughout the Mantaro Valley indicated that about 90% of the harvested tubers had some weevil damage. However, in about 60% of the cases, 20% or less of the tubers were damaged. According to farmers, this is approximately the level at which physical damage begins to have a significant economic impact.

Farmers actively attempted to control these weevils. The most commonly used insecticide was carbofuran, which was applied to the soil at planting to prevent weevil damage. This practice, promoted by chemical companies, was followed by over 80% of the farmers interviewed in the Mantaro Valley.

On-farm research in the late 1970's showed that soil application at the time of hilling were more effective than the current practice of applying carbofuran at planting (Franco *et al.*, 1981). However, farmers, even those that participated in the on-farm trials, did not adopt the recommended practice.

Peru's Coast: A Pesticide Treadmill

The Cañete valley on Peru's central coast presents a graphic illustration of the vulnerability of intensive production systems to pests, particularly when crops are grown outside their traditional agroclimatic range. Potatoes became an important commercial crop in Cañete starting in the early 1950's, with the rapid growth of demand for food in Lima, 150 kilometers to the north.

The agromyzid leafminer, *Liriomyza huidobrensis* (Blanchard), was present in Cañete as early as the 1940's, but it was not an important pest of any crop at that time. Insecticides were first introduced into the valley after World War II for use on cotton. They were applied to potato to control a lepidopterous pest, *Scrobipalpula absoluta* Meyrick. Consequently, natural enemies of *L. huidobrensis* were disrupted, and populations of *L. huidobrensis* increased dramatically. Farmers were subsequently led onto a "pesticide treadmill."

The pest management practices of a group of farmers were closely monitored for several seasons; damage levels and costs were estimated. Insecticides were found to be the single most costly input, and cost, in some cases, more than $1,000 U.S. per hectare. In years of serious attack, pest control accounted for over one-half of the total production costs.

Over the years, researchers and extension agents have repeatedly proposed to improve control of *L. huidobrensis* and to reduce pesticide use through regulation of planting dates and field monitoring. These proposals have never been adopted, however, because they require a coordinated group effort by the many small farmers of the valley, and these farmers have not been organized.

Potato Tuber Moth in Tunisia

The potato tuber moth, *P. operculella*, is the most important insect pest of potato in Tunisia and throughout North Africa. It attacks both above-ground plant parts and tubers and causes economic damage to tubers kept in unrefrigerated storage areas throughout the hot summer months.

CIP's regional office worked closely with Tunisia's national agricultural research institute, INRAT, to develop an improved IPM program for this moth. The program included the following components (von Arx *et al.*, 1987): frequent irrigation to prevent soil cracks through which larvae find their way to tubers; timely harvest, rapid handling, and careful selection to reduce initial infestation in storage; covering stored tubers with a thick layer of straw to serve as a mechanical barrier against subsequent infestation; and the use of the biological insecticide *Bacillus thuringiensis* Berliner. On-farm experiments, monitoring, and surveys were conducted to analyze farmers' control methods and to assess the potential value of the improved IPM program.

A survey in 1987 indicated that many farmers were aware of the recommended practices, although improvements in their management, timing, and combinations could be made. For example, most farmers apply insecticides both in the field and in storage heaps. Nearly 40% of the farmers recognized the importance of timely harvest. Nevertheless, because they were busy with other crops, nearly one-half of the farmers surveyed delayed their harvest until damage levels were high.

One reason why farmers did not improve control of this moth was that high damage levels are accepted in Tunisian markets. Farmers monitor the storage heaps, removing damaged potatoes and selling the lot if damage becomes severe. The burden is subsequently shifted to consumers, who pay higher prices for tubers of deteriorating quality from July through mid-October when new tubers begin to appear in the market. Samples taken in retail markets indicated that by October, buyers cut away and discarded 30 to 40% of the weight of tubers purchased before cooking.

While at present, farmers seem to have little incentive to improve the control of *P. operculella*, this will probably not last much longer because off-season potato production is increasing the supply of fresh, damage-free tubers in the market. This increased supply will probably lower the acceptable level of damage to tubers in storage, and subsequently force farmers to increase pesticide use, risking a pesticide treadmill, and/or to improve their control methods by other methods.

Lessons from Field Studies with Potato

Some important lessons for future IPM work on sweet potato can be drawn from the Peruvian and Tunisian experiences on potato. One lesson is that farmers who use traditional practices are often familiar with newer recommended practices for controlling pests. In all three cases described above, we found that many farmers knew more about pest problems than previously assumed, and many had experimented (in some cases years ago) with practices that researchers were now recommending as "new" and "improved." In many cases, after initial experimentation, farmers discontinued the use of some of these practices because they were too costly, or the effort was not adequately compensated for by additional economic return.

A second lesson is that perceptions of farmers and consumers about pest problems often differ substantially from those of researchers. In both the Mantaro Valley and in Tunisia, farmers were concerned with controlling the Andean weevil and the potato tuber moth, respectively, but they were much less concerned about these insects than local entomologists. Farmers and consumers had several other more important problems, and they have learned to live with pest problems. One implication of this is that IPM specialists must put their work in perspective by familiarizing themselves with the perceptions of farmers of the pests. A second implication is that field studies on sweet potato farming, marketing, and use are needed to complement information compiled on the major constraints of the crop.

A third lesson is that overcoming pest problems is an extremely complex task that requires more than technical research. In all three cases, technically sound solutions to pest problems were known. However, the recommended practices were not used by farmers for social and/or economic reasons. Thus, sweet potato IPM specialists need to maintain close links with farmers in order to understand their problems, and to monitor and/or modify the solutions offered. Effective IPM requires organized extension programs and group action by the producers. Social scientists can play important roles in IPM by obtaining information on the problems of farmers, and by participating in extension and group action.

CONCLUSIONS

Sweet potato is one of the world's most widely grown and valuable crops. Cultivated in more than 100 countries, it is especially important in the food system of Africa, Asia, and Latin America. Asia is the major producing region, but the relative importance of the sweet potato is greatest in Sub-Saharan Africa. The sweet potato is prized in many rural areas, where farmers rely on it to produce high yields on marginal land with few purchased inputs. It is an important component of food systems in many tropical and mild temperate areas, where it provides not only food for human consumption, but also feed for livestock, and starch for industry. In nutritional terms, in addition to energy, the sweet potato is a good source of quality protein, several vitamins, and minerals.

Despite the actual and potential value of this crop, the sweet potato has received little research attention, except in China. Consequently, little information is available on where and how the crop is grown and on the major constraints to its production and use.

An international survey of researchers indicated that the major constraints to sweet potato production and use are post-harvest and marketing problems; production problems were less important. According to the survey, the sweetpotato weevil, *C. formicarius*, is a major constraint in many areas and the single most important constraint in tropical areas. On a global scale, it would seem appropriate to focus the scarce resources available for sweet potato IPM on management of this weevil in tropical areas, where it is most damaging and where rural poverty and food needs are greatest. Other pests, however, may merit greater attention in specific countries and regions.

Field studies involving IPM specialists and social scientists are needed to provide additional information on the following: pest incidence and associated losses; the perceptions of farmers and consumers of their pest problems; the strengths and weaknesses of current pest management strategies; the efficacy of alternative control measures under field conditions; and on implementation and validation of IPM technologies through extension and farmer participation programs.

REFERENCES

Arx, R. von, J. Goueder, M. Cheikh & M. Ben Temime. 1987. Integrated control of potato tuber moth, *Phthorimaea operculella* (Zeller), in Tunisia. Insect Sci. Applic. 8:989-994.

Arx, R. von, P.T. Ewell, J. Goueder, M. Essamet, M. Cheikh & M. Ben Temime. 1988. Management of the potato tuber moth by Tunisian farmers. International Potato Center, Lima, Peru.

Ewell, P.T., H. Fano, K.V. Raman, J. Alcazar, M. Palacios, B. Eldridge & L. Carhuamaca. 1990. Management of insect pests by potato farmers in Peru: report of an interdisciplinary research project. International Potato Center, Lima, Peru. (In press).

FAO. 1984. FAO production yearbook, 1983. Food and Agricultural Organization, Rome.

Franco, E., D. Horton, R. Cortbaoui, F. Tardieu & L. Tomassini. 1981. Evaluacion agro-economica de ensayos conducidos en campos de agricultores en al Valle de Mantaro (Peru) Campana 1978/79. Documento de Trabajos 1981-1. Department of Social Science, International Potato Center, Lima, Peru (In Spanish).

Gregory, P., M. Iwanaga & D. Horton. 1990. Sweet potato research: global issues, pp. 462-468. *In* R.H. Howeler (ed.). Proceedings of the 8th international symposium for tropical root crops. CIAT, Bangkok, Thailand.

Horton, D. 1988a. Underground crops: long-term trends in production of roots and tubers. Winrock International Institute for Agricultural Development. Morrilton, Arkansas.

Horton, D.E. 1988b. World patterns and trends in sweet potato production. Trop. Agric. 65:268-270.

Horton, D.E. 1989. Constraints to sweet potato production and use, pp. 219-223. *In* Improvement of sweet potato (*Ipomoea batatas*) in Asia. International Potato Center, Lima, Peru.

Horton, D., G. Prain & P. Gregory. 1989. Sweet potato research and development: high level investment returns for international research and development. CIP Circ. 17:1-10.

Midmore, D. & R. Rhoades. 1987. Applications of agrometeorology to the production of potato in the warm tropics. Acta Horti. 214:103-136.

TAC (Technical Advisory Committee of the Consultative Group on International Agricultural Research). 1987. CGIAR priorities and future strategies. Food and Agricultural Organization, Rome, Italy.

Trewartha, G. & L. Horn. 1980. An introduction to climate. McGraw-Hill, New York.

U.S. Department of Agriculture. 1975. Composition of food. U.S. Department of Agriculture, Washington, D.C.

Woolfe, J. In press. Sweet potato: an untapped food resource. Cambridge University Press, Cambridge, United Kingdom.

23. Sweet Potato Pest Management: Future Outlook

Richard K. Jansson
Tropical Research and Education Center
University of Florida, I.F.A.S.
Homestead, Florida 33031 U.S.A.

Kandukuri V. Raman
International Potato Center
P.O. Box 5969
Lima, Peru

Oscar S. Malamud
International Potato Center
P.O. Box 25327
Santo Domingo, Dominican Republic

The preceding chapters covered the current knowledge world-wide and identified research needs and gaps in technology related to integrated pest management (IPM) programs for sweet potato. Most of the research in this area has been conducted in the United States, where the crop is of minor importance. In the developing world, sweet potato has received much less research and development attention than have other crops. Also, the high economic and nutritional value of the sweet potato has been underexploited (see Chapters 1 and 22).

Technological changes in research and development result in significant changes in crop improvement. Sweet potato IPM is no exception. Considerable advances have been achieved recently in several aspects of sweet potato IPM, including: an improved understanding of the taxonomy of *Cylas* weevils; the development of a synthetic sex pheromone for the sweetpotato weevil, *C. formicarius* (Fabricius); the development of IPM programs for *C. formicarius* in Asia; the use of entomopathogens for control of *C. formicarius* and the West Indian sweet potato weevil, *Euscepes postfasciatus* (Fairmaire); the potential use of *Agrobacterium*-mediated gene transfer to produce pest resistant sweet potato plants; the development of phytochemical methods to assess resistance of sweet potato to insects, especially *C. formicarius*; the development of several cultivars with resistance to pest complexes; and improved methods for managing vectors, viruses, and plant-parasitic nematodes of sweet potato. However, considerably more research and development is needed internationally to counteract the lack of research attention given to sweet potato IPM, especially in developing countries. A major cooperative program is needed to provide leadership in the following areas: (1) a global needs assessment for sweet potato IPM research and development; (2) preservation of sweet potato germplasm; (3) research in priority areas, such as host plant resistance to insects, nematodes, and diseases, and the use of sex pheromones and biological and cultural methods for managing these pests; (4) training, documentation, and information exchange; and (5) increased crop promotion and public awareness.

IPM programs for sweet potato pests have a promising future. The future of *C. formicarius* management is exciting and points towards a greater emphasis on the use of biological and parabiological approaches (see Chapters 6 and 9). These approaches are well suited for low input agricultural systems. For example, the synthetic sex pheromone of *C. formicarius* and entomopathogenic nematodes and fungi have good potential for integrating into current weevil management programs world-wide. The sex pheromone has already been integrated into weevil management programs in several developing countries and is currently being used by the International Potato Center (CIP) to develop monitoring systems for *C. formicarius* world-wide.

One of the major factors limiting the use of this technology in developing countries has been the availablity of the pheromone at a cost that developing countries can afford. In order to overcome this problem, synthetic chemists in certain developing countries have investigated the possibility of synthesizing the pheromone using locally available resources. Several countries are currently interested in producing this pheromone by the use of less expensive synthetic methods. Both Taiwan and India have synthesized this pheromone. However, as noted earlier (see Chapters 5 and 6), chemistry of pheromone synthesis is central to quality control of synthetic pheromone, and pheromone synthesized by alternative methods should be analyzed by HPLC, NMR, and mass spectroscopy to confirm its chemical structure and purity. In addition, field studies that use standardized protocols are needed to compare the efficacy and attractiveness of pheromone produced by alternative methods with that produced in the U.S. (see Chapters 5 and 6).

In most developing countries, sweet potato is grown in small fields rarely exceeding 1 ha. This pheromone also has good potential for managing *C. formicarius* populations by mass trapping and/or mating disruption. Studies are needed to assess its potential for managing this insect. Determination of the management potential of this pheromone is central for integrating the pheromone with other management tactics, especially cultural practices and the use of biological control agents, such as entomopathogenic nematodes and fungi, or insect parasitoids.

The two most efficacious and efficient traps at collecting *C. formicarius*, the plastic funnel trap, which was developed in the U.S. Virgin Islands and later modified in Florida, and the commercially available Universal moth trap (available from AgriSense, Fresno, California or Great Lakes IPM, Vestaburg, Michigan), are too costly for many developing countries. The cost of a trapping system may also be prohibitive in certain regions of the United States. For example, the North Carolina Department of Agriculture continues to use and recommend the screen-cone boll weevil trap in its monitoring program for *C. formicarius* in North Carolina despite the poor efficacy and efficiency of this trap. Although this trap is inferior to other traps that were tested, it is considerably less expensive than the Universal moth trap. For these reasons, alternative, inexpensive traps that are efficient at collecting this weevil are needed in both developing and developed countries. Additionally, in developing countries, traps are needed that can be constructed from locally available materials at minimal cost. Prototypes are currently being tested by CIP and cooperating national agricultural scientists in several countries.

Considering current domestic and international restrictions on trade and movement of sweet potato and associated plant parts, this pheromone also has great potential for use in regulatory entomology. Studies are needed to modify the existing monitoring system for regulatory purposes and develop a detection system for this weevil in sweet potato storage facilities.

Sex pheromones of other weevil species, such as *C. puncticollis* (Boheman), *C. brunneus* (Fabricius) and *E. postfasciatus*, have not been recognized to date. Preliminary studies are currently underway in Kenya to determine if *C. puncticollis* produces a sex pheromone. The availability of pheromone for *C. puncticollis* would help to improve the management of this weevil in Africa.

Preliminary work demonstrated that *C. formicarius* is attracted to host plant volatiles from a variety of *Ipomoea* species (see Chapter 13). More work is needed to characterize and identify the volatile(s) that are important in attracting these weevils to their host plants. Such an attractant might ultimately be useful for attracting female *C. formicarius* to traps. Currently, only male *C. formicarius* are caught in traps baited with a synthetic sex pheromone similar to that produced by female weevils. Integration of a female attractant in traps might help to manage this weevil in the future.

Concerning biological control of *Cylas* spp., there are indications that locally-adapted, tropical or subtropical strains and/or species of entomopathogenic nematodes may be more virulent to *C. formicarius* than those from temperate zones. Surveys are needed to isolate these nematodes from tropical soils. A survey for these nematodes is currently underway in the Caribbean basin, and several isolates have been found in Puerto Rico (W. Figueroa, R.K. Jansson, R.R. Gaugler, C. Cruz, unpubl. data). Also, the potential of these nematodes during the dry season in the tropics needs to be assessed. Other factors that may affect the success of these nematodes as biological control agents of *Cylas* spp. were presented earlier (see Chapter 9) and need considerably more research. The potential of these nematodes at managing populations of *E. postfasciatus* also needs to be determined. In addition to entomopathogenic nematodes, insect parasitoids and fungal pathogens also have potential. Insect parasitoids that achieve high levels of parasitism in the center of origin of *Cylas* species (i.e., Africa) and of *C. formicarius* (i.e., India) should be introduced into other regions for use in classical biological control programs (see Chapter 9). Research on identification, virulence, and efficacy of improved strains of the fungi *Beauveria bassiana* (Bals.) Vuill. and *Metarhizium anisopliae* (Metchnikoff) Sorikin and the bacteria *Bacillus thuringiensis* Berliner (var. *tenebrionis*) should be accorded high priority (see Chapter 9).

Progress in the area of host-plant resistance to sweet potato weevils has been slow. In spite of over 50 years of research, no cultivars with high levels of resistance are available. Low levels of resistance to weevils have been reported in several pools of sweet potato germplasm. The International Institute for Tropical Agriculture (IITA) and the Asian Vegetable Research Development Center (AVRDC) tried to increase levels of resistance to weevils by using conventional plant breeding techniques; however, they had little success.

There is potential to develop cultivars with higher levels of resistance in the future. Numerous reports indicate that different levels of weevil damage occur in certain populations of sweet potato. However, several problems have complicated interpretations of the results from some of the studies. For example, the experimental design used to assess levels of resistance in the field has often consisted of the use of small, single-row plots with few replications. These tests relied on either natural infestations and/or artificial infestations of weevils and lacked consistency in their methodology. Additionally, various methods have been used to assess weevil damage in the field thereby reducing the possibilities for comparative analyses among tests. For these reasons, standardized protocols for evaluating germplasm for resistance in the field are needed world-wide. The use of such methods may enhance the ability to select for resistant clones, especially for a trait of low heritability.

Many of the wild species of *Ipomoea* and primitive cultivars maintained in the world collection of sweet potato germplasm at CIP need to be evaluated for levels of resistance. Preliminary studies at CIP and AVRDC showed that a wild species, *Ipomoea trifida* (HBK) Don, may contain valuable genes for resistance to *C. formicarius*. Some success has been achieved in crossing *I. trifida* (4X form) with cultivated sweet potato (6X), and in producing *I. trifida* that are more compatible with *I. batatas*. Other forms of *I. trifida* (2X, 4X) are also known. Further studies are needed to make genes of *I. trifida* more available for conventional plant breeding programs so that the potential value of this and other wild species of *Ipomoea* can be assessed. Based on previous studies, resistance to sweet potato weevils is a quantitatively inherited trait with a low heritability (see Chapters 20 and 21). This type of trait is usually difficult to evaluate in a selection and breeding program. Restriction fragment length polymorphism (RFLP) markers have been associated with quantitative trait loci (QTL's) in other crops. Specific RFLP patterns must be determined in sweet potato to link genes which control QTL's for weevil resistance. Direct selection for RFLP marker loci associated with weevil resistance may result in a correlated increase in the frequency of favorable alleles for resistance (see Chapter 21).

Biochemical and biotechnological techniques which may lead to a better understanding of the weevil and the resistance mechanisms of sweet potato are improving rapidly. Oviposition and other volatile stimulants/attractants have been studied (see Chapters 11 and 12). Studies are needed to determine the quantitative and/or qualitative relationships between phytochemicals in storage roots and the expression of resistance to weevils in the field. An analytical approach based on the levels of kairomones in storage roots, which affect oviposition of *C. formicarius*, could be an important tool for plant breeders.

A method to transform sweet potato plants and transfer genes that govern toxicity to pests, such as those from *B. thuringiensis*, protease inhibitors, or chitinases, by using *Agrobacterium tumefaciens* plasmids as a gene vector system has potential for improving pest management programs on sweet potato in the future (see Chapter 10). Caution is advised in this area of research, however, because recently, several insects, such as Indianmeal moth, *Plodia interpunctella* Hubner, diamondback moth, *Plutella xylostella* (L.), Colorado potato beetle, *Leptinotarsa decemlineata* (Say), almond moth, *Cadra cautella* (Walker), tobacco budworm, *Heliothis virescens* (Fabricius), and others, developed resistance to *B. thuringiensis*.

There also is an increasing need to solve the systematic problems involving the members of *C. formicarius* and *C. puncticollis* species complexes. Future systematic work on these species should include biochemical, karyological, and DNA-based analyses for more conclusive answers. Examination of mitochondrial DNA may help to elucidate parental lines and similarity of different weevil populations. Such information may help plant breeders target development and subsequent distribution of resistant plant varieties.

The IPM program for *C. formicarius* described in this book (see Chapter 7) is simple and is readily adaptable for controlling other major sweet potato weevil species, such as *C. puncticollis* and *E. postfasciatus*. Cultural practices are essential to the success of this program. Both international and national programs should encourage farmer participation in the evaluation and use of these approaches.

Progress in developing resistant cultivars for other pests, such as *Conoderus* spp., *Diabrotica* spp., and *Systena* spp., has promise. Several cultivars with multiple resistance to these pests are now available in the United States. These pests are of major importance in many countries of South America. For this reason, studies are needed to transfer genes from these cultivars to germplasm that is adapted to the tropics. More studies are also needed to better understand the biology of these pests on sweet potato in the tropics. The use of the sex pheromone of *D. balteata* LeConte should help to improve IPM programs for this pest world-wide. Biological control agents, such as the new strains of bacteria, entomopathogenic fungi, and entomogenous nematodes, which are effective against coleopterous pests, may have potential for managing these pests as part of an integrated control program in the future. Because of reductions in the numbers of chemical insecticides available for use against these pests, cultural and biological control strategies need to be developed and implemented to provide adequate control of these pests in the future.

Vine borers, *Omphisa anastomasalis* (Guenee) and *Megastes grandalis* (Guenee), are important pests of sweet potato in Asia and the Caribbean (see Chapter 17). Studies have identified resistance to both of these species in sweet potato. These sources of resistance should be incorporated into an active breeding program. There are also indications that adult females of *O. anastomasalis* produce a sex pheromone that attracts males. Research is needed to identify, bioassay, and synthesize the active component(s) of this pheromone.

The diverse complex of egg and larval parasitoids of these pests should also be investigated more fully. Studies are needed to develop and determine the impact of classical biological control programs for these pests in regions where they occur. Additional studies should determine if parasitoids of one vine borer species may also attack the other species. If crossover of parasitoids occurs and high levels of parasitism can be achieved, then exchange and introduction of these parasitoids for management of these vine borer species is encouraged.

A continuing problem facing many sweet potato improvement programs world-wide is the production and movement of pathogen-tested germplasm. CIP has concentrated on establishing phytosanitary standards for pests and pathogens for the distribution of both true seed and *in vitro* cultures of *Ipomoea*. The major constraints for developing such standards are the space and labor needed to index germplasm for virus detection. Newer methods of virus detection and testing, which require fewer resources, such as *in vitro* assay procedures which are now being used on potato, need to be examined for sweet potato. The status of the sweetpotato whitefly, *Bemisia tabaci* (Gennadius), as a major vector of several viruses needs to be assessed. This pest status of this insect recently accelerated in the Western Hemisphere. Because *B. tabaci* is a very polyphagous insect, holistic management programs that consider the surrounding, previous, and subsequent cropping systems and other surrounding vegetation are needed to successfully manage this insect and the diseases that it vectors. Studies are also needed to determine the importance of various alternate hosts as reservoirs for vectors, their associated viruses, and their natural enemies. Additional studies are needed to develop improved IPM programs for virus vectors and viruses on sweet potato.

More research is also needed to improve the management of plant parasitic nematodes. Host plant resistance to plant-parasitic nematodes has been identified in wild and cultivated accessions, and is currently maintained in the world collection of sweet potato germplasm at CIP. Studies are needed to incorporate resistance to these nematodes with resistance to other important pests and pathogens. Improved IPM programs need to be developed for these nematodes and integrated with management programs for other pests.

It is important to realize that pest problems on sweet potato are dynamic and many problems are usually not predictable. Season-long monitoring is needed for most pests. Improved monitoring programs are needed for most of these important pests, especially in developing countries, so that IPM programs can be effectively implemented and assessed.

Although this treatise focused primarily on the major insect, nematode, and virus pests of sweet potato world-wide, we recognize the need for further research on other pests that limit sweet potato production regionally. In addition, more research is needed on certain pests which were once considered a minor pest problem, but recently achieved major pest status, and on new pests which have recently appeared. The recent population buildup of a stem borer, *Ptericoptus* sp. probably *sinuatus* Berg., in Paraguay and the sweet potato leaf beetle, *Typophorus nigritus viridicyaneus* (Crotch), in Argentina demonstrate the need for such research. Continuous interactions with farmers are needed to recognize new pest problems such as these.

Successful development and implementation of IPM technologies requires an integrated program of research with an interdisciplinary perspective. Central to this approach is the need for an improved understanding of the relationships between cultural practices, such as those described in Chapter 4, and pest severity. In addition, the needs of the producer (farmer), the market, and the consumer must be considered when designing, testing, and implementing IPM technologies. Lastly, the socioeconomic constraints of the farming system in which the technologies will operate must be understood in order for IPM to operate effectively (see Chapter 22). More information is needed on each of these aspects so that appropriate IPM systems are developed and utilized by growers world-wide.

CIP is currently developing an international network for sweet potato IPM in developing countries. The objective of this network is to focus research on the components of IPM (host plant resistance, cultural control practices, biological control agents, sex pheromones, and the judiscious use of chemical insecticides), and develop more holistic and integrated approaches for managing pests on sweet potato. Sweet potato IPM programs will become a reality in developing countries when these technologies are developed, implemented, adapted to local conditions, and accepted by farmers. Pest surveys and on-farm demonstration research will help to facilitate this process. National and international agricultural research and development programs in these countries should encourage such activities.

In the United States, IPM programs for pests of sweet potato are more advanced; however, considerably more research and extension efforts are needed to develop and implement improved IPM programs. For example, although cultivars with resistance to multiple pests have been developed for the United States, they are not widely grown. The sweet potato industry is reluctant to change to new cultivars despite their potential benefits (i.e., reduced pesticide use, reduced production costs). Efforts are needed to educate the producers, marketing specialists, and consumers to increase acceptance of cultivars with pest resistance.

The development of new conventional, chemical insecticides to replace those removed from use on sweet potato by the Environmental Protection Agency, and those that have lost their effectiveness over time is unlikely. Sweet potato is considered a minor crop in the United States, and thus, it lacks the importance (i.e., market size) needed to justify development of new insecticides.

The reduced levels of acceptance of cultivars with resistance to pests and the reduced availability of effective chemical pesticides will undoubtedly trigger additional research on developing other alternative tactics for managing pests of sweet potato. The recent surge in research on agroecology and sustainable agriculture in other crops will probably be well represented in sweet potato in the future. Also, considerable research is expected to occur on the development and application of biotechnology for use in sweet potato pest management programs as we enter the 21st century.

Index